Biochar as Soil Amendment

Biochar as Soil Amendment

Impact on Soil Properties and Sustainable Resource Management

Special Issue Editor
José María De la Rosa

MDPI • Basel • Beijing • Wuhan • Barcelona • Belgrade • Manchester • Tokyo • Cluj • Tianjin

Special Issue Editor
José María De la Rosa
Consejo Superior de
Investigaciones Científicas
(IRNAS-CSIC)
Spain

Editorial Office
MDPI
St. Alban-Anlage 66
4052 Basel, Switzerland

This is a reprint of articles from the Special Issue published online in the open access journal *Agronomy* (ISSN 2073-4395) (available at: https://www.mdpi.com/journal/agronomy/special_issues/soil_biochar_amendment).

For citation purposes, cite each article independently as indicated on the article page online and as indicated below:

LastName, A.A.; LastName, B.B.; LastName, C.C. Article Title. *Journal Name* **Year**, *Article Number*, Page Range.

ISBN 978-3-03928-274-6 (Pbk)
ISBN 978-3-03928-275-3 (PDF)

© 2020 by the authors. Articles in this book are Open Access and distributed under the Creative Commons Attribution (CC BY) license, which allows users to download, copy and build upon published articles, as long as the author and publisher are properly credited, which ensures maximum dissemination and a wider impact of our publications.

The book as a whole is distributed by MDPI under the terms and conditions of the Creative Commons license CC BY-NC-ND.

Contents

About the Special Issue Editor vii

Preface to "Biochar as Soil Amendment" ix

Marco Racioppi, Maria Tartaglia, José María De la Rosa, Mauro Marra, Elisa Lopez-Capel and Mariapina Rocco
Response of Ancient and Modern Wheat Varieties to Biochar Application: Effect on Hormone and Gene Expression Involved in Germination and Growth
Reprinted from: *Agronomy* 2020, 10, 5, doi:10.3390/agronomy10010005 1

Prakriti Bista, Rajan Ghimire, Stephen Machado and Larry Pritchett
Biochar Effects on Soil Properties and Wheat Biomass vary with Fertility Management
Reprinted from: *Agronomy* 2019, 9, 623, doi:10.3390/agronomy9100623 13

Beatriz Gámiz, Kathleen Hall, Kurt A. Spokas and Lucia Cox
Understanding Activation Effects on Low-Temperature Biochar for Optimization of Herbicide Sorption
Reprinted from: *Agronomy* 2019, 9, 588, doi:10.3390/agronomy9100588 23

Chen-Chi Tsai and Yu-Fang Chang
Carbon Dynamics and Fertility in Biochar-Amended Soils with Excessive Compost Application
Reprinted from: *Agronomy* 2019, 9, 511, doi:10.3390/agronomy9090511 39

Domenico Ronga, Enrico Francia, Giulio Allesina, Simone Pedrazzi, Massimo Zaccardelli, Catello Pane, Aldo Tava and Cristina Bignami
Valorization of Vineyard By-Products to Obtain Composted Digestate and Biochar Suitable for Nursery Grapevine (*Vitis vinifera* L.) Production
Reprinted from: *Agronomy* 2019, 9, 420, doi:10.3390/agronomy9080420 55

Otávio dos Anjos Leal, Deborah Pinheiro Dick, José María de la Rosa, Daniela Piaz Barbosa Leal, José A. González-Pérez, Gabriel Soares Campos and Heike Knicker
Charcoal Fine Residues Effects on Soil Organic Matter Humic Substances, Composition, and Biodegradability
Reprinted from: *Agronomy* 2019, 9, 384, doi:10.3390/agronomy9070384 77

Marina Paneque, Heike Knicker, Jürgen Kern and José María De la Rosa
Hydrothermal Carbonization and Pyrolysis of Sewage Sludge: Effects on *Lolium perenne* Germination and Growth
Reprinted from: *Agronomy* 2019, 9, 363, doi:10.3390/agronomy9070363 93

Samieh Eskandari, Ali Mohammadi, Maria Sandberg, Rolf Lutz Eckstein, Kjell Hedberg and Karin Granström
Hydrochar-Amended Substrates for Production of Containerized Pine Tree Seedlings under Different Fertilization Regimes
Reprinted from: *Agronomy* 2019, 9, 350, doi:10.3390/agronomy9070350 105

Muhammad Zafar-ul-Hye, Subhan Danish, Mazhar Abbas, Maqshoof Ahmad and Tariq Muhammad Munir
ACC Deaminase Producing PGPR *Bacillus amyloliquefaciens* and *Agrobacterium fabrum* along with Biochar Improve Wheat Productivity under Drought Stress
Reprinted from: *Agronomy* 2019, 9, 343, doi:10.3390/agronomy9070343 123

Khaled D. Alotaibi and Jeff J. Schoenau
Addition of Biochar to a Sandy Desert Soil: Effect on Crop Growth, Water Retention and Selected Properties
Reprinted from: *Agronomy* **2019**, *9*, 327, doi:10.3390/agronomy9060327 139

Alessandro Calamai, Enrico Palchetti, Alberto Masoni, Lorenzo Marini, David Chiaramonti, Camilla Dibari and Lorenzo Brilli
The Influence of Biochar and Solid Digestate on Rose-Scented Geranium (*Pelargonium graveolens* L'Hér.) Productivity and Essential Oil Quality
Reprinted from: *Agronomy* **2019**, *9*, 260, doi:10.3390/agronomy9050260 153

Takafumi Konaka, Shin Yabuta, Charles Mazereku, Yoshinobu Kawamitsu, Hisashi Tsujimoto, Masami Ueno and Kinya Akashi
Use of Carbonized Fallen Leaves of *Jatropha Curcas* L. as a Soil Conditioner for Acidic and Undernourished Soil
Reprinted from: *Agronomy* **2019**, *9*, 236, doi:10.3390/agronomy9050236 167

Supitrada Kumputa, Patma Vityakon, Patcharee Saenjan and Phrueksa Lawongsa
Carbonaceous Greenhouse Gases and Microbial Abundance in Paddy Soil under Combined Biochar and Rice Straw Amendment
Reprinted from: *Agronomy* **2019**, *9*, 228, doi:10.3390/agronomy9050228 181

Sara de Jesus Duarte, Bruno Glaser and Carlos Eduardo Pellegrino Cerri
Effect of Biochar Particle Size on Physical, Hydrological and Chemical Properties of Loamy and Sandy Tropical Soils
Reprinted from: *Agronomy* **2019**, *9*, 165, doi:10.3390/agronomy9040165 193

Cosmas Wacal, Naoki Ogata, Daniel Basalirwa, Takuo Handa, Daisuke Sasagawa, Robert Acidri, Tadashi Ishigaki, Masako Kato, Tsugiyuki Masunaga, Sadahiro Yamamoto and Eiji Nishihara
Growth, Seed Yield, Mineral Nutrients and Soil Properties of Sesame (*Sesamum indicum* L.) as Influenced by Biochar Addition on Upland Field Converted from Paddy
Reprinted from: *Agronomy* **2019**, *9*, 55, doi:10.3390/agronomy9020055 209

Wen-Tien Tsai, Po-Cheng Huang and Yu-Quan Lin
Characterization of Biochars Produced from Dairy Manure at High Pyrolysis Temperatures
Reprinted from: *Agronomy* **2019**, *9*, 634, doi:10.3390/agronomy9100634 229

About the Special Issue Editor

José María De la Rosa, (Dr.), is a researcher at The Institute of Natural Resources and Agrobiology of Seville (IRNAS) from the Spanish National Research Council (Consejo Superior de Investigaciones Científicas, CSIC). He obtained his B.S. and a Ph.D. degrees in Chemistry, both from the University of Seville, receiving the best Ph.D. thesis award (2007) from the same university. He is an expert in soil chemistry, focusing on the properties and mechanisms of stable forms of organic matter. His research on the application of biochar and other forms of pyrogenic organic matter as soil amendment is widely recognized. In the past, he held a postdoctoral position at the Technical University of Munich (TUM, Germany) and at the Instituto Superior Técnico of the University of Lisbon (IST-ITN, Portugal). Since 2012, he has worked in the Biogeochemistry Department at IRNAS-CSIC, where he has been a principal researcher in 7 projects and created a research line dedicated to the design and application of novel organic amendments for the improvement of agricultural soils and carbon stabilization. Dr. De la Rosa has published over 65 scientific papers, 25 book chapters, and 4 books and has given numerous presentations at international conferences and meetings. He has received several awards for his research on sustainable agriculture and on the mechanisms of C and N stabilization in pyrogenic organic matter.

Preface to "Biochar as Soil Amendment"

The role of biochar in improving soil fertility is increasingly being recognized and is leading to recommendations of biochar amendment of degraded soils. In addition, biochars offer a sustainable tool for managing organic wastes and to produce added-value products. The benefits of biochar use in agriculture and forestry can span enhanced plant productivity, an increase in soil C stocks, and a reduction of nutrient losses from soil and non-CO_2 greenhouse gas emissions. Nevertheless, biochar composition and properties and, therefore, its performance as a soil amendment are highly dependent on the feedstock and pyrolysis conditions. In addition, due to its characteristics, such as high porosity, water retention, and adsorption capacity, there are other applications for biochar that still need to be properly tested. Thus, the 16 original articles contained in this book, which were selected and evaluated for this Special Issue, provide a comprehensive overview of the biological, chemicophysical, biochemical, and environmental aspects of the application of biochar as soil amendment. Specifically, they address the applicability of biochar for nursery growth, its effects on the productivity of various food crops under contrasting conditions, biochar capacity for pesticide retention, assessment of greenhouse gas emissions, and soil carbon dynamics. I would like to thank the contributors, reviewers, and the support of the Agronomy editorial staff, whose professionalism and dedication have made this issue possible.

José María De la Rosa
Special Issue Editor

Article

Response of Ancient and Modern Wheat Varieties to Biochar Application: Effect on Hormone and Gene Expression Involved in Germination and Growth

Marco Racioppi [1], Maria Tartaglia [1], José María De la Rosa [2], Mauro Marra [3], Elisa Lopez-Capel [4] and Mariapina Rocco [1,*]

1. Department of Science and Technology, University of Sannio, 82100 Benevento, Italy; marco.racioppi91@gmail.com (M.R.); marytrt@live.it (M.T.)
2. Instituto de Recursos Naturales y Agrobiología de Sevilla (IRNAS-CSIC), Reina Mercedes Av., 10, 41012 Seville, Spain; jmrosa@irnase.csic.es
3. Department of Biology, University of Tor Vergata, 00133 Rome, Italy; mauro.marra@uniroma2.it
4. School of Natural and Environmental Sciences, Newcastle University, Newcastle upon Tyne NE1 7RU, UK; elisa.lopez-capel@newcastle.ac.uk
* Correspondence: rocco@unisannio.it; Tel.: +39-0824-305168

Received: 27 October 2019; Accepted: 15 December 2019; Published: 18 December 2019

Abstract: Agriculture has changed dramatically due to mechanization, new technologies, and the increased use of chemical fertilizers. These factors maximize production and reduce food prices, but may also enhance soil degradation. Sustainable agricultural practices include altering crop varieties and the use of soil amendments to increase production, improve irrigation, and more effectively use fertilizers. Ancient and modern durum wheat varieties have been shown to be tolerant to conditions caused by climate change and increase production. Biochar soil amendments have been reported to increase crop yields, soil fertility, and to promote plant growth. However, results are variable depending on biomass source, application conditions, and crop species. This study evaluates the crop response of two contrasting durum wheat varieties on an Eutric Cambisol amended with beech wood biochar. Wheat varieties used are Saragolla, an ancient variety traditionally used in Southern Italy, and Svevo, a widely used commercial variety. The effect of biochar soil amendment on the expression of genes involved in the germination of these two varieties of wheat was determined using RT-PCR. The content of hormones such as gibberellins (GAs), auxins (IAA), and abscisic acid (ABA) was determined. Results demonstrate that biochar had a stimulatory effect on the growth performances of Svevo and Saragolla cultivars at the molecular level. This correlated to the promoted transcription of genes involved in the control of plant development. Overall, the presence of biochar as soil amendment improved the germination rates of both varieties, but the ancient wheat cultivar was better suited to the Eutric Cambisol than the commercial variety. This trend was also observed in un-amended pots, which may indicate better adaptability of the ancient wheat cultivar to withstand environmental stress than the commercial variety.

Keywords: wheat; biochar; germination; hormone; gene expression

1. Introduction

Cereals are one of the most popular sources of food for humans and animals. According to FAOStat, the global wheat production reached 715.9 million tons in 2013 [1], and durum wheat (*Triticum durum*) is the most widespread crop in the Mediterranean area. It is the exclusive raw material used for pasta production, and a basic product of the Mediterranean diet. The area cultivated with durum wheat in Italy remained on an annual average of 1.6–1.8 million hectares [1], but the increased risk of land degradation due to unsustainable land practices and climate change could affect productivity [2].

The high yields of "modern" wheat cultivars, such as Svevo, require the use of a large amount of mineral fertilizers, chemical herbicides, and fungicides, leading to a greater risk of environmental pollution [3,4]. Increased public interest in this problem and growing consumer demand for healthier products, have led to a greater emphasis on crops grown under an integrated farm management approach to sustainable agricultural systems. Although ancient wheat cultivars provide high quality semolina, they have been progressively abandoned in favour of genetically uniform high-yield commercial varieties, as they do not adapt to intensive cultivation parameters. This reduction in biodiversity of modern varieties may enhance susceptibility to pathogens and disease; and decrease grain quality and productivity under adverse environmental conditions [4]. Traditional wheat cultivars, such as Saragolla, are expected to demonstrate greater adaptability and resilience as a result of better tolerance to diseases and more efficient capacity for using soil resources [3]. Therefore, ancient varieties could be the best choice to low-input and organic growing systems.

There is a need to develop more sustainable agricultural practices to avoid a decrease in soil fertility and organic carbon contents, and an increase in soil erosion [4]. Soil degradation in Mediterranean regions is particularly critical due to the climatic peculiarities of these areas, which can lead to permanent and irreversible degradation of soil quality and productivity. In restoration processes, the application of organic amendments is used for the creation of soil substrates with incipient structure and stable aggregates and to improve the biological functionality of the soil [5]. The use of biochar as soil amendment has been proposed as a sustainable strategy to improve crop productivity [6,7], control soil salinity [8], and mitigate global warming of degraded soils. Biochar has been reported to enhance plant physiological response [9] and improve plant adaptation to dry periods in biochar amended soils [6].

Biochar is rich in stable forms of C that decompose in soil at slower rate than untreated biomass residues [10]. The composition and properties of biochars are largely dependent on pyrolysis conditions and on the nature of the feedstock [11]. In general, the addition of biochar to a degraded soil improves its structure, increases porosity, decreases bulk density and enhances aggregation and water retention, reducing irrigation demand [12]. The biochar effect on soil properties influences crop growth [13], being more effective in degraded soils than in healthy ones [14]. For this reason, there is an increased interest in biochar applications for ecological restoration and carbon sequestration in nutrient-poor soils. Land management adaptations to enhance crop production in response to land degradation and potential climate change include the use of varieties resistant to heat shock and drought [7]. Crop germination, growth and yield depend on both genetic and environmental factors [15]. The balance of plant hormones is of fundamental importance in germination. There is evidence that biochar application can stimulate gibberellin, auxin, and brassinosteroid regulation, promoting plant growth [16]. Gibberellins are growth hormones that stimulate plant elongation, germination and flowering in cereal grains. French and Yyer-Pascuzzi [16] observed that biochar promotes growth partially through stimulation of the Gibberellins (GA) pathway. Although these is some evidence of the impact of biochar on crop production, the impact of biochar on plant growth and on the activity of enzymes and hormones involved in plant growth has not been fully investigated [17,18]. The mechanism of whether and how biochar soil amendment influences hormones and gene expression involved in crop germination remains largely unknown. Therefore, further research is needed to understand the signaling pathways on different plant species and cultivars within species [19]. The aim of this study is to investigate the effect of beech-wood biochar on germination performances, hormone level variations, and transcription of growth-related genes of a commercial (Svevo) and an ancient (Saragolla) wheat cultivar from Southern Italy in an irrigated soil from the Campania region in Italy. We hypothesized that biochar may play a key role on the stimulation of gibberellins in wheat. This knowledge is crucial for optimizing the use of biochar in wheat production and for breeding biochar-responsive wheat varieties.

2. Materials and Methods

2.1. Description and Analysis of Samples

Surface samples (0–25 cm depth) were collected from a traditional agricultural soil in Avellino province, within the Campania region (Southern Italy) classified as Eutric Cambisol (FAO classification). After decades of intensive cultivation of cereal crops the land was fallowed for a ten-year period (with no cultivation or fertilizer treatment). Wheat cultivation was recently reintroduced at this area under organic management practices. Soil samples were dried at 40 °C for 24 h and sieved (<2 mm) to remove roots, small branches and mosses. The biochar used in this experiment was produced by Verforfood (GREEN BIOCHAR S.C.A.R.L., Turin, Italy). It is a fine grain char (<0.251 mm) produced from hard beech (*Fagus sylvatica*) wood at a pyrolysis temperature of 550 °C. Soil and biochar pH was measured in triplicate in H_2O (1/2.5 v/v) using a pH meter (MM40 CRISON S.A, Barcelona, Spain)) after shaking for 60 min and overnight sedimentation as reported by Obia et al. [19]. Total carbon (C) and nitrogen (N) concentrations of soil and biochar samples were determined in triplicate by dry combustion (1000 °C) using a Perkin-Elmer 2400 series 2 elemental analyzer. Water Holding Capacity (WHC) and ash content of soil, biochar and 5% biochar amended soils were measured following the procedure described by De la Rosa et al. [20]. Briefly, samples were dried at 105 °C and subsequently heated in a muffle furnace (750 °C for 5 h). The ash percentage is the proportional weight of the remaining ash from the oven-dried weight sample. Soil, biochar, and biochar amended soils properties are shown in Table 1.

Table 1. Properties of biochar, soil and 5% biochar amended soils.

Sample	pH	% C	% N	% WHC	% Ash
Soil	5.62	3.1	0.16	35.9	88.0
Biochar	8.21	81.1	0.91	364.4	7.7
Soil + Biochar (5%)	7.10	n.a.	n.a.	53.4	n.a.

C: Total Carbon content (%); N: Total Nitrogen content (%); WHC: Water holding capacity (%); Ash: Ash content (%); n.a. = not analysed.

2.2. Seed Germination

Seeds of the T. durum cultivars Svevo (Agrisemi Minicozzi, Benevento, Italy) and Saragolla (SYNGENTA, Milan, Italy) were sterilized in 1% w/v sodium hypochlorite for 30 min and rinsed in distilled water to remove the excess of chemicals.

Svevo Control (SVC), Saragolla Control (SRC), and biochar amended soil treatments (SVB and SRB) were prepared by placing 100 seeds of Svevo and Saragolla seeds in 1 kg soil (dry weight and 1 kg soil mixed with 50 g biochar in pots (23 cm diameter and 20 cm height), respectively. The dose of biochar applied was 5% w/w, equivalent to about 25 t ha^{-1}, and within the range used for pot biochar studies [21]. Three pots per treatment were prepared, randomly placed, wetted with deionized water to 50% water holding capacity (WHC) [22], and kept at 4 °C for 24 h. Germination was monitored and recorded every 24 h for 7 days (after 24, 48, 72, 96, 120, 144 and 168 h) at a controlled environment chamber at 30 °C [23]. Seeds were rewetted and water content maintained at 50% WHC during the 7-day germination experiment.

2.3. Hormone Extraction and Analysis

In order to correlate the hormone content of the Svevo and Saragolla varieties with the effect of biochar treatment on their germination performances, seedlings of the two wheat cultivars, grown in control soils and soils amended with 5% biochar were collected three days after sowing. Abscisic acid (ABA), indoleacetic acid (IAA) and gibberellin A3 and A4 (GA) content in wheat seedlings was measured by HPLC [24]. Five wheat seedlings were collected from each pot and frozen at −80 °C. Wheat samples (2 g) were grinded to a fine powder and diluted in methanol (2.5 mL g^{-1} of fresh tissue). Each extract was centrifuged (16,000 rpm for 10 min at 4 °C). The supernatant was concentrated under vacuum, a volume of deionized water was added to each sample and extracted with an equal volume of ethyl acetate. Aqueous and organic phases were separated by centrifugation (16,000 rpm for 2 min). The lower aqueous phase was transferred to a new tube, and then the pH of the solution was adjusted below 3 to keep all the hormones in protonated form. The upper organic phase was recovered, dried under vacuum and dissolved in 30 µL of methanol before analysis by reversed phase-HPLC. HPLC analysis was performed on a LC-20 Prominence HPLC system (Shimadzu, Kyoto, Japan) equipped with LC-20AT quaternary gradient pump, SPD-M20A photo diode array detector (PDAD) and SIL-20 AH autosampler (20 µL injection vol). Plant hormones were separated on a Gemini-NX C18 column (250 × 4.5 mm, 5 µm; Phenomenex, Torrance, CA, USA), which was assembled with a Security Guard® (Phenomenex) pre-column and eluted with a gradient of acetonitrile (ACN) containing 0.1% v/v trifluoroacetic acid (TFA) in aqueous 0.1% v/v TFA at 45 °C; ACN ramped from 15% to 30% in 5 min, from 30% to 50% in 5 min, from 50% to 80% in 2 min, with a flow rate of 1.5 mL min^{-1}. Separated compounds were identified through their retention times, UV spectra and literature data by comparison with IAA (12886, Sigma-Aldrich, St Louis, MO, USA), GA3 (G7645, Sigma-Aldrich) and GA4 (G7276, Sigma-Aldrich) standards. These standard compounds were also used to build up calibration curves (in the range 5–200 µg/mL) at specific wavelengths (λIAA = 254 nm; λGAs = 205 nm; λABA = 254 nm). Gibberellin concentrations are reported as the sum of GA3 and GA4 content and results are shown as µg of hormone per gram of fresh tissue.

2.4. Data Analysis

The results of germination assay and HPLC-based hormone analysis are expressed as means ±S.D. The identification of significant differences was performed by analysis of variance (ANOVA), followed by the Student-Newman-Keulus test with a minimum level of significance of $p < 0.05$.

2.5. RNA Extraction, cDNAs Synthesis and Reverse Transcription-Quantitative PCR (RT-qPCR)

The effect of biochar addition over especific key enzymes of the biosynthetic pathway of GAs was analyzed through their expression in three-days germinated wheat seedlings of the Svevo and Saragolla varieties, by RT-qPCR.

A Sigma mirPremier microRNA isolation Kit was used to extract RNA from wheat leaf samples. RNeasy/QIamp columns and RNase-Free DNase set (QUIAGEN) were used to degrade genomic DNA and obtain an eluate of pure RNA. The extracted RNA was retrotranscribed to cDNA and stored at −20 °C. "ImProm-II Reverse Transcription System Kit" (Promega) and the "Mj mini thermal cycler" (BioRad) were used for retrotranscription, and samples were stored at −20 °C. The expression of some genes involved in germination, gibberellins biosynthesis, cell expansion and growth were analyzed by RT-qPCR. Gene primers (as shown in Table 2) were designed using the NCBI Primer Blast tool. "EvaGreen 2X qPCR MasterMix-R" (Applied Biological Materials) kit was used for RT-qPCR. The thermal cycler used "7300 Real-Time PCR System" was set to perform an initial denaturation at 95 °C for 1 min, an annealing phase of 9 min at 95 °C and 40 successive cycles of denaturation (95 °C for 30 s), annealing (58 °C for 30 s) and extension (72 °C for 30 s). Experiments were carried out in triplicate and the relative quantification in gene expression was determined using the $2^{-\Delta\Delta Ct}$ method [25].

Table 2. List of primers used for real time PCR.

Primers	Sequences
β-ACTIN	F: TGGACTCTGGTGATGGTGTC R: CCTCCAATCCAAACACTGTA
BIN2	F: GAGATCTAAAGCCTCAAAATCTT R: TGGCTTCACCTTTAACGAGCT
PIN1	F: ATCATCTGGTACACGCTCAT R: GGGAACTGCTCGGTTGAT
TIP2	F: GATGACTCCTTCAGCTTGG R: GGCGAAGACGAAGATGAG
XTH	F: GCCCTTCGTCGCCTCCTAC R: CGGCACAACAACAACTAGTGGTAG
EXP A2	F: CCACCATGATGTGTTGTTCC R: AGTAGGAGTGGCCGTTGATG
TaCPS	F: GTATGCAAGCTTACCGCGTG R: ACCCCCACAAGAATGTCCTC
TaKS	F: CAGGCCGGGGAGAAATCTT R: TGAGACAGCTCATCTGGGGA
TaKO	F: CCGGCACCGAGATAGTCATC R: GAGCAAATCCAGCACCTCAT
TaGA20ox1a	F: CCATCCTCCACCAGGACAAC R: GAGCTCCATCCTCTGTCTGG

3. Results and Discussion

3.1. Effect of Biochar on the Germination of Svevo and Saragolla Seeds

Figure 1 shows germination results for Svevo and Saragolla seeds measured every 24 h during a period of 7 days at 30 °C. The application of 5% biochar in soil improved and accelerated the germination rate of both Svevo and Saragolla wheat varieties. The ancient cultivar, Saragolla, exhibited better germination rates than Svevo, both in germination percentage and precocity. The addition of biochar influenced the soil properties of this Eutric Cambisol from the Campania region of Italy by increasing its pH from 5.6 to 7.1 and the water retention from 35.9% to 53.4% (Table 1). The latter was probably due to biochar high WHC, porosity and organic carbon content. The liming effect of biochar probably facilitated the germination and development of wheat seedlings. This is in agreement with reported literature on biochar promotion of plant growth [16]. Therefore, biochar amendment on this agricultural Campanian soil could be associated with an improvement in soil structure, water retention, nutrient availability [12], and facilitated root development of wheat varieties.

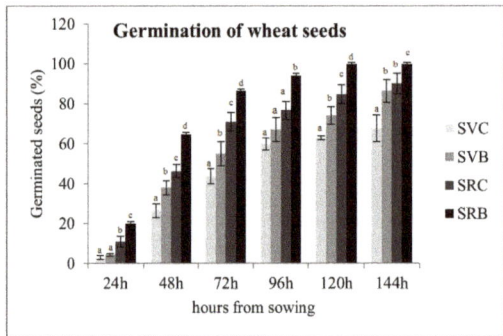

Figure 1. Effect of biochar on the germination of Svevo and Saragolla seeds. The histograms indicate the germination percentage for each sample; the vertical bars represent the standard deviation of mean of three replicates Svevo Control (SVC), Saragolla Control (SRC), Svevo Biochar (SVB) and Saragolla Biochar (SRB). Bars labeled with dissimilar letters are significantly different ($p < 0.05$).

3.2. Effect of Biochar on the Hormone Content of Svevo and Saragolla Seedlings

Figure 2 shows that the two varieties grown in soil without biochar (SVC and SRC) contained similar amounts of ABA while levels of 3-IAA increased significantly due to biochar addition. A significant increase of the GA content was also observed in biochar amended soils for both varieties. Overall, the total amount of GA was higher in Saragolla than Svevo cultivars, under both control and biochar amendment conditions. Germination and hormonal differences between the Saragolla and Svevo varieties support a model in which biochar application stimulates the GAs pathway in wheat, especially in Saragolla variety.

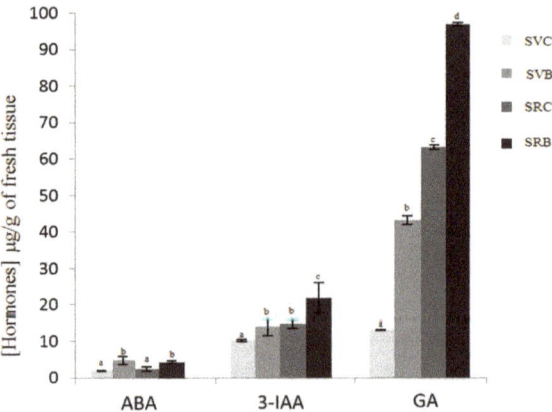

Figure 2. Hormone content of three-days germinated Svevo and Saragolla from sowing grown in soil amended with biochar. Histograms represent abscisic acid (ABA), auxins (3-IAA), and gibberellins (GA) contents ($\mu g\ g^{-1}$ fresh weight) for each sample. Vertical bars represent the standard deviation of mean of three replicates. Svevo Control (SVC), Saragolla Control (SRC), Svevo Biochar (SVB) and Saragolla Biochar (SRB). Bars labeled with dissimilar letters are significantly different ($p < 0.05$).

ABA and 3-IAA concentrations were significantly lower than those of GA. It is known that the stimulatory role of gibberellins (GA) in cereal grain have an interactive inhibitory effect on ABA transactivation activity, while the same does not affect GA activity [26,27]. These results demonstrate that, at a molecular level, the biochar soil amendment was able to markedly improve GA content of wheat seeds, thereby increasing their germination ability.

3.3. Effect of Biochar on Gene Transcription

GA Biosynthesis Genes

Figure 3 shows a higher expression of ent-copalyl diphosphate synthase (TaCPS), ent-kaurene synthase (TaKS), ent-kaurene oxidase (TaKO) and GA 20-oxidase (TaGA20ox1a) in Saragolla than Svevo varieties. Moreover, the application of 5% wood biochar significantly increased the transcription of these genes in both varieties. In general, results confirm that the positive effect of biochar on the germination performance of wheat varieties is due to an increase in GA synthesis, which in turn depends on the stimulation of transcription of GA biosynthetic enzymes. Furthermore, the higher germination ability of the "ancient" Saragolla breed than the "modern" Svevo can be attributed to constitutively higher expression levels of GA biosynthesis enzymes. These results are in agreement with those from French and Iyer-Pascuzzi [16]. Using the ga3ox1-3 Arabidopsis mutant for GA biosynthesis, they demonstrated that the GA pathway is involved in biochar-mediated plant growth promotion and hypothesized that the occurrence of hormone-like substances, such as karrikins, stimulate the GA pathway and stressed the importance of measuring GA levels in biochar-treated plants [28]. Our investigation demonstrates that biochar treatment stimulates wheat growth by increasing endogenous GAs concentration rather than by the release of hormone-like compounds. Figure 3 shows that biochar stimulates GA pathway by increasing endogenous gibberellins concentration, The increase of GA concentration (Figure 2), and GA biosynthetic enzymes (Figure 3); and accelerated germination (Figure 1) observed in biochar treated crops, provides conclusive evidence of the functional role of the GA pathway in biochar-induced plant growth promotion.

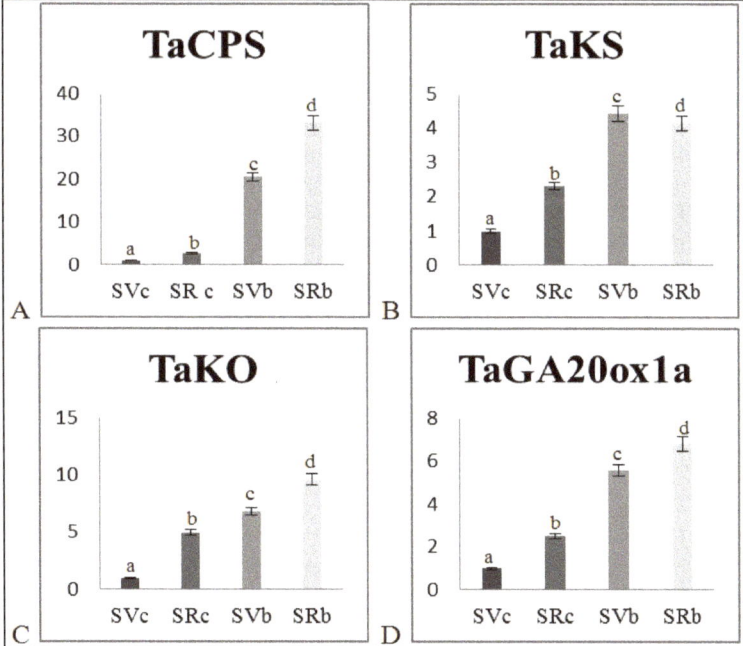

Figure 3. RT-qPCR analysis of the expression of the GA biosynthesis genes in Svevo and Saragolla cultivars upon biochar administration. (**A**) ent-copalyl diphosphate synthase (TaCPS); (**B**) ent-kaurene synthase (TaKS); (**C**) ent-kaurene oxidase (TaKO) and (**D**) GA 20-oxidase (TaGA20ox1a) for each treatment: Svevo Control (SVC), Saragolla Control (SRC), Svevo Biochar (SVB) and Saragolla Biochar (SRB). Bars labeled with dissimilar letters are significantly different ($p < 0.05$).3.3.2. Growth-Related Genes.

Viger et al. (2015) [29] reported the stimulation of the growth-promoting hormones auxin and brassinosteroid in Arabidopsis in plants treated with biochar. In order to determine whether the growth-promoting effect of biochar on Saragolla and Svevo cultivars could trigger auxin and/or brassinosteroid regulation, the expression of some key growth-related genes were analysed. Expansine A2 (EXP A2), xyloglucan endotransglicosylase (XTH), aquaporin 2 (TIP2), auxin efflux carrier (PIN 1) and brassinosteroid-insensitive 2 (BIN 2) gene expressions proteins were analysed by RT-qPCR on three-day old germinated samples. EXP A2, TIP2 and PIN 1 and XTH proteins participate at different levels of the growth-promoting pathway regulated by auxin; such as plant cell wall weakening (EXP A2), water transport (TIP2), auxin transport (PIN 1) and cell wall remodeling (XTH) [16,30]. Figure 4 shows that transcript levels of all of these proteins increased in both wheat cultivars in biochar amended soil treatments. This suggests that activation of the auxin responsive growth-promoting pathway can contribute to the stimulation of germination and growth observed in the wheat cultivars with biochar treatment. Both in dicot and monocot XTHs are encoded by a multigene family and the expression of the individual XTH genes is differentially regulated by environmental stimuli [31] and growth-promoting hormones such as GAs and IAA [32]. XTHs are involved in the control of cell wall extensibility during growth stimulated by GAs and IAA [33] and the observed increase for XTH gene expression in the Saragolla cultivar upon biochar treatment is in accordance with the reported stimulation of growth and enhanced GAs biosynthesis of the cultivar after biochar treatment. Reasons for the absence of a similar increase in the Svevo cultivar are unclear. The BIN 2 gene encodes for a negative regulator of the brassinosteroid (BR) pathway which in plant regulates growth and development [34], interacting with other hormone pathways such as that activated by IAA [35]. Remarkably, levels of BIN 2 transcripts greatly decreased in both wheat cultivars after biochar soil amendment and particularly in the Saragolla cultivar, which showed the best germination and growth performance. This finding strongly suggests that the brassinosteroid-responsive pathway becomes activated upon biochar treatment and cooperates with GA and possibly IAA pathways to stimulate growth and development. Viger et al. [29] proposed that pH modification and increased K^+ supply and availability in biochar-amended soil could activate Ca^{2+} and ROS-mediated cell signaling, leading in turn to the stimulation of IAA and BR growth-promoting pathways. Our results show increased levels of endogenous GAs and suggest the activation of IAA and BR pathways, which are in line with the hypothesis of Viger et al. [29]. This provides mechanistic evidence of how biochar soil amendment influences gene expression involved in wheat germination.

This study reports that the amendment of a typical agricultural Cambisol with 5% wood biochar enhanced seed germination and growth-hormone content of an "ancient" (Saragolla) and a commercial (Svevo) durum wheat variety. Biochar soil amendment had an effect on soil properties which could have facilitated wheat germination and root development. Agronomic data, hormone and gene expression analyses provided evidence of the growth performances and the stimulatory effect of biochar application on both varieties. At the molecular level, growth stimulation could be associated with an increase in GA levels and in transcripts of GAs biosynthesis genes. RT-qPCR results suggested that IAA and BRs pathways become activated upon biochar treatment, interacting with the GA pathway in the stimulation of growth and development of wheat cultivars. Although, the application of a 5% biochar soil amendment improved the germination and selected hormone content of both varieties, the "ancient" wheat cultivar achieved better results than the commercial variety for both untreated and to biochar amended cambisol. This study suggests that the use of biochar as a soil amendment and the use of traditional wheat varieties could improve wheat productivity in a sustainable way.

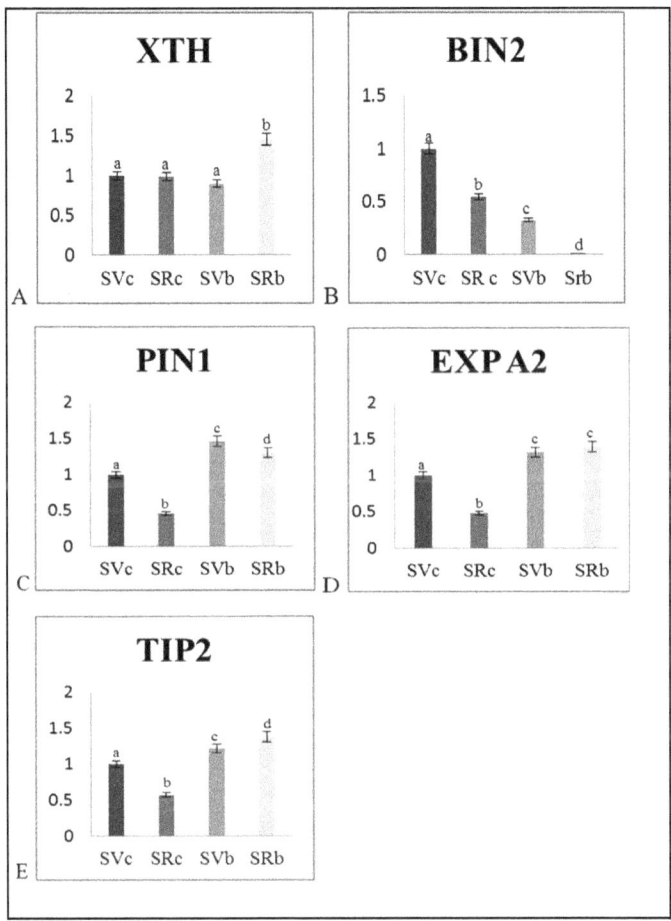

Figure 4. RT-qPCR analysis of the expression of growth-related genes in Svevo and Saragolla varieties upon biochar administration (**A**) xyloglucan endotransglicosylase (XTH), (**B**) brassinosteroid-insensitive 2 (BIN2), (**C**) auxin efflux carrier (PIN1); (**D**) expansine A2 (EXP A2), and (**E**) aquaporin 2 (TIP2) for the treatments Svevo Control (SVC), Saragolla Control (SRC), Svevo Biochar (SVB) and Saragolla Biochar (SRB). Bars labeled with dissimilar letters are significantly different ($p < 0.05$).4. Conclusions.

Author Contributions: M.R. (Mariapina Rocco), M.R. (Marco Racioppi) and J.M.D.l.R. conceived the work. Conceptualization: M.R. (Mariapina Rocco); formal analysis: M.R. (Marco Racioppi) and M.T.; investigation: M.R. (Mariapina Rocco), M.R. (Marco Racioppi), M.T.; resources: M.R. (Mariapina Rocco) and M.M.; data curation: M.R. (Mariapina Rocco), M.R. (Marco Racioppi), M.T. and M.M.; Writing—original draft preparation: M.R. (Mariapina Rocco), M.R. (Marco Racioppi), M.T., J.M.D.l.R., E.L.-C. and M.M.; Edition of the manuscript: M.R., E.L.-C. and J.M.D.l.R.; Visualization: M.M. All authors have read and agreed to the published version of the manuscript.

Funding: This research received no external funding.

Conflicts of Interest: The authors declare no conflict of interest.

References

1. Călinoiu, L.F.; Vodnar, D. Whole grains and phenolic acids: A review on bioactivity, functionality, health benefits and bioavailability. *Nutrients* **2018**, *10*, 1615. [CrossRef]
2. Hamdi, L.; Suleiman, A.; Hoogenboom, G.; Shelia, V. Response of the Durum Wheat Cultivar Um Qais (*Triticum turgidum* subsp. *durum*) to Salinity. *Agriculture* **2019**, *9*, 135. [CrossRef]
3. Guarda, G.; Padovan, S.; Delogu, G. Grain yield, nitrogen-use efficiency and baking quality of old and modern Italian bread-wheat cultivars grown at different nitrogen levels. *Eur. J. Agron.* **2004**, *21*, 181–192. [CrossRef]
4. Martini, D.; Taddei, F.; Ciccoritti, R.; Pasquini, M.; Nicoletti, I.; Corradini, D.; D'Egidio, M.G. Variation of Total Antioxidant Activity and of Phenolic Acid, Total Phenolics and Yellow Coloured Pigments in Durum Wheat (*Triticum turgidum* L. Var. *durum*) as a Function of Genotype, Crop Year and Growing Area. *J. Cereal Sci.* **2015**, *65*, 175–185. [CrossRef]
5. Fernández, J.M.; Hernández, D.; Plaza, C.; Polo, A. Organic matter in degraded agricultural soil amended with composted and thermally-dried sewage sludge. *Sci. Total Environ.* **2008**, *378*, 75–80. [CrossRef] [PubMed]
6. Paneque, M.; De la Rosa, J.M.; Franco-Navarro, J.; Colmenero-Flores, J.M.; Knicker, H. Effect of biochar amendment on morphology, productivity and water relations of sun-flower plants under non-irrigation conditions. *Catena* **2016**, *147*, 280–287. [CrossRef]
7. Howden, S.M.; Soussana, J.-F.; Tubiello, F.N.; Chhetri, N.; Dunlop, M.; Meinke, H. Adapting agriculture to climate change. *Proc. Natl. Acad. Sci. USA* **2007**, *104*, 19691–19696. [CrossRef]
8. Huang, M.; Zhang, Z.; Zhu, C.; Zhai, Y.; Lu, P. Effect of biochar on sweet corn and soil salinity under conjunctive irrigation with brackish water in coastal saline soil. *Sci. Hortic.* **2019**, *250*, 405–413. [CrossRef]
9. Álvarez, J.M.; Pasian, C.; Lala, R.; López, R.; Díaz, M.J. Morpho-physiological plant quality when biochar and vermicompost are used as growing media replacement in urban horticulture. *Urban For. Urban Green.* **2018**, *34*, 175–180. [CrossRef]
10. De la Rosa, J.M.; Rosado, M.; Paneque, M.; Miller, A.Z.; Knicker, H. Effects of aging under field conditions on biochar structure and composition: Implications for biochar stability in soils. *Sci. Total Environ.* **2018**, *613–614*, 969–976. [CrossRef]
11. Zhang, H.; Voroney, R.P.; Price, G.W. Effects of temperature and processing conditions on biochar chemical properties and their influence on soil C and N transformations. *Soil Biol. Biochem.* **2015**, *351*, 263–265. [CrossRef]
12. Schmidt, H.P.; Pandit, B.H.; Martinsen, V.; Cornelissen, G.; Conte, P.; Kammann, C.I. Fourfold increase in pumpkin yield in response to low-dosage root zone application of urine-enhanced biochar to a fertile tropical soil. *Agriculture* **2015**, *5*, 723–741. [CrossRef]
13. Dari, B.; Nair, V.D.; Harris, W.G.; Nair, P.K.R.; Sollenberger, L.; Mylavarapu, R. Relative influence of soil-vs. biochar properties on soil phosphorus retention. *Geoderma* **2016**, *280*, 82–87. [CrossRef]
14. Hussain, M.; Farooq, M.; Nawaz, A.; Al-Sadi, A.M.; Solaiman, Z.M.; Alghamdi, S.S.; Ammara, U.; Ok, Y.S.; Kadambot, H.; Siddique, M. Biochar for crop production: Potential benefits and risks. *J. Soils Sediments* **2016**, *17*, 1–32. [CrossRef]
15. Graziano, S.; Marando, S.; Prandi, B.; Boukid, F.; Marmiroli, N.; Francia, E.; Pecchioni, N.; Sforza, S.; Visioli, G.; Gullì, M. Technological Quality and Nutritional Value of Two Durum Wheat Varieties Depend on Both Genetic and Environmental Factors. *J. Agric. Food Chem.* **2019**, *67*, 2384–2395. [CrossRef]
16. French, E.A.; Iyer-Pascuzzi, A.S. A role for the gibberellin pathway in biochar-mediated growth promotion. *Sci. Rep.* **2018**, *8*, 1–10. [CrossRef]
17. Farhangi-Abriz, S.; Torabian, S. Biochar Increased Plant Growth-Promoting Hormones and Helped to Alleviates Salt Stress in Common Bean Seedlings. *J. Plant Growth Regul.* **2018**, *37*, 591–601. [CrossRef]
18. Egamberdieva, D.; Wirth, S.; Behrendt, U.; Abd-Allah, E.F.; Berg, G. Biochar treatment resulted in a combined effect on soybean growth promotion and a shift in plant growth promoting rhizobacteria. *Front. Microbiol.* **2016**, *7*, 209. [CrossRef]
19. Obia, A.; Cornelissen, G.; Mulder, J.; Dörsch, P. Effect of soil pH increase by biochar on NO, N_2O and N_2 production during denitrification in acid soils. *PLoS ONE* **2015**, *10*, e0138781. [CrossRef]

20. De la Rosa, J.M.; Paneque, M.; Miller, A.Z.; Knicker, H. Relating physical and chemical properties of four different biochars and their application rate to biomass production of Lolium perenne on a Calcic Cambisol during a pot experiment of 79 days. *Sci. Total Environ.* **2014**, *499*, 175–184. [CrossRef]
21. Jeffrey, S.; Verheijen, F.; van der Velde, M.; Bastos, A.C. A quantitative review of the effects of biochar application to soils on crop productivity using meta-analysis. 2011. *Agric. Ecosyst. Environ.* **2011**, *144*, 175–187. [CrossRef]
22. Bista, P.; Ghimire, R.; Machado, S.; Pritchett, L. Biochar Effects on Soil Properties and Wheat Biomass vary with Fertility Management. *Agronomy* **2019**, *9*, 623. [CrossRef]
23. Barros, A.C.; Freund, M.T.; Villavicencio, L.; Delincee, A.L.C.H.; Arthura, H. Identification of irradiated wheat by germination test, DNA comet assay and electron spin. *Radiat. Phys. Chem.* **2002**, *63*, 423–426. [CrossRef]
24. Rocco, M.; Tartaglia, M.; Izzo, F.P.; Varricchio, E.; Arena, S.; Scaloni, A.; Marra, M. Comparative proteomic analysis of durum wheat shoots from modern and ancient cultivars. *Plant Physiol. Biochem.* **2019**, *135*, 253–262. [CrossRef] [PubMed]
25. Livak, K.J.; Schmittgen, T.D. Analysis of relative gene expression data using real-time quantitative PCR and the 2(-Delta Delta C(T)). *Methods* **2001**, *25*, 402–408. [CrossRef]
26. Camoni, L.; Visconti, S.; Aducci, P.; Marra, M. 14-3-3 Proteins in Plant Hormone Signaling: Doing Several Things at Once. *Front. Plant Sci.* **2018**, *9*, 297. [CrossRef]
27. Guangwu, Z.; Xuwen, J. Roles of gibberellin and Auxin in promoting seed germination and seedling Vigor in Pinus massoniana. *For. Sci.* **2014**, *60*, 367–369. [CrossRef]
28. Vishal, B.; Kumar, P.P. Regulation of seed germination and abiotic stresses by gibberellins and abscisic acid. *Front. Plant Sci.* **2018**, *9*, 838. [CrossRef]
29. Viger, M.; Hancock, R.D.; Miglietta, F.; Taylor, G. More plant growth but less plant defence? First global gene expression data for plants grown in soil amended with biochar. *GCB Bioenergy* **2015**, *7*, 658–672. [CrossRef]
30. Zhang, Y.; Ni, Z.; Yao, Y.; Nie, X.; Sun, Q. Gibberellins and heterosis of plant height in wheat (*Triticum aestivum* L.). *BMC Genet.* **2007**, *8*, 40. [CrossRef]
31. Xu, W.; Campbell, P.; Vargheese, A.K.; Braam, J. The Arabidopsis XET-related gene family: Environmental and hormonal regulation of expression. *Plant J.* **1996**, *9*, 879–889. [CrossRef] [PubMed]
32. Vissenberg, K.; Oyama, M.; Osato, Y.; Yokoyama, R.; Verbelen, J.P.; Nishitani, K. Differential expression of AtXTH17, AtXTH18, AtXTH19 and AtXTH20 genes in Arabidopsis roots. Physiological roles in specification in cell wall construction. *Plant Cell Physiol.* **2005**, *46*, 192–200. [CrossRef] [PubMed]
33. Sánchez-Rodríguez, C.; Rubio-Somoza, I.; Sibout, R.; Persson, S. Phytohormones and the cell wall in Arabidopsis during seedling growth. *Trends Plant Sci.* **2010**, *15*, 291–301. [CrossRef] [PubMed]
34. Singla, B.; Tyagi, A.K.; Khurana, J.P.; Khurana, P. Analysis of expression profile of selected genes expressed during auxin-induced somatic embryogenesis in leaf base system of wheat (*Triticum aestivum*) and their possible interactions. *Plant Mol. Biol.* **2007**, *65*, 677–692. [CrossRef] [PubMed]
35. Kim, T.W.; Lee, S.M.; Joo, S.H.; Yun, H.S.; Lee, Y.; Kaufman, P.B.; Kirakosyan, A.; Kim, S.H.; Nam, K.H.; Lee, J.S.; et al. Elongation and gravitropic responses of Arabidopsis roots are regulated by brassinolide and IAA. *Plant Cell Environ.* **2007**, *30*, 679–689. [CrossRef]

© 2019 by the authors. Licensee MDPI, Basel, Switzerland. This article is an open access article distributed under the terms and conditions of the Creative Commons Attribution (CC BY) license (http://creativecommons.org/licenses/by/4.0/).

Article

Biochar Effects on Soil Properties and Wheat Biomass vary with Fertility Management

Prakriti Bista [1,*], Rajan Ghimire [2], Stephen Machado [1] and Larry Pritchett [1]

1. Columbia Basin Agricultural Research Center, Oregon State University, Pendleton, OR 97801, USA; Stephen.Machado@oregonstate.edu (S.M.); Larry.Pritchett@oregonstate.edu (L.P.)
2. Agriculture Science Center, New Mexico State University, Clovis, NM 88101, USA; rghimire@nmsu.edu
* Correspondence: pbistaghimire@gmail.com; Tel.: +1-575-985-2292

Received: 14 September 2019; Accepted: 6 October 2019; Published: 10 October 2019

Abstract: Biochar can improve soil health and crop productivity. We studied the response of soil properties and wheat growth to four rates of wood biochar (0, 11.2, 22.4, and 44.8 Mg ha^{-1}) and two fertilizer rates [no fertilizer and fertilizer (90 kg N ha^{-1}, 45 kg P ha^{-1}, and 20 kg S ha^{-1})]. Biochar application increased soil organic matter (SOM), soil pH, phosphorus (P), potassium (K), sulfur (S) contents, and the shoot and root biomass of wheat. However, these responses were observed at biochar rates below 22.4 Mg ha^{-1}, particularly in treatments without fertilizer. In fertilizer-applied treatments, soil nitrate levels decreased with an increase in biochar rates, mainly due to better crop growth and high nitrate uptake. However, without N addition, the high C:N ratio (500:1) possibly increased nutrient tie-up, reduced plant biomass, and SOM buildup at the highest biochar rate. Based on these results, we recommend biochar rates of about 22.4 Mg ha^{-1} and below for Walla Walla silt loams.

Keywords: nutrient cycling; soil health; soil organic matter

---

1. Introduction

Approximately 64 million dry tons of forest harvest residues are produced annually in the US, and an additional 87 million dry tons of wood remain as milling residues [1]. These residues are a potential feedstock for the production of biochar. Biochar is a charcoal-like product of thermal degradation of biomass in limited presence or absence of oxygen (pyrolysis) that could be used as a soil amendment to improve soil health and crop productivity [2]. However, biochar is not a uniform product. A characteristic of the feedstock type and pyrolysis conditions determines its structure, nutrient content, pH, and other properties [3–5]. Biochar has between 40% to 90% carbon (C) and, depending on the pyrolysis temperature, it could be either acidic or alkaline. Low-temperature pyrolysis (<400 °C) usually produces acidic biochar while high-temperature pyrolysis (>600 °C) produces alkaline biochar.

In the inland Pacific Northwest (iPNW) of the USA, there is growing interest in biochar to remedy deteriorating soil conditions particularly in regions where winter wheat-summer fallow (WW-SF) has been practiced for the last 80 to 100 years. Soils under the WW-SF system have lost more than 60% of soil organic matter (SOM) in the top 0–30 cm depth profile. In this system, only one crop is grown in two years, and the resultant crop residues are inadequate to maintain or increase SOC. Growing cover crops or annual cropping could restore SOC, which is rather challenging for low precipitation zones of iPNW where WW-SF is practiced. In addition, the soils in iPNW acidified over time due to the continual use of ammoniacal N fertilizers with some soils now showing pH values as low as 4.6 to 4.8 in the top 30-cm depth [6,7].

The presence of biochemically recalcitrant and predominantly aromatic carbon in biochar is often attributed to its long-term C storage potential [8]. Therefore, applying biochar can quickly increase th total C pool in iPNW soils. In addition, biochar amendments supply phosphorus (P), potassium

(K), sulfur (S), and other trace minerals to the soil. Adding lime (CaCO$_3$) to the soil is the most recommended method to reduce soil acidity. However, the application of agricultural lime results in carbon dioxide (CO$_2$) emissions that contribute to global warming. For example, in 2001, about 20–30 Tg of lime was applied to soils in the US, resulting in net CO$_2$ emissions of 4.4–6.6 Tg [9]. Alkaline biochar with high liming value can be a substitute for lime in reducing soil acidity observed in iPNW soils without releasing excess CO$_2$. Liming reduces soil acidity and alleviates aluminum (Al) and iron (Fe) toxicity [10]. Amending acidic and low SOM soils of iPNW with alkaline biochar can increase total C, reduce soil acidity, improve soil health, and contribute to climate change mitigation [11].

Studies of biochar impacts on soil health and crop productivity have shown varied responses across soil types and management systems. Biochar application rates from 0.5 to 135 t ha^{-1} have produced plant growth responses ranging from −29% to 324% [12]. Plant and soil responses of biochar application also varied with agricultural systems, crop type, climatic conditions [13,14], and fertilization status [15]. Lone et al. [16] showed that biochar could affect soil N cycling and several transformation mechanisms, such as reduced inorganic N leaching by increasing nutrient retention due to cation and anion exchange reactions and immobilization of inorganic N due to labile C fractions of biochar. Biochar could also prevent nitrification and denitrification losses by increasing adsorption of NH$_4^+$ and NO$_3^-$. Especially, in fertilized systems with biochar amendments, there is a greater reduction in N loss and consequently, higher fertilizer use efficiency [16]. Yet, biochar and fertilizer interactions on plant production and soil health are inconclusive. Some authors have reported that treatments receiving both biochar and fertilizer increased fertilizer use efficiency by enhancing plant growth and soil N mineralization than in treatments receiving either [17–19]. However, a recent meta-analysis using 371 independent studies indicated no additive or synergistic relation between biochar and fertilizer [20]. The study also reported differences in the efficiency of fertilizer with varying rates of biochar, and reduced efficiency, especially at higher doses [15]. Besides, information on application rates for different crops and soils is lacking [2]. Clearly, site-specific research on the effects of biochar on soil properties and crop production are needed. Studies conducted under diverse soil nutrient management practices will assist producers to find optimal biochar rates for their soils to improve agricultural sustainability [16,21].

A greenhouse experiment was conducted with the objective of investigating the effect of different rates of wood biochar on soil properties and winter wheat growth in an iPNW soil. We evaluated (1) soil chemical and biochemical properties, and (2) winter wheat shoot and root growth using different biochar application rates in the presence and absence of chemical fertilizers.

2. Materials and Methods

2.1. Experimental Setup

A Walla Walla silt loam soil (coarse silty, mixed, superactive, mesic Typic Haploxerolls) [22] with 18% Clay, 70% silt, and 12% fine sand was collected from the top 20 cm depth of a WW-SF field at the Columbia Basin Agriculture Center (CBARC) near Pendleton, Oregon (45°42' N, 118°36' W, Elev. 438 m) for the study. A factorial randomized block design experiment with four replications consisting of four biochar rates (0, 11.2, 22.4, 44.8 Mg ha^{-1}) with and without fertilizer was established in the greenhouse at CBARC. Fertilizer treatments were equivalent to 90 kg N ha^{-1}, 45kg P ha^{-1} and 20 kg S ha^{-1}, the rate typically used for winter wheat production in the region. Air-dried soil, fertilizers and biochar were evenly mixed in a custom-build portable rotary cement mixer for 5 min and packed into 4 L capacity plastic pots (14 cm i.d. by 14 cm tall). Soil moisture was measured using Stevens Hydra probe (Stevens Water Monitoring Systems Inc., Portland, OR, USA) and soil water content was adjusted by adding deionized water in pots when the moisture was below 70% of the field capacity.

The biochar used in this study was a co-product of energy production in Philomath, OR (Biological Carbon, LLC, Philomath, Oregon, USA). The pyrolysis temperature was about 900 °C, and Douglas fir (*Pseudotsuga menziesii*) was the main feedstock for the biochar production. Biochar chemical characteristics were determined using the standard method for wood charcoal analysis [23] at Control Laboratories Inc. (Watsonville, CA, USA). Soil samples collected for the experiment were also analyzed for basic soil properties in the AgSource Laboratory (Umatilla, OR, USA). Analyses of biochar and soil properties are reported in Table 1.

Table 1. Physiochemical characteristics of Douglas fir biochar and soil used in the study.

Characteristics	Biochar	Soil
C	900 g kg^{-1}	6.4 g kg^{-1}
N	1.8 g kg^{-1}	0.6 g kg^{-1}
C:N	500:1	11:1
pH	10.6	4.8
Volatile matter	51 g kg^{-1}	-
Ash content	188 g kg^{-1}	-
Moisture	48 g kg^{-1}	-
EC	-	2.8 dS m^{-1}
CEC	-	3.5 cmol kg^{-1}
NO$_3$-N	-	271 mg kg^{-1}
NH$_4$-N	-	28.0 mg kg^{-1}
P	-	40.3 mg kg^{-1}
K	-	1105 mg kg^{-1}
S	-	30.5 mg kg^{-1}

Wheat seeds were pre-germinated in Petri-dishes on moistened paper towels in growth champers set at 4 °C for 48 h, and six pre-germinated seeds were sown in each pot. Seedlings were thinned to four per pot two weeks after sowing. The pots were placed in shallow trays and watered with deionized water to maintain field capacity throughout the 10 weeks of the experiment.

2.2. Plant and Soil Analysis

The whole wheat shoot biomass was harvested by cutting shoots at the soil surface using stainless steel scissors and weight to determine fresh weight and dried at 70 °C for 72 h to estimate dry matter content. Roots in the bulk soil were separated by passing soil through a 2 mm sieve. Roots passing through 2 mm sieves were separated by wet sieving 500 g sub-sample through 250 μm sieves for 20 min [24] and dried for 24 h at 70 °C to estimate dry mass.

Soil from each pot was homogenized, passed through a 2 mm sieve to separate the root and shoot residue and a 500 g sub-sample was collected and stored at 4 °C until analysis. Soil inorganic N [sum of NO$_3^-$ and NH$_4^+$] was determined by extracting 10 g soil sub-samples in 50 mL 1 M potassium chloride solution and analyzed in an Astoria Analyzer with micro-segmented flow analysis system (Astoria-Pacific Inc., Clackamas, OR, USA). Soil pH and electrical conductivity (EC) were determined in 1:2 soil to deionized water ratio (w/v) using a pH/Conductivity meter (Thermo Scientific™ Orion™ Star A215 pH and Conductivity Benchtop Meter). The rest of the soil sub-samples were sent to a commercial lab (AgSource Laboratories, Umatilla, OR) for the analysis of other soil properties. Soil total C and N concentrations were determined by combustion analysis (Flash EA 1112 series, Thermo Finnigan, San Jose, CA, USA) of soil that had been oven-dried (60 °C, 72 h) and finely ground for 2 min in Shatter 1 Box 8530 ball mill (Spex Sample Prep., Metuchen, NJ, USA). Soil pH was less than 6.5 in all soil samples; hence total C from these samples were considered as SOC.

2.3. Statistical Analysis

Statistical significance ($\alpha = 0.05$) for mean comparisons of variables was determined using a mixed model analysis of variance (Proc. Mixed) with two-way analysis term for factorial experiments (SAS ver. 9.4, SAS Institute, Cary, NC, USA). Plant growth parameters and soil properties were response variables, fertilizer and biochar treatments were explanatory variables, and replication was considered

3. Results

Biochar and fertilizer application had significant effects on shoot biomass, and only biochar had a significant effect on root biomass (Table 2). Biochar increased wheat shoot biomass by 15% to 20% (control: 5.1 g^{-1} pot) when applied at the rate of 11.2 and 22.4 Mg ha^{-1} without fertilizer addition; and by 17% to 18% (control: 5.5 g^{-1} pot) with fertilizer application (Figure 1a). The highest biochar application rate (44.8 Mg ha^{-1}) increased shoot biomass of wheat by 18% (6.45 g^{-1} pot) compared to the control in fertilized treatments. However, without fertilizer addition, only two to three plants survived in two out of four replications resulting in reduced shoot and root biomass in treatments with the highest biochar rate. Root biomass increased by 39 and 45% compared to the control (1.3 g^{-1} pot) without fertilizer addition, and by 19% and 24% with fertilizer application at the rate of 11.2 and 22.4 Mg ha^{-1} biochar application (Figure 1b).

Table 2. Analysis of variance (p value) of biochar and fertilizer effects on soil properties (pH, EC, OM, P, K, S, NO$_3$-N and NH$_4$-N) and plant biomass (shoot and root).

	pH	EC *	SOM	P	K	S	NO$_3$-N	NH$_4$-N	Shoot	Root
Biochar (B)	<0.0001	0.6073	0.0453	<0.0001	<0.0001	<0.0001	0.0009	0.0792	0.0204	0.0032
Fertilizer (F)	0.0004	0.0009	0.0689	0.0010	0.3876	0.0002	0.0145	0.0037	0.0067	0.96
B × F	0.0020	0.0889	0.0637	<0.0001	0.0006	0.0011	0.0326	0.0333	0.0566	0.26

* EC = electrical conductivity, SOM = soil organic matter, P = phosphorus, K = potassium, S = sulfur.

There were significant biochar and fertilizer interactions on many soil properties (Table 2). Th soil pH, P, K, and S concentration, increased with biochar rates (Figure 2). Without fertilizer addition, the soil pH increased by 1.1 units in the 44.8 Mg ha^{-1} biochar treatment compared to the control (5.3). Soil pH increased by 1.2 units (control: 5.1) at the same biochar rate when the fertilizer was added. There was a strong positive correlation among soil pH and P, K, and S concentrations regardless of fertilizer application (Table 3). Soil P was 18% to 50% greater with 11.2 to 44.8 Mg ha^{-1} biochar rates, respectively, than controls (33 mg P kg^{-1} soil) without fertilizer addition while it was 14% to 36% greater, respectively, with biochar than control (34 mg P kg^{-1} soil) in fertilized treatments (Figure 2). Sulfur was 93% to 380% greater in biochar treatments than control (18.8 mg S kg^{-1} soil) without fertilizer and 66 to 297% more than control (29 mg S kg^{-1}) with fertilizer application. Soil K was 16 to 55% greater in biochar treatments than control (888 mg K kg^{-1} soil) without fertilizer and 12% to 40% greater than control (951 mg K kg^{-1} soil) with fertilizer addition. The soil pH and nutrients were negatively correlated with soil NO$_3$ N and NH$_4$-N in treatments receiving fertilizer application, whereas, without fertilizer, soil pH and nutrients were positively correlated with soil NO$_3$-N and NH$_4$-N concentrations (Table 3).

The NO$_3$-N and NH$_4$-N contents decreased with increasing biochar rates in fertilized treatments (Figure 2). Without fertilizer addition, both NO$_3$-N and NH$_4$-N did not respond to increasing biochar rates except at the highest biochar rates. There was no significant biochar and fertilizer interaction on SOM content, but increasing biochar rates significantly increased SOM compared to the control (Table 2 and Figure 1c). The biochar application increased SOM content by 13% to 20%, and the highest increase was observed at 22.4 Mg ha^{-1} biochar rate. With fertilizer application, SOM was positively correlated with nutrients (P, K, and S) and negatively correlated with NO$_3$-N and NH$_4$-N whereas without fertilizer application there was no correlation of SOM with the nutrients (Table 3).

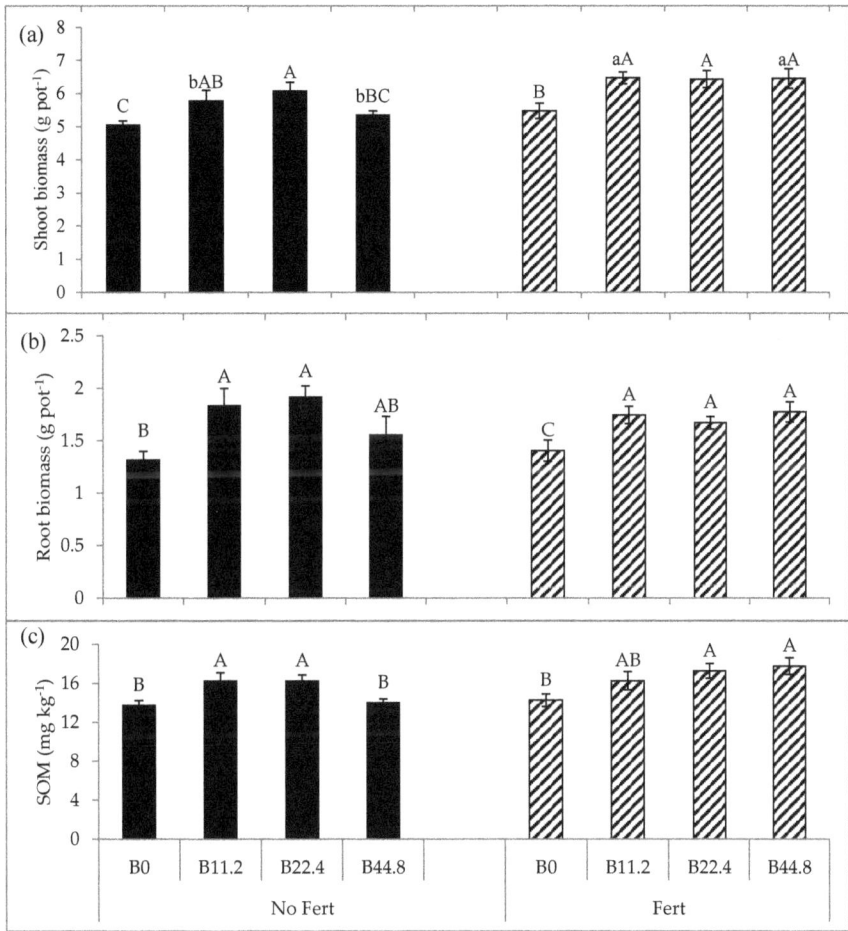

Figure 1. Effect of biochar and fertilizer treatments on (**a**) wheat shoot biomass, (**b**) wheat root biomass, and (**c**) soil organic matter (SOM). Upper case letters indicate significant difference among the different rate of biochar within fertilizer applied (Fert) and no fertilizer (No Fert) treatments, and lower case letters indicate significant difference within the same rate of biochar applied treatment with and without fertilizer.

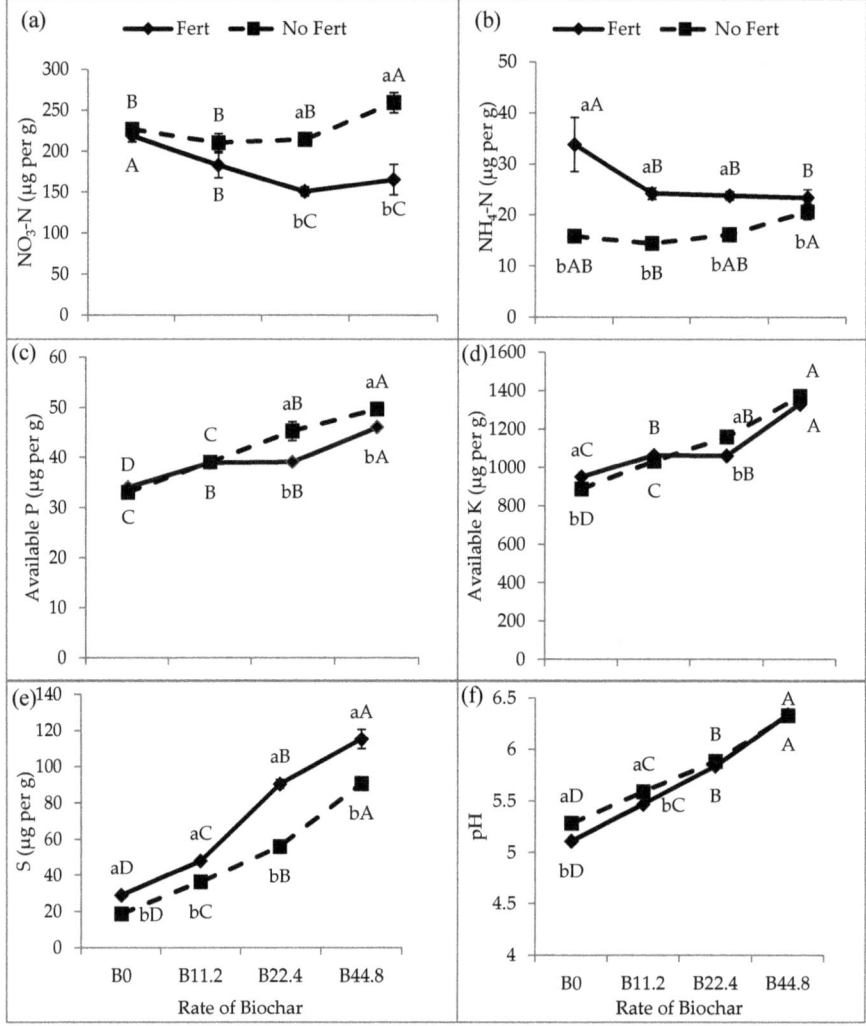

Figure 2. Effect of biochar and fertilizer treatments on soil properties (a) NO_3-N, (b) NH_4-N, (c) available P, (d) available K, (e) S and (f) pH. Upper case letters indicate significant differences among the different rate of biochar rates within fertilizer applied (Fert) and no fertilizer (No Fert) treatments, and lower case letters indicate significant differences within the same biochar treatment with and without fertilizer.

Table 3. Correlation among soil properties in biochar applied soils with and without fertilizer addition.

	EC	OM	P	K	S	NO$_3$-N	NH$_4$-N
			With Fertilizer application				
pH	−0.38	0.62	0.94 ***	0.93 ***	0.98 ***	−0.61 *	−0.51 *
EC		0.12	−0.32	−0.42	−0.38	0.06	−0.21
OM			0.62 **	0.54 *	0.62 **	−0.61 **	−0.70 **
P				0.96 ***	0.88 ***	−0.45	−0.53 *
K					0.86 ***	−0.40	−0.38
S						−0.64 **	−0.48
NO$_3$-N							0.56 *
			Without fertilizer application				
pH	−0.21	−0.02	0.96 ***	0.99 ***	0.99 ***	0.49 *	0.66 **
EC		−0.21	−0.24	−0.22	−0.20	0.13	0.28
OM			0.05	−0.01	−0.06	−0.37	−0.47
P				0.96 ***	0.94 ***	0.41	0.60 **
K					0.99 ***	0.56 *	0.61 **
S						0.59 **	0.63 **
NO$_3$-N							0.40

*, **, and *** indicate significant correlations at the 0.05, 0.01, and 0.001 probability levels, respectively.

4. Discussion

The alkaline forest biochar increased soil pH and SOM in all treatments compared to soil with no biochar, suggesting a potential for improving soil health with biochar application. Regardless of fertility, soil P, K, and S increased with increasing biochar rates indicating that either biochar contributed the nutrients or the increase in pH increased nutrient availability in the soil. Biochar is a source of several nutrients; its complex reaction with soil releases nutrients, making them available for plant uptake over time [20,25]. Availability of N, P, K, and S also increase with pH [12,26]. In this study, biochar increased soil pH by more than a unit from about 5 to 6 and there was a strong positive correlation among soil pH and P, K and S. The soil pH increase, however, was greater in treatments receiving fertilizer application than in the corresponding treatments with no fertilizer addition indicating that fertilizer application could enhance the biochar effect to reduce soil acidity. The increase in pH and the nutrients probably enhanced shoot and root growth.

Unlike other nutrients, application of biochar decreased soil NO$_3$-N and NH$_4$-N concentration. The reduction in NO$_3$-N and NH$_4$-N coincided with the increased wheat shoot and root biomass, suggesting that the reduction of these nutrients was due to increased plant uptake. The opposite was true when the soil was not fertilized. Soil NO$_3$-N and NH$_4$-N remained high at the highest biochar rates indicating the inability of plants to take up these nutrients. Subsequently, wheat shoot and root biomass were reduced under the highest biochar rate (44.8 Mg ha^{-1}) when the soil was not fertilized. Biochar used in this study had very high C:N ratio (500:1) that could have tied up some of N, thereby limiting its availability to plants. The reduced availability of inorganic N without fertilizer addition suggests the strong affinity of biochar for NH$_4$-N and NO$_3$-N [27,28]. Biochar N is mainly found in the heterocyclic compound, which is suggested to be resistant to microbial degradation hence not easily available for plant use [29,30]. Moreover, biochar also increases nutrient retention due to cation and anion exchange reactions and immobilizes inorganic N due to high C:N ratio (500:1) [16]. Therefore, in the treatment that received a high dose of biochar without fertilization, nutrient supply, especially soil available N was likely not enough to support plant survival.

The SOM was positively correlated with soil P, K, and S concentrations, and negatively correlated with NO$_3$-N and NH$_4$-N in fertilized soil; however there was no correlation in unfertilized treatments. These findings highlight the positive role of fertilization on the effectiveness of biochar in nutrient availability and plant uptake. Although SOM generally increased with an increase in biochar rates, it was lower in unfertilized than in fertilized treatments at the highest biochar rate. Death of plants at the highest biochar rates in treatments without fertilizer resulted in reduced shoot and root biomass which ultimately resulted in the low SOM accumulation. Barontia et al. [31] observed reduced durum wheat productivity biochar rates above 10 Mg ha^{-1}. Wheat death at a higher biochar rate implies that biochar application rates greater than 22.4 Mg ha^{-1} may not be beneficial for wheat biomass production in a Walla Walla silt loam soil. Overall, the alkaline biochar was effective in ameliorating soil acidity and increasing SOM, nutrient availability and uptake by plants, and plant shoot and root biomass. Further on-farm studies will enhance our understanding of the agronomic and environmental benefits of biochar on soil health and plant productivity.

5. Conclusions

Biochar amendment improved the soil chemical environment of the Walla Walla silt loam and wheat growth and has the potential to enhance agricultural sustainability in iPNW. Applying biochar at rates up to 22.4 Mg ha^{-1}, with and without fertilizer addition, increased wheat shoot and root biomass. However, at the highest rate of (44.8 Mg ha^{-1}), wheat biomass was reduced, especially when no fertilizer was added. Biochar positively influenced on soil properties, which ultimately affected plant production. Increasing biochar application rates significantly increased SOM, soil pH, availability of P, K, and S content; all factors critical for improving soil health. The increase in SOM was not solely due to C additions from the biochar but also from root biomass as indicated by the lowering of SOM at the highest biochar application rate without fertilization where plants died. It appears that the optimum biochar application rate for the Walla Walla silt loam is 22.4 Mg ha^{-1}. The reduction in soil NO_3-N with increase in biochar application rates was attributed to plant uptake under the fertilized treatment. Our study suggested that the addition of biochar could enhance soil health and wheat production. The biochar application can reduce soil acidity and improve wheat yield. However, because of the high C:N ratio, we recommend the addition of adequate fertilizer when amending the soil with biochar, particularly in low-fertility soil and at high biochar rates of application. Our greenhouse study generated positive preliminary results on soil and wheat responses to biochar application. However, field experiments are needed to measure and quantify long-term agronomic and environmental benefits of biochar on agricultural soils in the iPNW.

Author Contributions: P.B., R.G., and S.M. conceived and designed the experiment, P.B., R.G., and L.P. performed the experiment; P.B. analyzed the data and wrote the first draft of the manuscript; S.M. contributed reagents/materials/analysis tools and acquired funding; all authors revised the manuscript.

Funding: Please add: This research was funded by the Biological Carbon LLC, and the National Institute of Food and Agriculture, USA Department of Agriculture, under award number 2011-68002-30191.

Acknowledgments: The authors would like to acknowledge the support by staffs of Columbia Basin Agricultural Research Center.

Conflicts of Interest: The authors declare no conflict of interest. The funders had no role in the design of the study; in the collection, analyses, or interpretation of data; in the writing of the manuscript, or in the decision to publish the results.

References

1. White, E.M. *Woody Biomass for Bioenergy and Biofuels in the United States—A Briefing Paper*; General Technical Report PNW-GTR-825; U.S. Department of Agriculture, Forest Service, Pacific Northwest Research Station: Portland, OR, USA, 2010; p. 45.
2. Lehmann, J.; Joseph, S. *Biochar for Environmental Management: Science and Technology*; Earthscan: London, UK, 2009.
3. Bruun, E.W.; Hauggaardd-Nielsen, H.; Ibradhim, N.; Egsgaard, H.; Ambus, P.; Jensen, A.P.; Dam-Johansen, K. Influence of fast pyrolysis temperature on biochar labile fraction and short-term carbon loss in a loamy soil. *Biomass Bioenergy* **2011**, *35*, 1182–1189. [CrossRef]

4. De la Rosa, J.M.; Paneque, M.; Miller, A.Z.; Knicker, H. Relating physical and chemical properties of four different biochars and their application rate of biomass production of *Lolium perenne* on a Calcic Cambisol during a pot experiment of 79 days. *Sci. Total Environ.* **2014**, *499*, 175–189. [CrossRef]
5. Novak, J.M.; Busscher, W.J.; Laird, D.L.; Ahmedna, M.; Watts, D.W.; Niandou, M.A.S. Impact of biochar amendment on fertility of a Southeastern coastal plain soil. *Soil Sci.* **2009**, *174*, 105–112. [CrossRef]
6. Brown, T.T.; Huggins, D.R. Soil carbon sequestration in the dryland cropping region of the Pacific Northwest. *J. Soil Water Conserv.* **2012**, *67*, 406–415. [CrossRef]
7. Ghimire, R.; Machado, S.; Bista, P. Soil pH, soil organic matter, and crop yields in winter wheat-summer fallow systems. *Agron. J.* **2017**, *109*, 706–717. [CrossRef]
8. Keith, A.; Singh, B.; Singh, B.P. Interactive priming of biochar and labile organic matter mineraliztion in smecitie-rich soil. *Environ. Sci. Technol.* **2011**, *45*, 9611–9618. [CrossRef] [PubMed]
9. West, T.O.; McBride, A.C. The contribution of agricultural lime to carbon dioxide emissions in the United State: Dissolution, transport and net emissions. *Agric. Ecosyst. Environ.* **2005**, *108*, 145–154. [CrossRef]
10. Major, J.; Rondon, M.; Molina, D.; Riha, S.J.; Lehmann, J. Maize yield and nutrition during 4 years after biochar application to a Colombian savanna oxisol. *Plant Soil* **2010**, *333*, 117–128. [CrossRef]
11. Brown, S.; Carpenter, A.; Beecher, N. A calculator tool for determining greenhouse gas emissions for processing and end use. *Environ. Sci. Technol.* **2010**, *44*, 9509–9515. [CrossRef]
12. Glaser, B.; Lehmann, J.; Zech, W. Ameliorating physical and chemical properties of highly weathered soils in the tropics with charcoal—A review. *Biol. Fertil. Soils* **2002**, *35*, 219–230. [CrossRef]
13. Gundale, M.J.; DeLuca, T.H. Charcoal effects on soil solution chemistry and growth of Koeleria macrantha in the ponderosa pine/Douglas-fir ecosystem. *Biol. Fertil. Soils* **2007**, *43*, 303–311. [CrossRef]
14. Unger, R.; Killorn, R.; Brewer, C. Effects of soil application of different biochars on selected soil chemical properties. *Commun. Soil Sci. Plant* **2011**, *19*, 2310–2321. [CrossRef]
15. Haefele, S.M.; Konboon, Y.; Wongboon, W.; Amarante, S.; Maarifat, A.A.; Pfeiffer, E.M.; Knoblauch, C. Effects and fate of biochar from rice residues in rice-based systems. *Field Crops Res.* **2011**, *121*, 430–440. [CrossRef]
16. Lone, A.H.; Najar, G.R.; Ganie, M.A.; Sofi, J.A.; Ali, T. Biochar for Sustainable Soil Health: A Review of Prospects and Concerns. *Pedosphere* **2015**, *25*, 639–653. [CrossRef]
17. Albuquerque, J.A.; Salazar, P.; Barron, V.; Torrent, J.; del Campillo, M.D.; Gallardo, A.; Villar, R. Enhanced wheat yield by biochar addition under different mineral fertilization levels. *Agron. Sustain. Dev.* **2013**, *33*, 475–484. [CrossRef]
18. Chan, K.Y.; van Zwieten, L.; Meszaros, I.; Downie, A.; Joseph, S. Agronomic values of greenwaste biochar as a soil amendment. *Aust. J. Soil Res.* **2007**, *45*, 629–634. [CrossRef]
19. Steiner, C.; Teixeira, W.G.; Lehmann, J.; Nehls, T.; de Macedo, J.L.V.; Blum, W.E.H.; Zech, W. Long term effects of manure, charcoal and mineral fertilization on crop production and fertility on a highly weathered Central Amazonian upland soil. *Plant Soil* **2007**, *291*, 275–290. [CrossRef]
20. Biederman, L.A.; Harpole, W.S. Biochar and its effects on plant productivity and nutrient cycling: A meta-analysis. *GCB Bioenergy* **2013**, *5*, 202–214. [CrossRef]
21. Jeffery, S.; Verheijen, F.G.A.; van der Velde, M.; Bastos, A.C. A quantitative review of the effects of biochar application to soils on crop productivity using meta-analysis. *Agric. Ecosyst. Environ.* **2011**, *144*, 175–187. [CrossRef]
22. Soil Survey Staff. Web Soil Survey. Natural Resources Conserv, Serv., USDA, 2016. Available online: http://websoilsurvery.sc.gov.usda.gov (accessed on 5 December 2016).
23. ASTM International. *ASTM D1762-84. 2013. Standard Test Method for Chemical Analysis of Wood Charcoal*; ASTM International: West Conshohocken, PA, USA, 2013.
24. Ghimire, R.; Norton, J.B.; Pendall, E. Alfalfa-grass biomass, soil organic carbon, and total nitrogen under different management approaches in an irrigate agroecosystem. *Plant Soil* **2014**, *374*, 173–184. [CrossRef]
25. Lehmann, J.; Kuzyakov, Y.; Pan, G.; Ok, Y.S. Biochar and the plant-soil interface. *Plant Soil* **2015**, *395*, 1–5. [CrossRef]
26. Chintala, R.; Mollinedo, J.; Schumacher, T.E.; Malo, D.D.; Julson, J.L. Effect of biochar on chemical properties of acidic soil. *Arch. Agron. Soil Sci.* **2013**, *60*, 393–404. [CrossRef]
27. Atkinson, C.J.; Fitzgerald, J.D.; Hipps, N.A. Potential mechanisms for achieving agricultural benefits from biochar applications to temperate soils: A review. *Plant Soil* **2010**, *337*, 1–18. [CrossRef]

28. Lehmann, J.; da Silva, J.P.; Seiner, C.; Nehls, T.; Zech, W.; Glaser, B. Nutrient availability and leaching in an archaeological anthrosol and a Ferralsol of the central Amazon basin: Fertilizer, manure and charcoal amendments. *Plant Soil* **2003**, *249*, 343–357. [CrossRef]
29. Schmidt, M.W.I.; Noack, A.G. Black carbon in soils and sediments: Analysis, distribution, implications, and current challenges. *Glob. Biogeochem. Cycles* **2000**, *14*, 777–793. [CrossRef]
30. Knicker, H. How does fire affect the nature and stability of soil organic nitrogen and carbon? A review. *Biogeochemistry* **2007**, *85*, 91–118. [CrossRef]
31. Baronti, S.; Alberti, G.; Vedove, G.D.; Gennaro, F.D.; Fellet, G.; Genesio, L.; Migletta, F.; Peressotti, A.; Vaccari, F.P. The biochar option to improve plant yields: First results from some field and pot experiments in Italy. *Ital. J. Agron.* **2010**, *5*, 3–12. [CrossRef]

© 2019 by the authors. Licensee MDPI, Basel, Switzerland. This article is an open access article distributed under the terms and conditions of the Creative Commons Attribution (CC BY) license (http://creativecommons.org/licenses/by/4.0/).

Article

Understanding Activation Effects on Low-Temperature Biochar for Optimization of Herbicide Sorption

Beatriz Gámiz [1,*], Kathleen Hall [2], Kurt A. Spokas [3] and Lucia Cox [1]

1. Instituto de Recursos Naturales y Agrobiología de Sevilla, Consejo Superior de Investigaciones Científicas (IRNAS-CSIC), Reina Mercedes, Av. 10, 41012 Seville, Spain; lcox@irnase.csic.es
2. Department of Soil, Water, and Climate, University of Minnesota, 1991 Upper Buford Cir., St. Paul, MN 55101, USA; Kathleen.Hall@state.mn.us
3. United States Department of Agriculture—Agricultural Research Service, St. Paul, MN 55101, USA; kurt.spokas@usda.gov
* Correspondence: bgamiz@irnase.csic.es; Tel.: +34-954624711 (ext. 208197)

Received: 4 August 2019; Accepted: 22 September 2019; Published: 27 September 2019

Abstract: Activation treatments are often used as a means of increasing a biochar's sorption capacity for agrochemical compounds but can also provide valuable insight into sorption mechanisms. This work investigates the effects of H_2O_2 activation on a low-temperature (350 °C) grape wood biochar, evaluates subsequent changes to the removal efficiency (RE) of cyhalofop and clomazone, and elucidates potential sorption mechanisms. Activation by H_2O_2 decreased the biochar pH, ash content, and C content. Additionally, the biochar O content and surface area increased following activation, and Fourier transform infrared spectroscopy (FTIR) data suggested a slight increase in surface O groups and a decrease in aliphatic C. Cyhalofop RE significantly increased following activation, while clomazone RE was unchanged. The increased sorption of cyhalofop was attributed to pH effects and charge-based interactions with biochar O moieties. Results from this study suggest that H_2O_2 activation treatments on low-temperature biochars may improve the removal of organic acid herbicides but are of little value in optimizing the removal of polar, non-ionizable herbicides.

Keywords: activated charcoal; aging; pesticides

---

1. Introduction

There is a growing interest in biochar as sustainable soil amendment due to the agronomical benefits derived from its use. Lately, biochar has attracted attention for showing sorbent properties towards a wide variety of agrochemical compounds. By binding chemicals such as pesticides, biochar can help prevent and remediate contamination in soil and water [1,2] acting, among others, as a soil ameliorant tool. To further enhance the sorption capacity of biochars, activation processes akin to those used in the production of activated charcoal are increasingly being adopted [3,4]. These activation treatments may be particularly useful for improving herbicide sorption capacities of low-temperature biochar (< 400 °C), which are favored for soil fertility applications, but have relatively low sorption capacities. However, for biochar activation to be effective, it is necessary to understand which biochar physicochemical properties are affected, and whether or not these changes influence the binding of the target compound.

Activation treatments can be physical, chemical, or thermal in nature. For example, biochars can be activated through grinding or ball milling (physical activation), treating with acids, bases, and oxidizers [e.g., hydrogen peroxide (H_2O_2)] (chemical activation), or re-pyrolyzing at a higher temperature than the production temperature (thermal activation). Typically, the goal of these post-production treatments is to increase biochar's specific surface area (SSA) and introduce reactive functional groups which can increase the material's adsorption capacity [5].

Generally, changes in biochar SSA and O functionality are credited for the increases in adsorption following activation [6]. However, to predict whether such changes will increase sorption, it is key to understand which adsorption mechanisms dominate. A variety of adsorption mechanisms have been proposed to describe interactions between biochar and organic compounds including hydrophobic interactions, electrostatic surface complexation, ion exchange, hydrogen bonding, π–π interactions, co-precipitation, inner-sphere complexation, and the formation of charge-transfer metal complexes [7,8]. The degree of adsorption, as well as the mechanisms involved, depends on the interactions between the properties of the biochar, the chemical involved, as well as the solution chemistry. Biochar properties cited to influence the binding of organic chemicals, include porosity, SSA, pH, and the presence of specialized functional groups [2]. With regard to the sorbate, molecular size, solubility, and its potential to ionize are a few of the well-known influential characteristics [9].

Adsorption mechanisms are often inferred from correlations between biochar physicochemical properties (e.g., SSA; aromaticity; O:C; ash content) and sorption capacity. Many studies rely on the variability in the biochar physicochemical properties that arises from differences in feedstock materials or changes in pyrolysis temperatures [10]. However, it is difficult to discern the importance of individual parameters by this method, since each pyrolysis system possesses its own thermal transfer properties and thereby confers differential chemical and sorptive properties to the biochar [11]. Additionally, it is challenging to ascertain the influence of biochar surface moieties when the bulk chemical composition and physical structure differ (e.g., such as differential dissolution of carbonate/ash material from biochar). However, some activation treatments can function as an improved strategy to deduce adsorption mechanisms by altering limited properties of a single biochar and creating a more direct comparison [12]. Furthermore, few studies have used modified low-temperature biochars as a tool to elucidate biochar-herbicide binding mechanisms [13], and useful insight may be gained through the application of a chemical activation-based approach. Although many works have reported the sorption mechanisms of biochar prepared at low temperature, compared to high-temperature biochars, the effects of activation on low-temperature materials are less clearly defined. Changes to the biochar properties depend largely on what activation treatment is used, as well as the original biochar material. To our knowledge, little has been reported dealing with the activation of biochars produced at low temperature, i.e., 350 °C, with H_2O_2, since other authors studied different activation (chemical or physical) methods and other pyrolysis temperatures. For instance, Liu et al. [14] observed an increase in the SSA of a rice husk biochar (500 °C) following potassium hydroxide, KOH, activation at 65 °C, with π–π interactions responsible of the adsorption mechanism of tetracycline on biochar. More recently, Xiao and Pignatello [12] reported an increase in SSA with thermal air oxidation of a maple wood biochar (400 °C), together with an increase in carboxyl and phenolic groups, reporting that the main adsorption mechanisms for some ionizable compounds on activated chars were H-bonding and pore filling.

Finally, the activation has also been used to simulate and predict the long-term behavior of biochars when they are added to soil, since their surfaces undergo changes that can modify their sorption properties and H_2O_2 has been proposed to this aim [15]. Mia et al. [16] proposed the activation of a biochar prepared at 550 °C with H_2O_2 to predict the long-term behavior of biochars to sorb nutrients, phosphate, and ammonium. Coadsorption and pore filling were the mechanisms proposed after the chemical activation. Regarding pesticides, this knowledge is still inconsistent; in a previous work, we found that three highly persistent ionizable pesticides had a completely different sorption behavior over time on the same aged biochar and differences were attributed to the water-soluble fraction [17].

In this work, the activation of a grape biochar produced at 350 °C was performed by its treatment with H_2O_2. H_2O_2 is a desirable means of activation because it is relatively inexpensive, decomposes to H_2O and O_2, is effective at ambient temperatures, and does not introduce foreign elements into the biochar structure. We hypothesized that the activation of the biochar prepared at 350 °C with H_2O_2 could modify its surface affecting differently to its adsorption capacity towards ionizable and polar herbicides. Data reported here would be useful to prepare customized biochars to optimize the behavior of pesticides in agricultural soils, which is the critical step for the wide-spread use of biochar as soil amendment. Hence, the objectives of this study were to (1) assess the effects of H_2O_2 activation on a low-temperature grape wood biochar, (2) evaluate the impact of activation on the sorption of two herbicides, one anionic, cyhalofop, and other neutral and polar, clomazone, and (3) elucidate potential sorption mechanisms.

2. Materials and Methods

2.1. Chemicals and Biochars

Two herbicides with distinct chemistries were selected for comparison in this study (Figure 1). Cyhalofop-butyl (butyl(2R)-2-[4-(4-cyano-2-fluorophenoxy) phenoxy]propionate) is a post-emergence aryloxyphenoxy propionic acid herbicide used in dry-seeded, water-seeded, and transplanted rice [18]. In soil, cyhalofop-butyl is rapidly transformed into its acid form, which exhibits lower sorption and higher water solubility [19]. Therefore, cyhalofop acid ((2R)-2-[4-(4-cyano-2 fluorophenoxy) phenoxy] propanoic acid) (pKa = 3.9 at 25 °C) was used in this study, and is henceforth referred to simply as cyhalofop. Clomazone [2-(2-chlorophenyl) methyl-4,4-dimethyl-3-isoxazolidinone] is a nonionizable pre- or post-emergence isoxazolidinone herbicide used in rice production, as well as in soybean, peas, and cotton cropping systems.

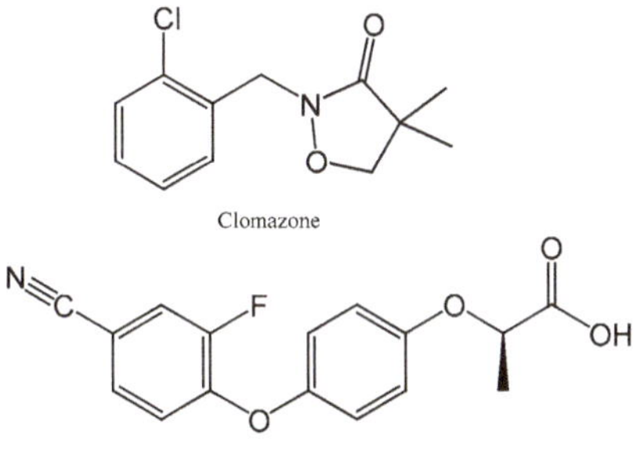

Figure 1. Chemical structures of clomazone and cyhalofop herbicides.

Analytical standards of cyhalofop and clomazone were purchased from Sigma-Aldrich (St. Louis, MO) for use in the sorption experiments. Solutions were prepared at a concentration of 1 mg L^{-1} in deionized (DI) water for each herbicide. Two additional cyhalofop solutions were prepared at equivalent concentrations (1 mg L^{-1}) with 0.01 M $CaCl_2$, and 0.01 M KCl to evaluate the influence of different salts on the sorption of this ionic compound.

Biochars used in this study were produced from grape wood pruning waste. Pyrolysis was carried out at temperatures of 350 °C (G-350), 500 °C (G-500), and 900 °C (G-900) using a Lindberg bench furnace equipped with a gas-tight retort (Lindberg/MPH, Riverside, MI) under a constant nitrogen inert gas flow. All biochars were held at their maximum temperature for two hours. Biochar was then cooled in the system (still under nitrogen purge) before being exposed to the atmosphere. The biochars were stored in sealed plastic bags until used. Biochar properties appear in Table 1.

Table 1. Biochars properties.

Feedstock	Pyrolysis Temperature (°C)	Moisture %	Ash %	VM [1] %	C %	H %	N %	O %	SSA [2] ($m^2 g^{-1}$)
Grape wood	350	3.5	10.9	39.5	66.6	4.0	1.1	17.5	<1
Grape wood	500	4.0	16.8	19.3	70.4	2.3	0.9	9.6	<1
Grape wood	900	1.3	22.2	6.6	71.6	0.1	1.0	4.9	124

[1] Volatile matter; [2] Specific surface area determined by N_2 sorption.

Activation of the G-350 biochar was carried out by treatment with H_2O_2. Approximately 10 g of G-350 biochar was immersed in 50 mL of a 3% H_2O_2 solution and allowed to react for a period of 24 h at room temperature (22 °C). This biochar was selected for being prepared at low temperature, which was expected to maintain the agronomic benefits described for these types of biochars as compared to those prepared at higher pyrolysis temperature [20]. The 3% H_2O_2 solution was used to minimize the amount of oxidizing agent while the properties of the biochar are modified, according to previously published data [21]. The biochar was then rinsed with DI water and oven dried (110 °C) prior to use. Activation was carried out shortly after the biochar was received (<2 weeks after production). Henceforth, the activated material will be referred to as the G-350 H_2O_2.

2.2. Biochar Characterization-G-350 Versus G-350 H_2O_2

Bulk biochar properties including C, N, H, and ash content were analyzed by Micro-analysis, Inc. (Wilmington, DE). All values were reported on a % dry weight, ash-free basis. Oxygen content was calculated by difference (O=100–C–N–H–ash).

Biochars were imaged using a Hitachi S3500N Variable Pressure Scanning Electron Microscope (SEM) (Toyko, Japan) to visually inspect their physical structure. To reduce sample charging and improve image resolution, biochar samples were coated with a 2 nm layer of metal (60% Au, 40% Pd) using a Cressington 108 Auto Sputter Coater (Watford, England).

Fourier transform infrared spectroscopy (FTIR) was used to assess the biochar surface functional groups. FTIR was performed using a Jasco FT/IR 6300 spectrometer (Jasco Europe s.r.l.) equipped with Mercury-cadmium-telluride (MCT) detector fitted with an attenuated total reflection (ATR) accessory (MIRacle™ Single Reflection ATR from Pike Technologies, WI, USA). The spectral range used ranges between 4000 and 580 cm^{-1}, with nominal resolution of 2 cm^{-1} and 15–20 numbers of scans. Biochar samples were oven dried and ground prior to analysis.

Zeta potential was measured to evaluate electrostatic characteristics of biochar particles suspended in solution. Analysis was performed on G350 biochar with and without activation in suspensions of DI water and 0.01 M $CaCl_2$ (40 mg: 8 mL) by dynamic light scattering using a Zetasizer Nano ZS Analyzer (Malvern Instrument, Malvern, United Kingdom).

2.3. Water Vapor and Herbicide Sorption

Water vapor sorption on biochars was measured following the methods of McDermot and Arnell [22] and Medic et al. [23] to evaluate the accessible surface area and surface energy. Biochars were placed in sealed chambers held at different relative humidities using salt solutions [24]. The biochars were periodically removed and weighed until equilibrium was established, then total water sorption was determined by mass difference. Water vapor sorption data were fit to a linear Brunauer-Emmett-Teller (BET) model from which SSA [1] and the isoteric heat of adsorption (Ea) [2] were calculated.

The specific surface area, SSA; is estimated from the following relationship:

$$SSA = \left(\frac{n_m N \sigma}{m}\right) \qquad (1)$$

where n_m = moles H_2O in monolayer, N = Avogadro's number (6.02 × 10^{23}), σ = surface area per H_2O molecule (10.8 × 10^{-20} m^2), and m = mass of sorbent (g). The isoteric heat of adsorption (E_a) is calculated by the following:

$$E_a = \ln(c)RT \qquad (2)$$

where c = BET constant, R = gas constant (8.314 × 10^{-3} kJ K^{-1} mol^{-1}), and T = temperature (K).

Herbicide sorption measurements were performed in triplicate using the batch equilibration method [25]. Biochar (40 mg) and 1 mg L^{-1} herbicide solution (8 mL) were combined in glass centrifuge tubes with Teflon lined caps (preliminary studies showed neither herbicide sorbed on the centrifuge tubes) [26,27]. After shaking in an end-over-end shaker (30 r.p.m.) at 20 ± 2 °C for 24 h, samples were centrifuged (5000 rpm, 10 min), and 4 mL of supernatant were removed and filtered (0.45 µm), discarding the first 2 mL. The remaining 2 mL of filtered solution were transferred to 2 mL amber screw top vials for analysis by high performance liquid chromatography (HPLC), as detailed below.

Herbicide sorption on the biochar materials was evaluated based on removal efficiency (RE), calculated as:

$$RE = \frac{C_i - C_e}{C_i} \times 100 \qquad (3)$$

where C_i = initial herbicide solution concentration (mg L^{-1}) and C_e = solution concentration at equilibrium (mg L^{-1}). The pH of the supernatant was measured directly in the sample tube at the end of the experiment. The solution pH was not adjusted or buffered during sorption measurements to fully account for the effects of activation and avoid further changes to the surface chemistries of the materials.

The sorption-desorption isotherms were also obtained for clomazone and cyhalofop on untreated and H_2O_2 activated G-350 biochar. For this purpose, in triplicate, 40 mg of each biochar were equilibrated with 8 mL of different solution of each herbicide with initial concentration ranging between 0.2 and 5 mg/L for 24 h at 20 ± 2 °C. Then, the suspensions were centrifuged and the supernatants were removed, filtered, and analyzed by HPLC. Sorption isotherms were fitted to the Freundlich equation:

$$C_s = K_f C_e^{N_f} \qquad (4)$$

where C_s = amount of herbicide sorbed (mg kg^{-1}) and C_e = solution concentration at equilibrium (mg L^{-1}). K_f and N_f are the empirical Freundlich constants calculated from its linearized form.

2.4. HPLC Analysis

Both cyhalofop and clomazone were measured by HPLC using a C-18 column (Kinetex C-18) in a Waters 600E chromatograph (Milford, MA) coupled to a diode array detector (Waters 996). The UV absorbance was monitored at 233 and 230 nm for cyhalofop and clomazone, respectively. The mobile phase was 50:50 (v/v) acetonitrile/water adjusted to pH 2.0 with phosphoric acid (H_3PO_4) for cyhalofop, and 35:65 (v/v) methanol/water for clomazone, with a 1 mL min^{-1} flow and 25 µL injection volume. Instrumental limit of detection (LOD) was calculated as the lowest observable concentration giving a signal-to-noise (S/N) ratio of 3:1, while instrumental limit of quantification (LOQ) was calculated as the concentration resulting in an S/N ratio of 10:1. The LOD and LOQ were 0.01 mg L^{-1} and 0.03 mg L^{-1}, respectively, for both herbicides.

2.5. Statistical Analyses

Statistical evaluations of sorption based on RE values are detailed below for the various comparisons. All data are the means of triplicate samples with the exception of cyhalofop sorption in

0.01 M KCl which was conducted in duplicate. The effects of H_2O_2 activation on RE were evaluated for each herbicide using t-tests ($p < 0.05$) comparing individual herbicide-solution pairs (e.g., RE of cyhalofop on G350 versus G-350 H_2O_2 in H_2O). The effects of variable background solutions on cyhalofop RE were evaluated by analysis of), variance (ANOVA and means were compared using Tukey's honest significant difference (HSD) test ($p < 0.05$). The effects of biochar production temperature on sorption for each herbicide were also evaluated based on ANOVA, and if statistically significant differences existed, the means were further compared using Tukey's HSD test ($p < 0.05$). Additionally, mean RE values of the two herbicides were analyzed pair-wise for each temperature using a series of t-tests to determine statistically significant differences ($p < 0.05$). All statistical analyses were performed in R (Version 3.3.2). Figures were plotted using Sigma Plot® (Version 13.0 for Windows, 2014, Systat Software Inc., Point Richmond, CA, USA).

3. Results and Discussion

3.1. Biochar Activation Observations

Vigorous, rapid bubbling was observed upon the addition of H_2O_2 to the biochar, consistent with observations of effervescence by Lawrinenko et al. [21]. It is unclear whether the bubbling was due to reactions with the carbon structure of the material [28], or simply the decomposition of H_2O_2 catalyzed by metal oxides in the biochar ash component [29], or a combination of the two mechanisms. A variety of reactions have been reported following the exposure of biochar to H_2O_2, depending on the biochar properties [30] and activation conditions (e.g., pH, presence of ferrous salts [31]).

Once dry, the H_2O_2 activated G-350 biochar was noticeably lighter in color, shifting from a dark gray to a brownish hue (not shown). Alkaline H_2O_2 is commonly used in the bleaching of wood pulp and appears to have similar effects on low-temperature biochar. The bleaching effects of H_2O_2 on wood pulp have been attributed to the removal of chromophores and the breakdown of lignin [32]. However, the presence of transition metal ions is known to diminish the bleaching process by directly degrading the H_2O_2. Because bleaching effects are curtailed by metal-catalyzed degradation of H_2O_2 [32], the observed lightening of the biochar itself suggests that the H_2O_2 was not solely degraded by metals associated with the ash component.

The bulk composition of the G-350 biochar showed the following changes with activation: carbon content decreased from 62.7% to 59.2%, whereas nitrogen content increased from 1.06% to 1.15%, and oxygen content increased from 32.4% to 36.6% following H_2O_2 activation. The ash content of the biochar decreased from 10.5% to 3.5% following H_2O_2 activation. While other studies have observed a similar decrease in biochar carbon content with H_2O_2 treatments [16,30], changes may be affected by the source material. Huff and Lee [21] reported no changes in the bulk structure of a pinewood biochar (400 °C) with H_2O_2 treatments ranging from 1% to 30% H_2O_2. In the present study, the measured decrease in carbon could be due to either a loss of inorganic carbonates or organic carbon in the form of labile aliphatic groups. The loss of carbon can lead to the concentration of other elements. Because no new nitrogen atoms should be directly introduced by the H_2O_2 treatment, the increase in N suggests some degree of concentration due to mass loss. Furthermore, the reduction in the ash content could contribute to the concentration of N. In contrast, external O may be introduced on the biochar by activation, and the loss of ash may decrease the O content through the loss of oxides and carbonates. Therefore, bulk elemental analyses alone are not enough to confirm changes in the oxygen functionality of the material.

SEM images of G-350 (a and b) and G-350 H_2O_2 (c and d) are shown in Figure 2. Compared to the untreated biochar, the G-350 H_2O_2 has more open pores. The uncovering or opening of these pores likely resulted from the degradation of the more labile carbon (i.e., aliphatic groups) at the biochar surface by the H_2O_2. The oxidative degradation of carbon structures by H_2O_2 is a well-known reaction; for example, soils are often treated with H_2O_2 to remove organic matter [33]. The opening of pores via the removal of labile C is also supported by the decrease in C content following activation (62.7% to 59.2%).

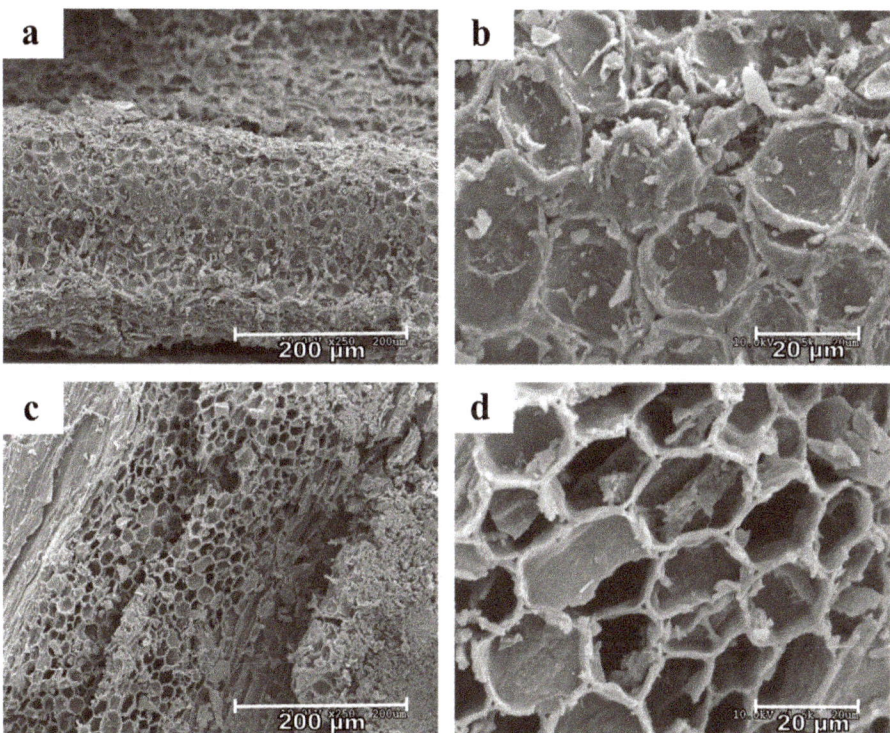

Figure 2. Scanning Electron Microscope (SEM) images of (**a**) the untreated G-350 (250×); (**b**) the untreated G-350 (1.5k×); (**c**) G-350 H_2O_2 (250×); (**d**) G-350 H_2O_2 (1.5k×.) biochars.

ATR-FTIR spectra of H_2O_2 activated G-350 biochar, G-350, G-500 and G-900 are shown in Figure 3. The following peak identifications correspond with those reported by Li et al. [34]. The broad band between 3200 and 3600 cm^{-1} represents the O–H stretch from adsorbed water, while the peaks at 2921, 2855, and 1419 cm^{-1} result from aliphatic C–H. The presence of C=C and C=O stretching in aromatic rings are responsible for the peak at 1616 cm^{-1}, and the peak at 1317 cm^{-1} was attributed to aromatic CO– and phenolic –OH groups. The peaks at 1018, 872, and 777 cm^{-1} likely result from the presence of silicate minerals (Si–O–Si) in the biochar samples, which is not surprising given that the G-350 biochar is over 10% ash. Alternatively, the peak at 777 cm^{-1} has been attributed to aromatic C–H stretching in biochar samples [21], and the peak at 872 cm^{-1} has been attributed to aromatic C–H as well as C–O from carbonates [35].

While the overall differences between the G-350 spectra before and after activation appear to be minor, slight changes in O functionality and aliphatic groups are evident. The G350 H_2O_2 spectrum suggests an increase in O-containing functional groups, most notably at 1310 and 1616 cm^{-1}. A similar increase in O-groups was reported by Huff and Lee [27] following H_2O_2 activation of a 400 °C pinewood biochar at room temperature, as was an increase in the 775 cm^{-1} peak. Huff and Lee [27] proposed that the increase at 775 cm^{-1} represented an increase in C–H stretching, possibly resulting from the opening of aromatic rings. In contrast, the decreased intensity of the aliphatic bands near 2900 cm^{-1} observed in the present data suggests a decrease in aliphatic functional groups, which is consistent with their degradation by reactions with H_2O_2. Regarding the biochars prepared at 500 °C and 900 °C, there was a decreased in the intensity of the O–H stretching of the hydroxyl groups at 3424 cm^{-1}, due to the loss of hydration, and the C–H stretching of the aliphatic vibration groups (2921–2855 cm^{-1}). The disappearance of the phenolic –OH and aromatic CO also suggest a loss of functionality in these materials, as has been published previously [36,37], which would explain the observed sorption behavior.

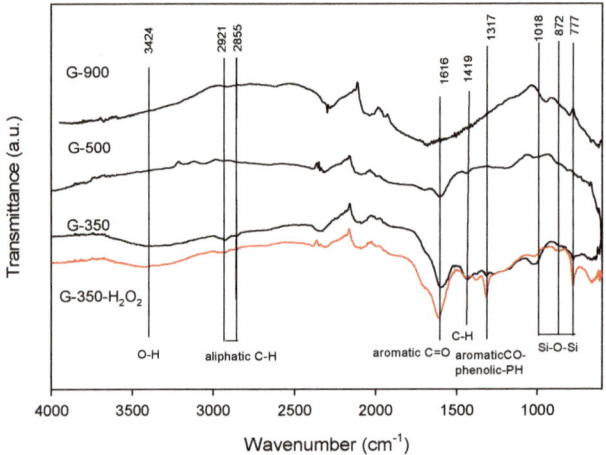

Figure 3. Fourier transform infrared spectroscopy (FTIR)-attenuated total reflection (ATR) spectra of H$_2$O$_2$ activated G-350 biochar, G-350, G-500, and G-900.

Biochar produced at low temperatures generally have a variety of surface functional groups compared to high-temperature biochars, which more closely resemble aromatic graphitic carbon [35]. Limited decarbonylation and decarboxylation reactions take place below 350 °C, resulting in the presence of more polar surface groups, aliphatic groups, and a more amorphous character [10]. Typically, oxidative activation treatments add O to higher temperature biochars or charcoals in the case of activated carbon, which have limited functionality (i.e., flat FTIR spectra); therefore, the increase in O is clearly observable. However, it is unclear whether or not a similar oxygenation process is taking place on this low-temperature material with many pre-existing O groups. The FTIR data do not provide adequate evidence to determine whether the observed increase in the O-group peaks is from the addition of O (i.e., chemisorption of O on defect sites [30]), or if these peaks sharpen due to the removal of the aliphatic C groups (i.e., uncovering pre-existing atoms or concentrating their signal).

A key challenge in interpreting biochar FTIR spectra is that the peaks are the sum of interactions of difference groups, and the information is not often quantitative [38]; the magnitude of changes to biochar that were observed are not known. Furthermore, bands of mineral components can overlap the typical C–O and C–H peaks associated with the organic phase [39], which can lead to misinterpretations of the data. This may be particularly problematic for biochars with high ash contents, and in comparing biochars with different ash content or composition.

Biochar pH was strongly influenced by the treatment with H$_2$O$_2$, decreasing from 7.9 to 4.8 in H$_2$O (7.5 to 4.5 in 0.01 M CaCl$_2$). The pH decrease may be related to the changes in functionality (i.e., increased surface acidity) [21,40]. However, given the drastic nature of this decrease, it is likely other factors also play a role and the importance of metal oxides in the ash cannot be ruled out. In comparison, Huff and Lee [27] reported a 1.5 unit decrease in pH with 30% H$_2$O$_2$ treatment, but only a 0.1 unit decrease with 3% H$_2$O$_2$ for different wood biochars. The decrease in ash content with H$_2$O$_2$ treatment likely facilitated the observed decrease in pH.

Chemical activation had no measurable effect on the zeta potential of the G-350 biochar in DI water. The original G-350 and its activated counterpart had measured zeta potential values of −27.4 ± 2.2 mV and −26.4 ± 1.2 mV, respectively, in DI water, indicating a net negative surface charge. The biochar zeta potentials were less negative when measured in 0.01 M $CaCl_2$ from the screening effects of the ions in solution (−10.4 ± 1.7 mV and −14.2 ± 1.6 mV for G-350 and G-350 H_2O_2, respectively), and G-350 H_2O_2 was significantly less negative than G-350 (t-test, $p = 0.04$).

The zeta potential was measured at the natural pH level of each biochar. Because zeta potential correlates with pH, if the variable-charge surface groups were unaltered, the lower pH of G-350 H_2O_2 would be expected to have a more positive zeta potential value. However, the unchanged value in DI water suggests more charge groups are present on the activated biochar, which is in agreement with the FTIR data.

H_2O_2 activation increased the water vapor SSA of G-350 by nearly 3-fold, from 47 to 140 $m^2\ g^{-1}$. The increase in SSA with activation is consistent with the opening of pores visible in the SEM images (Figure 2). Though only macropores are clearly distinguishable on the scale imaged, the opening of additional micropores may have also contributed to the rise in SSA.

Isoteric heats of adsorption of water vapor (E_a values) were 23 and −43 kJ mol^{-1} for G-350 and G350 H_2O_2 biochars, respectively. Values that are more negative indicate more energetically favorable adsorption. Activation of G-350 by H_2O_2 increased the biochar surface's affinity for water vapor, based on the large decrease in E_a. Water is a polar molecule; therefore, E_a values can be influenced by site-specific electrostatic interactions. It is possible that the newly developed (or uncovered) O-group nucleation sites are responsible for the observed decreased E_a with chemical activation. The decrease in aliphatic C-H observed in the FTIR spectra following activation, also supports this observed decrease in the G-350 biochar's hydrophobicity [41].

3.2. Herbicide Sorption

Activation of the G-350 biochar by H_2O_2 increased cyhalofop RE but did not significantly change clomazone RE as compared to unactivated biochar (Figure 4). In H_2O, cyhalofop RE increased from 6.3% to 35.4%, while clomazone RE did not significantly increase (65.0% to 70.3%). Though a different activation method was employed (i.e., post-pyrolysis air oxidation), Xiao and Pignatello [12] similarly found that activation had a greater effect on organic acid adsorption than on neutral compounds.

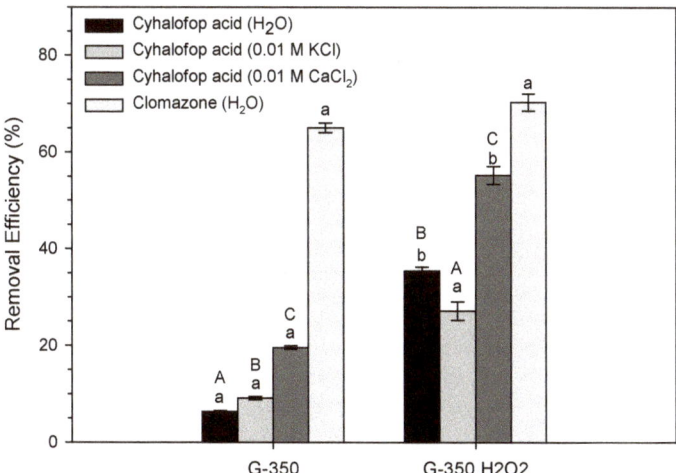

Figure 4. Removal efficiency (RE) of G-350 biochar with and without H_2O_2 activation for cyhalofop and clomazone. Error bars represent the standard error of the mean (n = 3). Lowercase letters indicate significant differences between RE means with and without activation for individual chemical/solution pairs ($p < 0.05$). Uppercase letters indicate significant differences among background solutions for each biochar ($p < 0.05$).

Clomazone RE was greater than cyhalofop RE on both the G-350 (65.0% versus 6.3%) and G-350 H_2O_2 (70.3% versus 35.4%) biochars. This finding is consistent with the sorptive behavior of clomazone versus cyhalofop in soil, where average distribution coefficient normalized to total organic carbon, K_{oc}, values are 300 and 186, respectively [42]. In previous work, clomazone was also sorbed greater than other anionic pesticide, byspiribac sodium, when the soil was amended with biochars prepared at 350 °C [27]. This was attributed to the presence of more amorphous organic matter within the carbonaceous matrix in biochar and further abundance of surface functional groups, according to higher O/C ratios, could favor specific chemical interactions, as reported for others polar organic compounds [27]. The scarce sorption of cyhalofop can be explained by its low pK_a (3.9) which favors the predominance of the anionic form at most biochar pH levels (pH > 7). In its anionic form there would be reduced attraction to negatively charged surfaces. This electrostatic repulsion along with its low SSA values (<1 m^2/g) has been the key factors cited for the low sorption of anionic pesticides on biochars [27].

The sorption isotherms were well-fitted to the Freudlich equation with R^2 > 0.970 for all cases (Table 2) and they confirmed the trend observed for the RE (Figure 5). That is, the activation was only satisfactory for cyhalofop whereas no changes were registered for clomazone after the treatment of of G-350 with H_2O_2 (Figure 5). The N_f values < 1 registered for the activated biochar was an indication of the limited availability of sorption sites whereas for G350 closer to 1 suggesting more partitioning medium (Table 2). This fact has been associated with adsorption mechanism for others organic compounds in biochars [27].

Table 2. Freundlich Coefficients for cyhalofop and clomazone sorption isotherms on G-350 and G-350-H_2O_2.

	G-350			G-350-H_2O_2		
Herbicide	K_f	N_f	R^2	K_f	N_f	R^2
Cyhalofop	13.9 (12.6–15.4) [1]	0.92 ± 0.09 [2]	0.970	48.3 (44.6–52.4)	0.75 ± 0.07	0.976
Clomazone	284 (264–306)	0.81 ± 0.05	0.990	284 (265–305)	0.74 ± 0.04	0.990

[1] Values in parentheses correspond to the range in the values of the Freundlich coefficient; [2] values ± standard error.

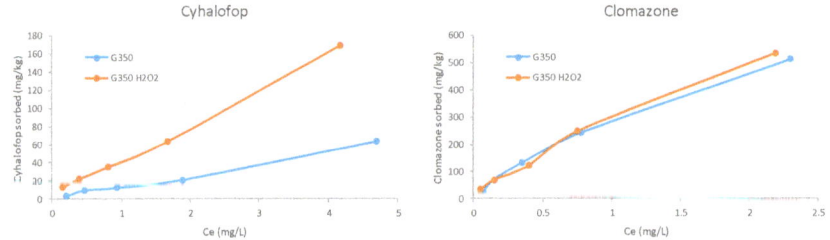

Figure 5. Sorption isotherms for cyhalofop and clomazone on G-350 and activated G-350 with H_2O_2.

To compare the effects of chemical activation to as opposed to thermal activation, REs of the untreated grape wood biochars (350, 500, and 900 °C) are shown in Figure 6. As seen with the activated biochar, clomazone sorption was significantly greater than cyhalofop sorption on both G-350 and G-500. Within the temperature series, the G-900 biochar had the greatest affinity for both herbicides, removing > 99% from solution, while the REs of G-350 and G-500 did not significantly differ from one another for either compound. The high RE of G-900 is consistent with the commonly reported trend of increased sorption with biochar production temperature, which is often attributed to increase in aromaticity and SSA with temperature (e.g., Chen et al. [10]). While the H_2O_2 activation of G-350 increased the RE of cyhalofop, the REs for both cyhalofop and clomazone remained much lower than with the G-900 biochar and were likely controlled by different sorption mechanisms.

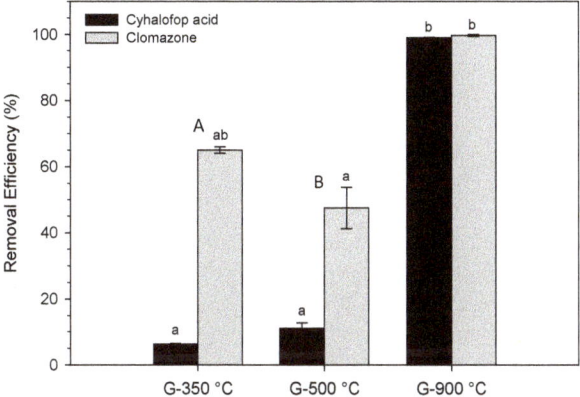

Figure 6. Removal efficiency (RE) of grape wood biochars for cyhalofop and clomazone in deionized (DI) water. Error bars represent the standard error of the mean (n = 3). Different lowercase letters indicate significant differences among temperatures for individual herbicides ($p < 0.05$). Uppercase letters indicate a significant difference between cyhalofop and clomazone REs for each biochar ($p < 0.05$).

The background solution had a significant impact on cyhalofop removal by both the untreated and H_2O_2 activated biochars; however, the effects were not consistent across the two materials (Figure 4). It is anticipated that adsorption will vary with solution ionic strength [9], consistent with that observed for cyhalofop on untreated G-350. Previous studies of aromatic acid sorption on carbonized sorbents have reported similar trends of increasing sorption with ionic strength [43], suggesting electrostatic-based mechanisms; however, other studies have reported no effect of ionic strength on organic acid sorption on biochar [44]. In contrast, cyhalofop sorption on G-350 H_2O_2 did not increase with ionic strength or follow any clear trend. The lower pH of the G-350 H_2O_2 biochar may have also dampened the screening or bridging effect of cations, as suggested by the lower magnitude of RE variability.

The effects of activation on G-350 biochar could be the results of a variety of mechanisms, i.e., changes in ash/mineral fraction, O functionality, aliphatic carbon, SSA, pH, zeta potential, π-interactions, or combinations thereof. If the mineral fraction of the biochar were actively involved in adsorption, we would expect to see a decrease in RE correlating to the decrease in % ash; however, this was not the case for either herbicide. Clomazone RE did not change, suggesting that the ash content has no effect. Cyhalofop RE increased, suggesting that the ash may have been blocking active binding sites [9], or otherwise altering properties that impact cyhalofop adsorption, specifically pH [45].

Both the bulk analysis and FTIR data suggest a slight increase in O functionality of the biochar surface following chemical activation. Because clomazone RE did not change with activation, there is no evidence to support direct interactions with these newly formed O groups. In contrast, the increase in cyhalofop RE on G-350 H_2O_2 suggests the new, or newly accessible, groups were influential; therefore, site-specific mechanisms such as charge-assisted hydrogen bonding (CAHB) may be critical for cyhalofop sorption.

The change in aliphatic C with activation can help in discerning the role of hydrophobic interactions. Unlike activated charcoals and high-temperature biochars where hydrophobicity is related to its large aromatic C structure, the hydrophobicity of low-temperature biochars arises from the presence of hydrophobic aliphatic functionalities [46,47]. The H_2O_2 activation appears to have degraded some of the aliphatic groups present on the original G-350 biochar, which should theoretically decrease the hydrophobicity of a low-temperature biochar surface. This was confirmed in our study by the decrease in water vapor E_a. Therefore, if the labile hydrophobic C were responsible for the removal of the herbicides, we would expect to see a decrease in RE. The unchanged RE of clomazone instead does not support hydrophobic effects with these non-aromatic groups. Similarly, the aliphatic groups did not appear to contribute to the removal of cyhalofop, and instead may have been preventing site-specific interactions with O groups.

The decrease in water vapor E_a with activation supports the decrease in biochar surface hydrophobicity; therefore, if hydrophobic effects were key in the sorption of either herbicide, decreased REs should have been observed. Water vapor adsorption data also identified an increase in SSA with activation. The increase in SSA corresponded to an increase in cyhalofop RE, but not clomazone RE, suggesting the newly exposed surfaces favored interactions with the ionizable compounds, or a removal of pore size restrictions for the slightly larger cyhalofop molecule.

Though porosity was not explicitly measured in this study, the increased SSA and SEM images suggest an increase in porosity with activation. While the SSA measurable by water vapor indicates a large increase, it is important to keep in mind that the larger clomazone and cyhalofop molecules cannot necessarily enter newly developed micropores accessible to the smaller water molecules. For all porous materials, pore filling is a potential removal mechanism; however, the data in this study do not support it as the dominant mechanism controlling adsorption on biochar. If pore filling were a dominant mechanism, we would expect to see both cyhalofop and clomazone adsorption increase following the H_2O_2 treatment, as both molecules are of similar size. Instead, only an increase in cyhalofop adsorption was observed. The treatment with 3% H_2O_2 at room temperature was relatively mild compared to other activation methods, such as thermal oxidation, where porosity has been shown to play a larger role in the increased organic chemical sorption (e.g., [7]).

Changes in pH arise from changes in the previously discussed biochar properties (i.e., ash content and surface functionality), but their effects must also be considered independently. Because pH influences the relative abundance of charged species for ionizable compounds such as cyhalofop, it is crucial to take into account changes in this distribution. As previously stated, pH was not adjusted in this experiment in order to fully account for the effect of activation and prevent further changes to the material. Based on the solution pH of the untreated G-350 biochar (pH 7.9), 100% of the cyhalofop would be present in its anionic form, while 89% is expected in the anionic form with the activated biochar (pH 4.8). Though important to note, this shift in species distribution does not appear to fully account for the >450% increase in cyhalofop RE with activation. Similarly, the slightly lower pH of the biochars in 0.01 M $CaCl_2$ may contribute in part to the higher cyhalofop RE (compared to H_2O), but other mechanisms such as cation bridging are also likely influential [48].

Zeta potential, likewise, is related to the aforementioned changes in biochar properties, as well as pH. For a negatively charged biochar surface, the general trend is for zeta potential to become less negative as pH decreases. Therefore, if we assume no change to the biochar surface with activation, the three-unit decrease in pH would be expected to result in a more positive zeta potential. However, we observed either no change (H_2O) or a more negative (0.01 M $CaCl_2$) zeta potential following activation, which is in agreement with the observed changes to the biochar surface chemistry. Because the net charge did not become more positive, non-site-specific electrostatic forces of the material would not be expected to favor a greater attraction of the anionic cyhalofop, again suggesting the importance of mechanisms such as CAHB with selected functional groups [44,49].

π-interactions (e.g., π–π overlap, polar –π) are often cited as adsorption mechanisms for the removal of organic compounds by biochar (e.g., antibiotics [48,50]), but most often are associated with high temperature biochars and activated charcoals. While many aromatic π-orbitals are present in the G-350 biochar, only small ring clusters exist within biochars produced at these low temperatures, and the structure is more amorphous than high temperature biochars [36]. Additionally, the presence of many functional groups, such as long aliphatic side chains, as have been registered by FTIR (Figure 3), can sterically hinder these interactions and influence the distribution of delocalized electrons.

Sorption mechanisms for organic compounds are often cited to shift from partitioning in the noncarbonized organic fraction to adsorption on the carbonized material as biochar pyrolysis temperatures increase [10]. However, partitioning is not specific mechanism that can be determined, but rather describes the removal of a sorbate from solution that follows a linear isotherm. Theoretically, the sorbate is dissolving into the organic matter; in the case of biochar, the noncarbonized (amorphous) fraction. Because partitioning is not a clearly defined mechanism, it is challenging to comment on its importance from the data in this study. However, if partitioning were occurring in the labile C fraction of the original biochar, we would have expected to see a decrease in RE with activation.

Numerous other mechanisms appear in the biochar literature (e.g., Van der Waals forces, hydrogen bonding, etc.); however, we cannot elucidate the individual contributions of each specific mechanism. It is also important to recall that multiple mechanisms simultaneously contribute to the observed net sorption. We can, however, begin to prioritize the mechanisms.

4. Conclusions

Characterization of a low temperature grape wood biochar before and after activation revealed that 3% H_2O_2 at room temperature, a relatively mild treatment, was sufficient to alter the surface chemistry of the biochar as well as the bulk composition. The activation induced changes improved the sorption of cyhalofop, but not clomazone, which suggests that H_2O_2 activation treatments may be of use for sorbing organic acid herbicides but is of little value in optimizing the removal of polar, non-ionizable compounds. However, the ability to target weak acid herbicides is particularly beneficial, as their anionic nature makes them particularly susceptible to leaching and contaminating groundwater. Furthermore, the demonstrated ability to improve sorption on a low temperature biochar is promising for agricultural use.

With regard to sorption mechanisms, it was found that neither ash, nor labile C (aliphatic groups) were key in the removal of cyhalofop and clomazone. Instead, cyhalofop removal appeared to be favored by the decreased pH with activation and the availability of O functionality; therefore, charge-based mechanisms such as CAHB are hypothesized to dominate its sorption by biochar. In contrast, the activation effects in this study suggest hydrophobic effects did not appear to be the driving mechanism. However, whether the similar mechanisms persist in the soil environment will need to be examined in future work. In particular, the localized effects of the H_2O_2-induced pH decrease should be studied to determine if this sorption-enhancing property remains effective in case of the other weak acid herbicides.

Author Contributions: The following summarize the contribution of the authors: conceptualization, K.H., B.G., K.A.S., and L.C.; methodology K.H., B.G., L.C., and K.A.S.; experimental investigation, K.H. and B.G.; experimental resources, K.A.S. and L.C.; writing—original draft preparation, K.H. and B.G.; writing—review and editing, B.G., K.H., L.C., and K.A.S.; visualization, K.H. and B.G.; supervision, K.A.S. and L.C.; project administration, L.C. and K.A.S.; funding acquisition, L.C., K.A.S., and K.H.

Funding: The Spanish Ministry of Science, Innovation and Universities (MICINN Project AGL2016-77821-R).

Acknowledgments: USDA is an equal opportunity provider and employer. K.H. would like to thank the University of Minnesota for a Thesis Research Travel Grant allowing for a visit to IRNAS-CSIC (Seville, Spain) and additional funding through a University of Minnesota Doctoral Dissertation Fellowship. Parts of this work were carried out in the Characterization Facility, University of Minnesota, which receives partial support from the National Science Foundation (NSF) through the Materials Research Science and Engineering Center (MRSEC) program. Special thanks to Jeff Novak and Don Watts for the biochar production from the USDA-ARS Florence, SC location. Special thanks also to William Koskinen for his assistance in reviewing an initial draft of this manuscript.

Conflicts of Interest: The authors declare no conflict of interest.

References

1. Tan, X.; Liu, Y.; Zeng, G.; Wang, X.; Hu, X.; Gu, Y.; Yang, Z. Application of biochar for the removal of pollutants from aqueous solutions. *Chemosphere* **2015**, *125*, 70–85. [CrossRef] [PubMed]
2. Ahmad, M.; Rajapaksha, A.U.; Lim, J.E.; Zhang, M.; Bolan, N.; Mohan, D.; Vithanage, M.; Lee, S.S.; Ok, Y.S. Biochar as a sorbent for contaminant management in soil and water: A review. *Chemosphere* **2014**, *99*, 19–23. [CrossRef] [PubMed]
3. Hagemann, N.; Spokas, K.; Schmidt, H.P.; Kägi, R.; Böhler, M.A.; Bucheli, T.D. Activated carbon, biochar and charcoal: Linkages and synergies across pyrogenic carbon's ABCs. *Water* **2018**, *10*, 1–19. [CrossRef]

4. Rajapaksha, A.U.; Chen, S.S.; Tsang, D.C.W.; Zhang, M.; Vithanage, M.; Mandal, S.; Gao, B.; Bolan, N.S.; Sik, Y. Engineered/designer biochar for contaminant removal/immobilization from soil and water: Potential and implication of biochar modification. *Chemosphere* **2016**, *148*, 276–291. [CrossRef] [PubMed]
5. Azargohar, R.; Dalai, A.K. Steam and KOH activation of biochar: Experimental and modeling studies. *Microporous Mesoporous Mater.* **2008**, *110*, 413–421. [CrossRef]
6. Benzigar, M.R.; Talapaneni, S.N.; Joseph, S.; Ramadass, K.; Singh, G.; Scaranto, J.; Ravon, U.; Al-Bahily, K.; Vinu, A. Recent advances in functionalized micro and mesoporous carbon materials: synthesis and applications. *Chem. Soc. Rev.* **2018**, *47*, 2680–2721. [CrossRef] [PubMed]
7. Pignatello, J.J.; Mitch, W.A.; Xu, W. Activity and reactivity of pyrogenic carbonaceous matter toward organic compounds. *Environ. Sci. Technol.* **2017**, *51*, 8893–8908. [CrossRef] [PubMed]
8. Ifthikar, J.; Wang, J.; Wang, Q.; Wang, T.; Wang, H.; Khan, A.; Jawad, A.; Sun, T.; Jiao, X.; Chen, Z. Highly efficient lead distribution by magnetic sewage sludge biochar: Sorption mechanisms and bench applications. *Bioresour. Technol.* **2017**, *238*, 399–406. [CrossRef]
9. Moreno-Castilla, C. Adsorption of organic molecules from aqueous solutions on carbon materials. *Carbon N. Y.* **2004**, *42*, 83–94. [CrossRef]
10. Chen, B.; Zhou, D.; Zhu, L. Transitional adsorption and partition of nonpolar and polar aromatic contaminants by biochars of pine needles with different pyrolytic temperatures. *Environ. Sci. Technol.* **2008**, *42*, 5137–5143. [CrossRef]
11. Spokas, K.A.; Cantrell, K.B.; Novak, J.M.; Archer, D.W.; Ippolito, J.A.; Collins, H.P.; Boateng, A.A.; Lima, I.M.; Lamb, M.C.; McAloon, A.J. Biochar: A synthesis of its agronomic impact beyond carbon sequestration. *J. Environ. Qual.* **2012**, *41*, 973–989. [CrossRef] [PubMed]
12. Xiao, F.; Pignatello, J.J. Effects of post-pyrolysis air oxidation of biomass chars on adsorption of neutral and ionic compounds. *Environ. Sci. Technol.* **2016**, *50*, 6276–6283. [CrossRef]
13. Kupryianchyk, D.; Hale, S.; Zimmerman, A.R.; Harvey, O.; Rutherford, D.; Abiven, S.; Knicker, H.; Schmidt, H.P.; Rumpel, C.; Cornelissen, G. Sorption of hydrophobic organic compounds to a diverse suite of carbonaceous materials with emphasis on biochar. *Chemosphere* **2016**, *144*, 879–887. [CrossRef]
14. Liu, P.; Liu, W.J.; Jiang, H.; Chen, J.J.; Li, W.W.; Yu, H.Q. Modification of bio-char derived from fast pyrolysis of biomass and its application in removal of tetracycline from aqueous solution. *Bioresour. Technol.* **2012**, *121*, 235–240. [CrossRef]
15. Lawrinenko, M.; Laird, D.A.; Johnson, R.L.; Jing, D. Accelerated aging of biochars: Impact on anion exchange capacity. *Carbon N. Y.* **2016**, *103*, 217–227. [CrossRef]
16. Mia, S.; Dijkstra, F.A.; Singh, B. Aging induced changes in biochar's functionality and adsorption behavior for phosphate and ammonium. *Environ. Sci. Technol.* **2017**, *51*, 8359–8367. [CrossRef] [PubMed]
17. Gámiz, B.; Velarde, P.; Spokas, K.A.; Celis, R.; Cox, L. Changes in sorption and bioavailability of herbicides in soil amended with fresh and aged biochar. *Geoderma* **2019**, *337*, 341–349. [CrossRef]
18. Ray, P.G.; Pews, R.G.; Flake, J.; Secor, J.; Hamburg, A. Cyhalofop butyl: A new graminicide for use in rice. In Proceedings of the 10th Australian Weeds Conference and 14th Asian Pacific Weed Science Society Conference, Brishane, Australia, 6 August–10 September 1993.
19. Jackson, R.; Douglas, M. An aquatic risk assessment for cyhalofop-butyl: A new herbicide for control of barnyard grass in rice. In Proceedings of the Human and Environmental Exposure to Xenobiotics, Proceedings of the XI Symposium Pesticide Chemistry, Cremona, Italy, 11–15 September 1999; Volume 1999, pp. 345–354.
20. Domingues, R.R.; Trugilho, P.F.; Silva, C.A.; De Melo, I.C.N.A.; Melo, L.C.A.; Magriotis, Z.M.; Sánchez-Monedero, M.A. Properties of biochar derived from wood and high-nutrient biomasses with the aim of agronomic and environmental benefits. *PLoS ONE* **2017**, *12*, 1–19. [CrossRef] [PubMed]
21. Huff, M.D.; Lee, J.W. Biochar-surface oxygenation with hydrogen peroxide. *J. Environ. Manage.* **2016**, *165*, 17–21. [CrossRef]
22. McDermot, H.L.; Arnell, J.C. Charcoal sorption studies. II The sorption of water by hydrogen treated charcoals. *J. Phys. Chem.* **1954**, *58*, 492–498. [CrossRef]
23. Medic, D.; Darr, M.; Shah, A.; Rahn, S. Effect of torrefaction on water vapor adsorption properties and resistance to microbial degradation of corn stover. *Energy & Fuels* **2012**, *26*, 2386–2393.
24. Rockland, L.B. Saturated salt solutions for static control of relative humidity between 5o and 40C. *Anal. Chem.* **1960**, *32*, 1375–1376. [CrossRef]

25. OECD (Economic Co-operation and Development). *Guideline 106: Adsorption—Desorption Using a Batch Equilibrium Method*; OECD Publishing: Paris, France, 2000; ISBN 9789264069602.
26. Wu, J.; Wang, K.; Zhang, Y.; Zhang, H. Determination and study on dissipation and residue determination of cyhalofop-butyl and its metabolite using HPLC-MS/MS in a rice ecosystem. *Environ. Monit. Assess.* **2014**, *186*, 6959–6967. [CrossRef]
27. Gámiz, B.; Velarde, P.; Spokas, K.A.; Hermosín, M.C.; Cox, L. Biochar soil additions affect herbicide fate: Importance of application timing and feedstock species. *J. Agric. Food Chem.* **2017**, *65*, 3109–3117. [CrossRef]
28. Yang, J.; Pignatello, J.J.; Pan, B.; Xing, B. Degradation of p-nitrophenol by lignin and cellulose chars: H_2O_2-mediated reaction and direct reaction with the char. *Environ. Sci. Technol.* **2017**, *51*, 8972–8980. [CrossRef] [PubMed]
29. Lousada, C.M.; Yang, M.; Nilsson, K.; Jonsson, M. Catalytic decomposition of hydrogen peroxide on transition metal and lanthanide oxides. *J. Mol. Catal. A Chem.* **2013**, *379*, 178–184. [CrossRef]
30. Fry, H.S.; Milstead, K.L. The action of hydrogen peroxide upon simple carbon compounds. III. Glycolic acid. *J. Am. Chem. Soc.* **1935**, *57*, 2269–2272. [CrossRef]
31. Gellerstedt, G.; Pettersson, I. Chemical aspects of hydrogen peroxide bleaching-Part II the bleaching of kraft pulps. *J. Wood Chem. Technol.* **1982**, *23*, 231–250. [CrossRef]
32. Robinson, W.O. The determination of organic matter in soils by means of hydrogen peroxide. *J. Agric. Res.* **1927**, *34*, 339–356.
33. Li, F.; Shen, K.; Long, X.; Wen, J.; Xie, X.; Zeng, X.; Liang, Y.; Wei, Y.; Lin, Z.; Huang, W.; et al. Preparation and characterization of biochars from Eichornia crassipes for cadmium removal in aqueous solutions. *PLoS ONE* **2016**, *11*, e0148132. [CrossRef]
34. Kloss, S.; Zehetner, F.; Dellantonio, A.; Hamid, R.; Ottner, F.; Liedtke, V.; Schwanninger, M.; Gerzabek, M.H.; Soja, G. Characterization of slow pyrolysis biochars: Effects of feedstocks and pyrolysis temperature on biochar properties. *J. Environ. Qual.* **2012**, *41*, 990. [CrossRef]
35. Keiluweit, M.; Nico, P.S.; Johnson, M.G.; Kleber, M. Dynamic molecular structure of plant biomass-derived black carbon (biochar). *Environ. Sci. Technol.* **2010**, *44*, 1247–1253. [CrossRef]
36. Trigo, C.; Cox, L.; Spokas, K. Influence of pyrolysis temperature and hardwood species on resulting biochar properties and their effect on azimsulfuron sorption as compared to other sorbents. *Sci. Total Environ.* **2016**, *566–567*, 1454–1464. [CrossRef]
37. Shafeeyan, M.S.; Daud, W.M.A.W.; Houshmand, A.; Shamiri, A. A review on surface modification of activated carbon for carbon dioxide adsorption. *J. Anal. Appl. Pyrolysis* **2010**, *89*, 143–151. [CrossRef]
38. Rhoads, C.A.; Senftle, J.T.; Coleman, M.M.; Davis, A.; Painter, P.C. Further studies of coal oxidation. *Fuel* **1983**, *62*, 1387–1392. [CrossRef]
39. Li, X.; Shen, Q.; Zhang, D.; Mei, X.; Ran, W.; Xu, Y.; Yu, G. Functional groups determine biochar properties (pH and EC) as studied by two-dimensional 13C NMR correlation spectroscopy. *PLoS ONE* **2013**, *8*, e65949. [CrossRef]
40. Kinney, T.J.; Masiello, C.A.; Dugan, B.; Hockaday, W.C.; Dean, M.R.; Zygourakis, K.; Barnes, R.T. Hydrologic properties of biochars produced at different temperatures. *Biomass and Bioenergy* **2012**, *41*, 34–43. [CrossRef]
41. PPDB Pesticides Properties Database. Available online: http://sitem.herts.ac.uk/aeru/bpdb/index.htm (accessed on 25 September 2019).
42. Sigmund, G.; Sun, H.; Hofmann, T.; Kah, M. Predicting the sorption of aromatic acids to noncarbonized and carbonized sorbents. *Environ. Sci. Technol.* **2016**, *50*, 3641–3648. [CrossRef]
43. Ni, J.Z.; Pignatello, J.J.; Xing, B.S. Adsorption of aromatic carboxylate ions to black carbon (biochar) is accompanied by proton exchange with water. *Environ. Sci. Technol.* **2011**, *45*, 9240–9248. [CrossRef]
44. Cao, X.; Harris, W. Properties of dairy-manure-derived biochar pertinent to its potential use in remediation. *Bioresour. Technol.* **2010**, *101*, 5222–5228. [CrossRef]
45. Gray, M.; Johnson, M.G.; Dragila, M.I.; Kleber, M. Water uptake in biochars: The roles of porosity and hydrophobicity. *Biomass Bioenergy* **2014**, *61*, 196–205. [CrossRef]
46. Zornoza, R.; Moreno-Barriga, F.; Acosta, J.A.; Muñoz, M.A.; Faz, A. Stability, nutrient availability and hydrophobicity of biochars derived from manure, crop residues, and municipal solid waste for their use as soil amendments. *Chemosphere* **2016**, *144*, 122–130. [CrossRef]
47. Jia, M.; Wang, F.; Bian, Y.; Jin, X.; Song, Y.; Kengara, F.O.; Xu, R.; Jiang, X. Effects of pH and metal ions on oxytetracycline sorption to maize-straw-derived biochar. *Bioresour. Technol.* **2013**, *136*, 87–93. [CrossRef]

48. Li, X.; Gámiz, B.; Wang, Y.; Pignatello, J.J.; Xing, B. Competitive sorption used to probe strong hydrogen bonding sites for weak organic acids on carbon nanotubes. *Environ. Sci. Technol.* **2015**, *49*, 1409–1417. [CrossRef]
49. Peng, B.; Chen, L.; Que, C.; Yang, K.; Deng, F.; Deng, X.; Shi, G.; Xu, G.; Wu, M. Adsorption of antibiotics on graphene and biochar in aqueous solutions induced by π-π interactions. *Sci. Rep.* **2016**, *6*, 31920. [CrossRef]
50. Sun, K.; Jin, J.; Keiluweit, M.; Kleber, M.; Wang, Z.; Pan, Z.; Xing, B. Polar and aliphatic domains regulate sorption of phthalic acid esters (PAEs) to biochars. *Bioresour. Technol.* **2012**, *118*, 120–127. [CrossRef]

© 2019 by the authors. Licensee MDPI, Basel, Switzerland. This article is an open access article distributed under the terms and conditions of the Creative Commons Attribution (CC BY) license (http://creativecommons.org/licenses/by/4.0/).

Article

Carbon Dynamics and Fertility in Biochar-Amended Soils with Excessive Compost Application

Chen-Chi Tsai * and Yu-Fang Chang

Department of Forestry and Natural Resources, National Ilan University, Ilan 26047, Taiwan
* Correspondence: cctsai@niu.edu.tw; Tel.: +886-3-931-7683

Received: 6 August 2019; Accepted: 2 September 2019; Published: 5 September 2019

Abstract: In Taiwan, farmers often apply excessive compost to ensure adequate crop yield in frequent tillage, highly weathered, and lower fertility soils. The potential of biochar (BC) to decrease soil C mineralization and improve soil nutrient availability in excessive compost application soil is promising, but under-examined. To test this, a 434-day incubation experiment of in vitro C mineralization kinetics was conducted. We added 0%, 0.5%, 1.0%, and 2.0% (w/w) woody BC composed of lead tree (*Leucaena leucocephala* (Lam.) de. Wit) to one Oxisol and two Inceptisols in Taiwan. In each treatment, 5% swine manure compost was added to serve as excessive application. The results indicated that soil type strongly influences the impact of BC addition on soil carbon mineralization potential. Respiration per unit of total organic carbon (total mineralization coefficient) of the three studied soils significantly decreased with increase in BC addition. Principal component analysis suggested that to retain more plant nutrients in addition to the effects of carbon sequestration, farmers could use locally produced biochars and composts in highly weathered and highly frequent tillage soil. Adding 0.5% woody BC to Taiwan rural soils should be reasonable and appropriate.

Keywords: biochar; soil carbon dynamics; soil fertility; excessive compost application; Ultisols

1. Introduction

Soil degradation due to erosion, salinization, depletion of soil organic matter (SOM), and nutrient imbalance is the most serious bio-physical constraint limiting agricultural productivity in many parts of the world [1]. Maintaining an appropriate level of SOM and ensuring the efficient biological cycling of nutrients are crucial to the success of soil management and agricultural productivity strategies [2,3], including the application of organic and inorganic fertilizers, combined with the knowledge of how to adapt these practices to local conditions, aiming to maximize the agronomic use efficiency of the applied nutrients and thus crop productivity [3]. In soils with low nutrient retention capacity, strong rains rapidly and easily leach available and mobile nutrients into the subsoil, where they are unavailable for most crops [4], rendering conventional fertilization highly inefficient [5]. SOM has been reduced in the arable lands of Taiwan over the last several decades due to highly frequent tillage, in association with high air temperature and rainfall; in addition, farmers often apply excess compost to ensure adequate crop yield.

Depending on the mineralization rate, organic fertilizers, such as compost, mulch, or manure, release nutrients in a gradual manner [6], and may therefore be more appropriate than inorganic fertilizers for nutrient retention under high-leaching conditions. Due to the relatively low levels of nutrients (10–20 g N/kg and less than 10 g P/kg) in compost, compared to a complete fertilizer, as well as the low plant availability of compost N and P, a large amount of compost is needed to meet the N and P crop requirements [7], and farmers often apply excess compost to ensure adequate crop yield, leading to excessive N and P loading into the environment. In the tropics, however, natural rapid mineralization of SOM is a limitation of the practical application of organic fertilizers; in addition to repeated application at high doses and the cost of application of organic materials, their rapid decomposition and mineralization may significantly contribute to global warming [8–10]. Excessive manure application often causes heavy metal accumulation (Cu, Pb, Zn, etc.) in the soil, and the soluble fraction of these metals tends to increase due to desorption and remobilization of metals previously bound to the soil matrix, leading to enhanced crop uptake of heavy metals [11]. In acidic and highly-weathered tropical soils, the application of organic fertilizers and charcoal increases nutrient stocks in the rooting zone of crops, reduces nutrient leaching, and thus improves crop production [5]. Biochar could be a key input in raising and sustaining production and simultaneously reducing pollution and dependence on fertilizers, and could also improve soil moisture availability and sequester carbon [12]. Biochar (BC) studies have mainly focused on the effects of pure BC addition or artificial fertilizers; however, pure BC does not provide a high amount of nutrients in most cases [13]. Incorporation of BC-compost into poor soil is considered a promising approach to produce a substrate like *terra preta*; the study demonstrated a synergistic positive effect of compost and BC mixtures on soil organic matter content, nutrients levels, and water-storage capacity of sandy soil under field conditions [13]. BC either helped stabilize manure C, or the presence of manure reduced the effect of BC on the mineralization of soil organic carbon (SOC) [14]. Trupiano et al. [15] showed that both BC amendment (65 g/kg) and compost (50 g/kg) addition to a moderately subalkaline (pH 7.1) and clayey soil poor in nutrients had a positive effect on lettuce plant growth and physiology, and on soil chemical and microbiological characteristics; however, no positive synergic or summative effects exerted by compost and BC in combination were observed compared to compost treatment alone. BC, compost, and the BC-compost blend have fewer environmental impacts than mineral fertilizer from a systems perspective [16].

In Taiwan, annual precipitation is about 2500 mm (ranging from 1500 mm to 4500 mm) and annual mean air temperature is about 23 °C (28~29 °C in summer, 16–19 °C in winter). Higher precipitation always results in enormous soil erosion and nutrient leaching, and warm temperatures cause rapid decomposition of soil organic matter. In addition, intensive and highly frequent tillage has resulted in obvious decrease of soil organic matter. These are major setbacks to Taiwan's agricultural soils. Harris et al. [17] suggested that proper soil organic carbon (SOC) content should be 4% to 6%. Wang et al. [18] indicated that the addition of 37 tons/ha organic fertilizer (carbon content was 58%) per year in rural Taiwan is necessary, and soil organic carbon can be maintained within an appropriate range. This is equivalent to 2% compost in general rural soil, considering 1800 Mg of soil per hectare (soil bulk density equal to 1.2 Mg/m and an arable soil layer of 15 cm). The carbon content of swine manure compost in this study is only 23.3%, much lower than the 58% mentioned above (Table 1). This suggests that [18] farmers should add at least 5% compost/ha/year to maintain appropriate soil organic carbon content. The 5% addition rate means adding 90 tons/ha compost to soil, as well as a large amount of 1800 kg N/ha and 900 kg P/ha to the soil. Taking into consideration economic viability, the doses of manure compost in Taiwan are recommended as 1% to 2%; however, some farmers apply more than 2% to 5% in intensive cultivation periods for short-term leafy crops, in an effort to add more N. However, in excessive compost application soils, little is known about the impact of BC application rates on the carbon mineralization and soil fertility of mixed-soils (BC, compost, and soil) in highly frequent tillage soil systems. The in vitro C mineralization kinetics of various BC addition rates in three selected soils were examined in this study. We hypothesized that the addition of BC may stabilize compost organic matter, diminish mixed-soil C mineralization, and improve soil nutrient status. Farmers can gradually reduce the addition of compost over the next few years by adding biochar to maintain appropriate SOC, reduce N, and prevent loss of nutrients. The aims of our research were: (1) to quantify the effects of woody BC additions on C mineralization and soil fertility, and (2) to evaluate the sustainability of woody BC additions in terms of maintaining high SOM content and nutrient availability.

Table 1. Characteristics of biochar, compost, and three studied soils.

Characteristics	Biochar (BC)	Compost	Pc Soil (SAO)	Eh Soil (MAI)	An Soil (SAI)
pH	9.9 [1]	8.41 [1]	6.1/5.0 [3]	7.5/7.2 [3]	6.5/6.2 [3]
EC (dS/m)	0.77 [1]/1.36 [2]	3.79 [1]	0.45	2.21	0.81
Sand (%)			11	24	33
Silt (%)			30	36	33
Clay (%)			59	39	34
Soil Texture			Clay	Clay loam	Clay loam
Total C (%)	82.5	23.3	2.03	1.11 (0.81) [4]	0.94
Total N (g/kg)	6.99	22.6	2.71	2.32	1.58
Total P (g/kg)	0.55	10.2	1.16	0.98	0.77
Ex. K (cmol(+)/kg soil)	1.91	6.43	0.32	0.29	0.21
Ex. Na (cmol(+)/kg soil)	1.26	1.09	0.31	0.26	0.37
Ex. Ca (cmol(+)/kg soil)	3.62	2.70	4.85	2.94	2.24
Ex. Mg (cmol(+)/kg soil)	0.40	2.72	0.64	0.80	0.36
CEC (cmol(+)/kg soil)	5.20	19.7	8.58	11.5	14.2
BS (%)	100	69	71	37	22
M3-P (mg/kg)	96.6	6874	163	236	94.0
M3-K (mg/kg)	616	8911	68.4	108	94.1
M3-Ca (g/kg)	4.09	14.5	2.03	8.22	2.99
M3-Mg (mg/kg)	278	3972	143	344	401
M3-Fe (mg/kg)	65.5	396	524	589	1199
M3-Mn (mg/kg)	20.9	188	29.0	213	185
M3-Cu (mg/kg)	0.02	6.22	9.77	9.95	3.17
M3-Pb (mg/kg)	ND [5]	1.23	10.8	11.7	1.54
M3-Zn (mg/kg)	0.35	62.4	20.4	7.98	5.28

[1] The pH and electrical conductivity (EC) of biochar and compost were measured using 1:5 solid: solution ratio after shaking for 30 min in deionized water; [2] Biochar EC was measured after shaking biochar-water mixtures (1:5 solid: solution ratio) for 24 h; [3] Soil pH was determined in soil-to-deionized water ratio of 1:1 (g mL^{-1}) and in soil-to-1N KCl ratio of 1:1 (g mL^{-1}); [4] carbonate content; [5] ND = not detected.

2. Materials and Methods

2.1. Soil Characterization

Three representative rural soils derived from different parent material in Taiwan were selected for the incubation experiment. The Pingchen (Pc) soil series is a relict tertiary Oxisol (slightly acidic Oxisol, SAO) in Northern Taiwan [19]. The Erhlin (Eh) soil series is an Inceptisol (mildly alkaline Inceptisol, MAI) developed from calcareous slate old alluvial parent material in Central Taiwan. The Annei (An) soil series is also an Inceptisol (slightly acid Inceptisol, SAI) developed from calcareous sandstone-shale new alluvial parent material in Southern Taiwan. Rice is the commonly grown crop in the sampled fields. Soil samples were collected in spring 2011 from the upper layers (0–15 cm) of three fields in Taiwan. Field moist soil samples were air-dried at room temperature (25–28 °C), gently crushed and passed through a 2-mm sieve, and then stored at room temperature for physiochemical analysis. The physical and chemical characteristics of the top soils (15 cm depth) are presented in Table 1.

Soil pH was determined in a soil-to-deionized water ratio of 1:1 (g/mL) and in soil-to-1 N KCl ratio of 1:1 (g/mL) [20], and electrical conductivity (EC) was determined by saturation extract of the soil sample [21]. Soil particle size was determined using the pipette method [22]. Soil total C (TC) content was determined by dry combustion [23], using an O.I. Analytical Solid Total Organic Carbon (TOC) (O.I. Corporation/Xylem, Inc., College Station, TX, USA). Soil TC was assumed to be organic in nature because the low or neutral soil pH precludes carbonates. Soil total nitrogen (TN) content was extracted by digesting a 1.0 g dried and powdered sample using concentrated H_2SO_4 in a Kjeldahl flask using K_2SO_4, $CuSO_4$, and Se powder as catalysts. TN concentration was determined via O.I. Analytical Aurora Model 1030W (O.I. Corporation/Xylem, Inc., College Station, TX, USA); content of soil total phosphorus (TP) in the digested solution was determined with inductively coupled plasma optical emission spectrometry (ICP-OES) (PerkinElmer, Inc., Optima 2100DV, Waltham, MA,

USA). The exchangeable bases (Ex-K, Na, Ca, and Mg), cation exchangeable capacity (CEC), and base saturation percentage (BS%) were measured using the ammonium acetate method at pH 7 [24]. Mehlich-3 extraction [25] was used for analysis of plant available nutrients. Mehlich-3 extractable (M3-) K, Na, Ca, Mg, Fe, Mn, Cu, Pb, Zn, and P values were measured with ICP-OES.

2.2. Studied BC

BC produced from lead tree (*Leucaena leucocephala* (Lam.) de. Wit) in an earth kiln was constructed by the Forest Utilization Division, Taiwan Forestry Research Institute, Taipei, Taiwan [26,27]. The charring for earth kilns typically requires several days and reaches temperatures of 500 °C to 700 °C. The highest temperature in the kiln at the end of carbonization was above 750 °C. The BCs were homogenized and ground to <2 mm mesh for analyses. The characterization of the studied BC was described in previous studies [28,29] (Table 1).

2.3. Incubation Experiment

To investigate the effect of biochar on C mineralization of compost excessive application soils, in this study, 5% commercially available swine manure compost was added as a soil fertilizer, which is twice the recommended amount of organic fertilizer in Taiwan. The economic viability of 5% manure compost is highly unlikely for most farmers, but that is not the objective of the present work.

In amended soils, laboratory incubation is generally used to obtain accurate information about C-mineralization dynamics [30], and the data can then be fitted to or with kinetic models to obtain complementary information, such as C-mineralization rates and the potentially mineralizable C. Therefore, a laboratory aerobic incubation experiment was conducted over 434 days to study and evaluate C-mineralization kinetics in a non-amended (no BC addition) soil sample (i.e., the control) and in three soils amended with three BC application rates. A total of 12 treatments were used in this study, and each treatment was set in five replicates. The application rate of BC, 0%, 0.5%, 1.0%, and 2.0% (w/w), equated to field applications of approximately 0, 9, 18, and 36 tons/ha, respectively, considering 1800 Mg of soil per hectare (soil bulk density equal to 1.2 Mg/m^3 and an arable soil layer of 15 cm). Twenty-five grams of mixed soil sample was placed in 30-mL plastic containers, which were subsequently put into 500-mL plastic jars containing a vessel with 10 mL of distilled water to avoid soil desiccation, and a vessel with 10 mL of 1 M NaOH solution to trap evolved CO_2. The jars were sealed and incubated at 25 °C. Soil moisture content was adjusted to 60% of field capacity before the incubation and was maintained throughout the experiment using repeated weighing. The incubation experiment was run for 434 days with 23 samples taken after 1, 3, 7, 14, 21, 28, 35, 42, 49, 56, 63, 77, 91, 105, 119, 133, 161, 189, 217, 245, 308, 371, and 434 days. After sampling, the vessel with 10 mL of 1 M NaOH solution was removed, resealed, and stored until analysis for CO_2 and replaced with fresh NaOH. A titrimetric determination method was used to quantify the evolved CO_2 [31]. The cumulative CO_2 released and C mineralization kinetics were calculated based on the amount of CO_2–C released during different intervals of time in each treatment. In addition, total mineralization coefficient (TMC) was calculated according to Díez et al. [32] and Méndez et al. [33], as follows:

$$\text{TMC (mg } CO_2\text{-C/g C)} = CO_2\text{-C evolved/initial TOC} \quad (1)$$

where CO_2-C evolved is expressed as mg CO_2-C/100 g soil and initial total organic carbon (TOC) is expressed as g C/100 g soil.

Samples of the BC-treated soil were collected after incubation for 434 days or analysis of plant available nutrients using Mehlich-3 extraction (M3-) [25]. M3-K, Na, Ca, Mg, Fe, Mn, Cu, Pb, Zn, and P values were measured with ICP-OES. To compare the changes and quantify the impacts of soil BC amendments on nutrients, soil pH, TC, TN, TP, exchangeable bases (Ex-K, Na, Ca, Mg), and CEC of the BC-treated soil on day 434 were also measured.

2.4. Statistical Analysis

The statistical analyses (calculation of means and standard deviations, differences of means) were performed using SAS 9.4 package (SAS Institute, Inc., SAS Campus Drive, Cary, NC, USA). The results were analyzed by analysis of variance (one-way ANOVA) to test the effects of each treatment. The statistical significance of the mean differences was determined using least-significant-difference (LSD) tests based on a t-test at a 0.05-probability level. The Pearson correlation coefficient (r) was calculated and principle component analysis (PCA) was performed using SAS 9.4 software. The multivariate statistical technique of PCA was used to investigate the most susceptible variances and to identify the important components explaining most of the variances in a large data set.

3. Results

3.1. Carbon Mineralization

The addition of woody BC showed significantly reduced CO_2 release in SAO soil, no significantly difference in MAI soil, and a significant increase in SAI soil (Figure 1 and Table 2). In SAO soil treatments, the CO_2-C release reduced by about 8.8%, 7.0%, and 9.4% for 0.5%, 1.0%, and 2.0% BC addition rates, respectively. No significant difference was observed in the MAI soil treatments; the CO_2-C release reduced by about 8.8%, 7.0%, and 9.4% for 0.5%, 1.0% and 2.0% BC addition rates, respectively. In contrast, in SAI soil treatments, the CO_2-C release increased by about 6.2%, 15.3%, and 7.9% for 0.5%, 1.0%, and 2.0% BC addition rates, respectively. The results of total mineralization coefficient (TMC) indicated a significantly reduced trend with increasing BC addition in SAO and MAI soil treatments; however, in SAI soil, only the 2% addition showed a significantly decrease, in comparison with the control. The value of TMC was higher in SAI soil treatments, followed by MAI soil treatments, and much lower in SAO soil treatments. The TMC value decreased by 16.5%, 24.0%, and 37.8% for 0.5%, 1.0%, and 2.0% BC additions to SAO soil, respectively. In MAI soil, TMC reduced by 19.6%, 20.7% and 32.5% for 0.5%, 1.0%, and 2.0% BC additions, respectively. In SAI soil, TMC reduced by 0.7% and 19.8 for 0.5% and 2.0% BC addition, respectively, but increased by 2.0% for 1.0% BC addition. We hypothesized that woody BC addition may stabilize compost organic matter and diminish C mineralization in soils with excessive compost application, and the results showed that addition of woody BC to SAO soil produced a favorable effect by decreasing the cumulative amount of CO_2–C evolution, but in SAI soil, it produced an unfavorable effect by increasing the cumulative amount of CO_2–C evolution. We observed no effect in MAI soil.

Table 2. CO_2-C evolved (mg C/100 g dry weight) and total mineralization coefficient (TMC) for control and amended soils after the incubation experiment [1].

Rate	CO_2 Evolved (mg C/100 g Dry Weight)	TMC (mg CO_2-C/g C)
SAO Soil		
0%	842 ± 8.7 A	333 ± 3.4 A
0.5%	768 ± 18 B	278 ± 6.4 B
1.0%	783 ± 15 B	253 ± 4.7 C
2.0%	763 ± 21 B	207 ± 5.7 D
MAI Soil		
0%	829 ± 30 A	526 ± 19 A
0.5%	782 ± 18 A	423 ± 9.6 B
1.0%	797 ± 17 A	417 ± 8.7 B
2.0%	803 ± 10 A	355 ± 4.5 C
SAI Soil		
0%	692 ± 20 C	455 ± 14 A
0.5%	735 ± 18 BC	452 ± 11 A
1.0%	798 ± 24 A	464 ± 14 A
2.0%	747 ± 10 B	365 ± 4.9 B

[1] Each value is the average ± standard deviation from three independent experiments. Means compared within a column, followed by a different uppercase letter, are significantly different at $p < 0.05$ using a one-way ANOVA (multiple comparisons vs. studied soil + 0% biochar as a control).

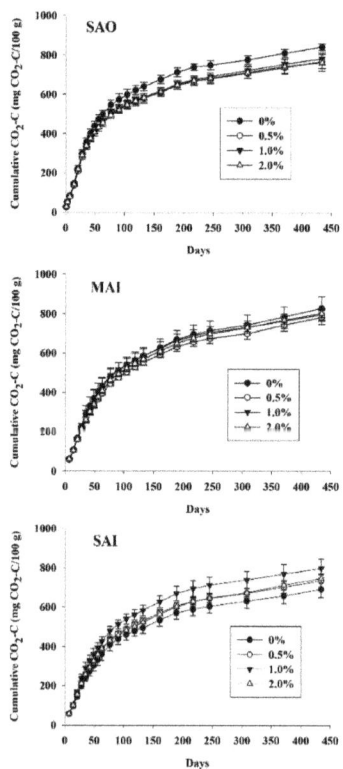

Figure 1. Cumulative CO_2-C (mg CO_2-C/100 g soil) from the three studied soils treated with 0%, 0.5%, 1.0%, and 2.0% woody biochar. Error bars indicated standard deviation of the mean.

3.2. Changes in Soil Properties and Fertility Characteristics

After 434 days of incubation, all treatments were analyzed to investigate if BC addition could result in increasing (enhancing) or decreasing (reducing) soil properties and fertility characteristics in compost over-applicated soils (Table 3). The enhancing effect on soil fertility characteristics suggests that adding BC can retain nutrients in compost over-applicated soils, even after one year of incubation. The high amount of nutrients retained in the soils at the end of the study period suggests that the farmer could apply less compost in the following year.

Table 3. Mean values of total soil carbon (TC), nitrogen (TN), and phosphorus (TP), soil pH, exchangeable bases (K, Na, Ca, and Mg), and cation exchangeable capacity (CEC) of four treatments of three soils after 434-day incubations [1].

Rate	pH	Ex-K	Ex-Na	Ex-Ca	Ex-Mg	CEC	TC	TN	TP	C/N
			-coml(+)/kg soil-					-g/kg-		
SAO Soil										
0%	5.66 b	2.55 b	0.72 b	14.9 a	3.58 a	16.4 a	23.9 c	4.37 ab	1.55 c	5.5 c
0.5%	5.75 b	2.87 a	0.91 a	15.0 a	3.73 a	16.4 a	28.0 b	4.43 a	1.77 b	6.3 b
1.0%	5.76 b	2.40 b	0.73 b	14.4 a	3.36 a	16.0 a	31.8 a	4.28 b	1.69 bc	7.4 a
2.0%	5.93 a	2.55 b	0.63 b	15.5 a	3.41 a	16.3 a	34.5 a	4.27 b	2.21 a	8.1 a
MAI Soil										
0%	7.53 c	2.64 b	0.66 a	22.9 a	3.37 a	9.7 b	18.2 c	3.64 b	0.88 bc	5.0 c
0.5%	7.58 b	2.92 b	0.68 a	25.5 a	3.78 a	10.1 b	21.9 b	3.62 b	0.75 c	6.0 bc
1.0%	7.58 bc	2.92 ab	0.68 a	25.5 a	3.78 a	10.7 a	22.2 b	4.06 a	1.05 ab	5.5 b
2.0%	7.65 a	3.24 a	0.76 a	25.9 a	3.75 a	10.0 b	32.4 a	4.15 a	1.18 a	7.8 a
SAI Soil										
0%	7.04 c	2.14 c	0.59 a	13.9 a	4.24 a	13.4 a	13.7 c	2.86 b	1.26 a	4.8 c
0.5%	7.11 b	2.30 bc	0.54 a	15.3 a	4.52 a	13.3 a	18.3 b	2.89 b	1.11 a	6.3 b
1.0%	7.14 b	2.61 a	0.62 a	15.6 a	4.56 a	13.6 a	21.4 b	3.06 a	0.88 b	7.0 b
2.0%	7.24 a	2.45 ab	0.54 a	14.9 a	4.23 a	12.8 b	26.6 a	3.07 a	0.64 c	8.7 a

[1] Each value is the average of three independent experiments. Means compared within a column followed by a different lowercase letter are significantly different at $p < 0.05$ using a one-way ANOVA (multiple comparisons vs. studied soil + 0% biochar as a control).

At the end of incubation, TC significantly increased with increase of BC addition in the three soils. The significant decreases in CO_2-C evolution and TMC with BC addition increase explain the soil carbon accumulation (sequence) in soils. That is, in this study, BC addition evidently reduced C-mineralization and TMC, and resulted in more soil C sequestrated in soils. TN content significantly increased with 1% and 2% BC addition in MAI and SAI soils, but slightly decreased in SAO soil. The application of woody BC with a high C/N ratio in three soils did not obviously result in soil nitrogen fixation, but in contrast, increased TN content. The TP content significantly increased with 0.5% and 2.0% BC addition in SAO soil and with 2.0% in MAI soil, but significantly decreased with 1.0% and 2.0% BC addition in SAI soil. The C/N ratio significantly increased with BC addition increase, the values of which were all less than 10:1 (Table 3).

Soil pH significantly increased with 2.0% BC addition of three soils—about 0.3 pH unit for SAO soil, 0.1 pH unit for MAI soil, and 0.2 pH unit for SAI soil (Table 3). Within the exchangeable bases, Ca and Mg showed insignificant difference from the control in the three soils, but obviously increased in the MAI and SAI soils. The addition of 0.5% BC resulted in a significant increase in the K and Na contents in SAO soil but a decrease with 1.0% and 2.0% additions. The 2% BC addition in MAI, and 1.0% and 2.0% BC additions in SAI soil significantly increased K content. CEC showed variable changes—significant increases occurred in 1.0% BC addition to MAI soil but significant decreases occurred with 2.0% addition to SAI soil.

In SAO soil, in terms of soil fertility characteristics, M3-P, K, Mg, Fe, and Mn content obviously and significantly decreased with increasing BC addition (Table 4). In contrast, Ca, Cu, Pb, and Zn content increased with increasing BC addition, especially with the 2.0% addition. The amount of Cu,

Pb, and Zn in SAO soil was about 8–9, 10–12, and 26–30 mg/kg, respectively. These values are not very high and cannot result in plant toxicity. However, we should pay more attention to SAO soil, to ensure that these metals are not fixed by BC, and their availability may increase after BC addition. In MAI soil, the amount of P, K, Ca, Mg, Fe, and Mn increased after BC addition, but only K content significantly increased with 1.0% and 2.0% BC addition. Significant decreases of Cu, Pb, and Zn occurred with 0.5%, 1.0%, and 2.0% BC addition (except for Zn with 2.0%). The application of woody BC in MAI soil can help retain some nutrients and significantly reduce heavy metal availability. Similar results for K, Cu and Pb were found for SAI soil. However, P content with 1.0% BC addition and Zn content with 0.5% and 1.0% addition was significantly increased in SAI soil. Ca, Mg, Fe, and Mn content decreased after BC addition. Adding BC to SAI soil could result in some nutrient decrease and reduce the availability of Cu and Pb, but we should pay attention to the risk of increased Zn availability.

Table 4. Mean values of soil fertility characteristics (Mehlich 3 extraction) (mg/kg) of four treatments of three soils after 434-day incubations [1].

Rate	P	K	Ca	Mg	Fe	Mn	Cu	Pb	Zn
SAO Soil									
0%	645 a	461 a	2701 b	533 a	953 a	5.7 ab	8.64 c	10.3 b	26.8 b
0.5%	653 a	467 a	3216 a	556 a	948 a	37.2 a	9.02 bc	10.2 b	28.6 b
1.0%	486 b	408 b	3118 a	444 b	739 b	31.6 c	9.36 ab	11.0 b	27.8 b
2.0%	537 b	457 a	3188 a	474 ab	777 b	3.5 bc	9.83 a	12.3 a	30.6 a
MAI Soil									
0%	769 ab	474 c	7594 a	636 ab	694 ab	286 a	8.73 a	12.9 a	12.2 a
0.5%	671 b	481 bc	7799 a	611 b	621 b	271 a	7.79 b	11.4 b	10.6 b
1.0%	832 a	545 a	7142 a	712 a	739 a	310 a	7.60 b	10.7 b	10.3 b
2.0%	795 a	534 ab	7697 a	660 ab	707 ab	301 a	7.66 b	11.3 b	11.1 ab
SAI Soil									
0%	476 b	384 c	3569 a	750 a	1257 a	197 a	1.70 a	0.84 a	9.48 c
0.5%	462 b	392 bc	3292 a	712 a	1147 a	186 a	1.66 a	0.67 ab	10.1 b
1.0%	564 a	474 a	3313 a	759 a	1200 a	196 a	1.54 b	0.57 bc	11.1 a
2.0%	470 b	437 ab	3648 a	726 a	1183 a	194 a	1.29 c	0.48 c	9.53 bc

[1] Each value is the average of three independent experiments. Means compared within a column followed by a different lowercase letter are significantly different at $p < 0.05$ using a one-way ANOVA (multiple comparisons vs. studied soil + 0% biochar as a control).

3.3. Principal Component Analysis

The PCA described substantial differences in soil physicochemical characteristics (pH, TC, TN, TP, M3-P, M3-K, M3-Cu, M3-Pb, and M3-Zn), and cumulative CO_2–C among the BCs (Figure 2). The PCA identified two primary components of SAO soil fertility, and PC1 and PC2 accounted for 49.1% and 21.0% of the total variance, respectively. AdditioPC1 and PC2 explained 43.0% and 19.8%, and 52.3% and 23.3% of the total variance in the MAI and SAI soil, respectively.

PCA showed two groupings for each of the three soils. The two grouping of SAO soil were: pH, TC, TP, M3-Pb, M3-Zn, and M3-Cu (Group 1); and TN, M3-P, M3-K, and cumulative CO_2-C (Group 2). The 2% BC addition was clustered near Group 1, whereas the 0.5% BC addition was clustered closer to Group 2. For the MAI soil, two groupings stood out: pH, TC, TN, TP, M3-P, and M3-K (Group 1); and M3-Cu, M3-Pb, M3-Zn, and cumulative CO_2-C (Group 2). The addition of 1% BC was clustered near Group 1. Lastly, the PCA for the SAI soil showed two main groupings: pH, TC, TN, M3-P, M3-K, M3-Zn, and cumulative CO_2-C (Group 1); and TP, M3-Cu, and M3-Pb (Group 2). Addition of 1% BC was clustered closer to Group 1, whereas 0.5% BC addition was clustered closely to Group 2.

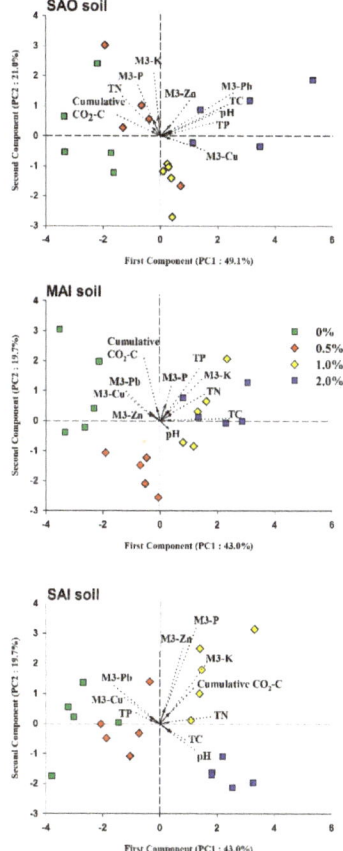

Figure 2. Principal component analysis based on soil chemical characteristics and cumulative CO_2-C (mg CO_2-C/100g soil) after 434-d incubation period in SAO, MAI, and SAI soils treated with 0%, 0.5%, 1.0%, and 2.0% woody BC.

4. Discussion

4.1. Effect of BC on Carbon Mineralization

While proper use of compost promotes soil productivity and improves soil quality, excess application degrades soil and water quality, and inhibits crop growth [34]. The net decrease in CO_2 emission with BC is clear, both directly through sequestration of BC-C and indirectly by altering the soil's physical, chemical, and microbiological properties [5,35]. The BC used in our study was a high-temperature pyrolysis product of wood with an accumulation of black C. This property makes it inert and resistant to microbial degradation [36]. In this study, we hypothesized that the addition of relatively small amounts of woody BC to soils with excess swine manure compost application could stabilize compost organic matter and decrease C mineralization. Decreasing C mineralization could contribute to reducing the decomposition of compost organic matter, enhance C sequestration, retain some nutrients, and may reduce the application rate of manure compost in the following year.

Carbon mineralization in each soil type was obviously greater in the initial days of incubation (Figure 1), especially on the first day, as reported in other studies [37–39]. Swine manure compost contains a significant amount of easily degradable organic C, and consequently, intense increases in soil microbial activity should occur after its application to soil, leading to high C mineralization. The BC treatments significantly reduced C mineralization in SAO soil, and showed insignificant difference in MAI soil (Table 2), but significantly increased C mineralization in SAI soil (1.0% and 2.0% BC treatments). Mukome et al. [40] showed that emissions of CO_2 from the interaction of BC with compost organic matter (COM) are dependent on the BC feedstock and pyrolysis temperature; however, net CO_2 emissions were less for the BC and compost mixtures compared to compost alone, suggesting that BC may stabilize COM and diminish C mineralization. The presence of easily metabolized organic C or additional labile organic carbon sources has been shown to accelerate BC decomposition (or increased soil CO_2 effluxes) [41–44], suggesting that co-metabolism contributes to BC decomposition in soils. Respiration per unit of TOC (TMC) of the three studied soils significantly decreased with increasing BC addition. The four treatments in SAO soil had significantly lower TMC than in MAI and SAI soils. Méndez et al. [33] suggested that a high TMC results in a more fragile humus and thus in a lower quality soil. In contrast, the lower TMC means that organic matter is conserved more efficiently and maintains activity of the microorganisms responsible for soil organic matter biodegradation.

BC amendments clearly have effects on soil CO_2 evolution, which vary with soil type. In coastal saline soil (pH 8.09), peanut-shell-derived BC addition increased the cumulative CO_2 emissions and the cumulative SOC mineralization due to labile C released from BC and enhanced microorganism proliferation [37]; however, the increased mineralized C only accounted for less than 2% in the 0.1%–3% BC treatments, indicating that BC may enhance C sequestration in saline soil. Rogovska et al. [14] indicated that BC additions sometimes increase soil respiration and CO_2 emissions, which could partially offset C credits associated with soil BC applications, and many uncertainties are related to estimation of mineralization rates of BC in soils. In this study, the result of CO_2 evolution and TMC both suggest that when adding excess swine manure compost to Oxisols, a higher BC application rate can stabilize and prevent rapid mineralization of compost. BC addition in mildly alkaline Inceptisols can stabilize compost organic matter, but only slightly decrease mineralization of the compost. In slightly acid Inceptisols, a higher BC application rate can stabilize compost organic matter but may significantly increase mineralization of the compost.

4.2. Effect of BC on Soil Properties and Fertility Characteristics

In the tropics, natural rapid mineralization of soil organic matter is a limitation of the practical application of organic fertilizers, despite it enhancing soil fertility [34]. Thus, the repeated application of organic materials at high doses can significantly contribute to global warming, plant toxicity, accumulation of heavy metals in plants, and ground and surface water pollution due to nutrient leaching. Some studies have indicated that the simultaneous application of BC and compost resulted in enhanced soil fertility, water holding capacity, crop yield, and C sequestration [45–48]. Schulz and Glaser [48] found that overall plant growth and soil fertility decreased in the order of compost > BC + compost > mineral fertilizer + BC > mineral fertilizer > control. The combination of BC with mineral fertilizer further increased plant growth during one vegetation period but also accelerated BC degradation during a second growth period. A combination of BC with compost showed the best plant growth and C sequestration, but had no effects on N and P retention. The blending of BC with compost has been suggested to enhance composting performance by adding more stable C and creating a value-added product (BC-compost blend) that can offset both the potential negative effects of the composting system and the pyrolysis BC system [16].

In addition to diminishing C mineralization in soils, we further examined the positive or negative effects of soil nutrients and heavy metals on mineralization and availability after 434 days of incubation. The results suggested that the effects of adding woody BC vary with soil type and element (Table 4). Without amendment with compost, the soils used in this study had low plant available contents of some nutrients, as well as low CEC. Soils with low CEC are often not fertile and are vulnerable to soil acidification [47]. The CEC of the studied soils followed the order: SAI soil > MAI soil > SAO soil. After incubation, the soil pH of the four treatments in SAO soil (Table 3) were lower than in bulk soil

(Table 1), suggesting low soil buffering capacity, and that soil acidification occurred after adding excess manure compost. In Dystric Cambisol with a loamy-sand texture, a maize (*Zea mays* L.) field trial with five treatments (control, compost, and three BC-compost mixtures with constant compost amount (32.5 Mg/ha) and increasing BC amount, ranging from 5–20 Mg/ha) was conducted [13]; the results demonstrated that total organic C content could be increased by a factor of 2.5 from 0.8% to 2% ($p < 0.01$) at the highest BC-compost level, compared to the control. TN content only slightly increased and plant-available Ca, K, P, and Na content increased by factors of 2.2, 2.5, 1.2, and 2.8, respectively. Trupiano et al. [15] indicated that, when compared to the addition of compost alone, the compost and BC combination did not improve soil chemical characteristics, except for an increase in total C and available P content. These increases could be related to BC capacity to enhance C accumulation and sequestration, and to retain and exchange phosphate ions by its positively charged surface sites. Oldfield et al. [16] suggested that BC recycles C and P; whereas compost recycles C, N, P, and K; and a blend of both resulted in the recycling of C, N, P, and K. Regional differences were found between BC, compost, and the BC-compost blend, and the BC-compost blend offered benefits in relation to available nutrients and sequestered C [16].

4.3. BC Addition Rate Effects on Soil Carbon Mineralization and Soil Fertility

Deteriorating soil fertility and the concomitant decline in agricultural productivity are major concerns in many parts of the world [46]. It is a critical problem in Taiwan. Biochar and biochar-compost applications positively impact soil fertility, for example, through their effect on SOC, CEC, and plant available nutrients [45]. Naeem et al. [49] suggested that the application of BC, in combination with compost and inorganic fertilizers, could be a good management strategy to enhance crop productivity and improve soil properties. Agegnehu et al. [46] indicated that as the plants grew, compost and biochar additions significantly reduced leaching of nutrients; separate or combined application of compost and biochar together with fertilizer increased soil fertility and plant growth. The application of compost and biochar improved water and nutrient retention in the soil, and thereby the uptake of water and nutrients by the plants [46]. The application of woody BC has potential for stabilizing compost organic matter, diminishing soil C mineralization, and improving soil nutrient availability in soil with excessive compost application, depending on soil type and application rate. Addition of BC in SAO soil and MAI soil led to substantial improvements in physicochemical properties, as well as significant and insignificant lower C mineralization, respectively (Figure 1 and Table 2). The 0.5% BC addition reduced the content of available P and K, and 2% addition could result in the risk of Cu, Pb, and Zn in SAO soil. In MAI soil, 1% addition increased pH and content of TC, TN, TP, M3-P, and M3-K. In contrast, BC addition in SAI soil resulted in significant higher C mineralization. The addition of 1% BC increased in soil pH and the contents of TC, TN, M3-P, M3-K, and M3-Zn, but 0.5% BC addition would reduce the contents of TP, M3-Cu, and M3-Pb.

PCA of the soil properties measured by Speratti et al. [50] found that both BC feedstocks had positive correlations between Ca, Fe, and Mn. Metals such as Fe and Mn, along with lower soil pH, can contribute to the formation of organo-mineral and/or organo-metallic associations that decrease BC mineralization [51]. This can increase BC-C stability in the soil, which may improve soil structure [52]. In this study, the free Fe oxide (dithionate-citrate-bicarbonate extractable) content was very high (43.1 g/kg) in SAO soil, followed by MAI soil, and SAI soil at 8.80 g/kg and 6.96 g/kg, respectively. Along with lower soil pH (<pH 6.0), BC, compost, and soil Fe oxide can contribute to the formation of organo-mineral and/or organo-metallic associations that improve soil structure, stabilize compost organic matter, and decrease mixed-soil C mineralization in SAO soil. The soil pH in MAI soil was highest. The potential of BC to reduce C mineralization in MAI soil was insignificant between the control and BC treatments but showed minor reductions after BC addition treatments. After BC addition, the mixed-soil C mineralization significantly increased, which could contribute to lesser formation of organo-mineral and/or organo-metallic associations due to the lower amount of Fe oxides and higher soil pH (7.1–7.2). Adding two biochars at 2% (*w/w*) composed of lac tree wood and mixed

wood (scrapped wood and tree trimmings), with and without vermicompost or thermocompost at 2% (w/w) in Hawaii in highly weathered soils (Ultisols and Oxisols), Berek et al. [53] indicated that soil acidity, nutrient in the soils, plant growth, and nutrient uptake improved with the amendments compared to the control. Nutrient increases and soil acidity reduction by additions of biochar combined with compost were the probable cause, and the use of locally produced biochars and composts was recommended to improve plant nutrient availability in highly weathered soils [53].

5. Conclusions

In this study, we assessed the capacity of woody BC in soils with excessive compost application to stabilize compost organic matter, diminish C mineralization, and improve nutrient availability in three highly frequent tillage soils in Taiwan (Oxisols, SAO; and Inceptisols, MAI and SAI). The effect of BC addition varied strongly according to soil type. Soil carbon mineralization significantly decreased with increasing BC addition in SAO soil, and produced insignificant changes in MAI soil, but significant increases in SAI soil. Respiration per unit of TOC (TMC) significantly decreased with increasing BC addition. In this study, a higher BC application rate stabilized and prevented the rapid mineralization of swine manure compost. Soil pH, exchangeable bases, and CEC only showed minor increases with increasing BC addition. BC addition had a positive effect on soil fertility, including TC, TN, TP, M3-P, K, Mg, Fe, Mn, Pb, and Zn, but had slightly positive effect on exchangeable Ca and negative effect on extractable Cu. To improve soil nutrient availability, adding BC generally increased the levels of plant macronutrients and reduced the concentrations of micronutrients. The results of PCA, even with low scores, indicated that adding BC has a positive impact on diminishing soil carbon mineralization (carbon sequestration), sustaining soil fertility, and preventing heavy metals contamination in compost over-applied soil. In this study, 1% biochar addition corresponds to 18 tons/ha. As adding a large amount of biochar in open fields would be unrealistic and not economically sustainable, we suggested that adding 0.5% woody BC to three studied soils should be reasonable and appropriate.

Author Contributions: Conceptualization, C.-C.T.; methodology, C.-C.T.; validation, C.-C.T., formal analysis, Y.-F.C.; investigation, C.-C.T. and Y.-F.C.; data curation, C.-C.T. and Y.-F.C.; writing—original draft preparation, C.-C.T.; writing—review and editing, C.-C.T.; supervision, C.-C.T.; funding acquisition, C.-C.T.

Funding: This research received no external funding.

Acknowledgments: The authors thank the National Science Council of the Republic of China for financially supporting this research under Contract No. NSC-100-2313-B-197-001. Special thanks to G. S. Hwang, Forest Utilization Division, Taiwan Forestry Research Institute for supplying the biochars.

Conflicts of Interest: The authors declare no conflict of interest.

References

1. Lal, R. Sequestering carbon and increasing productivity by conservation agriculture. *J. Soil Water Conserv.* **2015**, *70*, 55A–62A. [CrossRef]
2. Bationo, A.; Kihara, J.; Vanlauwe, B.; Waswa, B.; Kimetu, J. Soil organic carbon dynamics, functions and management in West African agro-ecosystems. *Agric. Syst.* **2007**, *94*, 13–25. [CrossRef]
3. Vanlauwe, B.; Bationo, A.; Chianu, J.; Giller, K.E.; Merckx, R.; Mokwunye, U.; Ohiokpehai, O.; Pypers, P.; Tabo, R.; Shepherd, K.D.; et al. Integrated soil fertility management: Operational definition and consequences for implementation and dissemination. *Outlook Agric.* **2010**, *39*, 17–24. [CrossRef]
4. Renck, A.; Lehmann, J. Rapid water flow and transport of inorganic and organic nitrogen in a highly aggregated tropical soil. *Soil Sci.* **2004**, *169*, 330–341. [CrossRef]
5. Steiner, C.; Teixeira, W.G.; Lehmann, J.; Nehls, T.; Vasconcelos de Macêdo, J.L.; Blum, W.E.H.; Zech, W. Long term effects of manure, charcoal and mineral fertilization on crop production and fertility on a highly weathered Central Amazonian upland soil. *Plant Soil* **2007**, *291*, 275–290. [CrossRef]
6. Burger, M.; Jackson, L.E. Microbial immobilization of ammonium and nitrate in relation to ammonification and nitrification rates in organic and conventional cropping systems. *Soil Biol. Biochem.* **2003**, *35*, 29–36. [CrossRef]

7. Hornick, S.B.; Sikora, L.J.; Sterrett, S.B.; Murray, J.J.; Millner, P.D.; Burge, W.D.; Colacicco, D.; Parr, J.F.; Chaney, R.L.; Willson, G.B. *Utilization of Sewage Sludge Compost as a Soil Conditioner and Fertilizer for Plant Growth*; United State Department of Agriculture: Washington, DC, USA, 1984.
8. Zech, W.; Senesi, N.; Guggenberger, G.; Kaiser, K.; Lehmann, J.; Miano, T.M.; Miltner, A.; Schroth, G. Factors controlling humification and mineralization of soil organic matter in the tropics. *Geoderma* **1997**, *79*, 117–161. [CrossRef]
9. Kaur, T.; Brar, B.S.; Dhillon, N.S. Soil organic matter dynamics as affected by longterm use of organic and inorganic fertilizers under maize–wheat cropping system. *Nutr. Cycl. Agroecosyst.* **2008**, *81*, 59–69. [CrossRef]
10. Srivastava, A.; Das, S.; Malhotra, S.; Majumdar, K. SSNM-based rationale of fertilizer use in perennial crops: A review. *Indian J. Agric. Sci.* **2014**, *84*, 3–17.
11. Leita, L.; De Nobili, M. Water-Soluble Fractions of Heavy Metals during Composting of Municipal Solid Waste. *J. Environ. Qual.* **1991**, *20*, 73–78. [CrossRef]
12. Barrow, C.J. Biochar: Potential for countering land degradation and for improving agriculture. *Appl. Geogr.* **2012**, *34*, 21–28. [CrossRef]
13. Liu, J.; Schulz, H.; Brandl, S.; Miehtke, H.; Huwe, B.; Glaser, B. Short-term effect of biochar and compost on soil fertility and water status of a Dystric Cambisol in NE Germany under field conditions. *J. Plant Nutr. Soil Sci.* **2012**, *175*, 698–707. [CrossRef]
14. Rogovska, N.; Laird, D.; Cruse, R.; Fleming, P.; Parkin, T.; Meek, D. Impact of biochar on manure carbon stabilization and greenhouse gas emissin. *Soil Sci. Soc. Am. J.* **2011**, *75*, 871–879. [CrossRef]
15. Trupiano, D.; Cocozza, C.; Baronti, S.; Amendola, C.; Vaccari, F.P.; Lustrato, G.; Di Lonardo, S.; Fantasma, F.; Tognetti, R.; Scippa, G.S. The effects of biochar and its combination with compost on lettuce (*Lactuca sativa* L.) growth, soil properties, and soil microbial activity and abundance. *Int. J. Agron.* **2017**, *2017*, 3158207. [CrossRef]
16. Oldfield, T.L.; Sikirica, N.; Mondini, C.; López, G.; Kuikman, P.J.; Holden, N.M. Biochar, compost and biochar-compost blend as options to recover nutrients and sequester carbon. *J. Environ. Manag.* **2018**, *218*, 465–476. [CrossRef] [PubMed]
17. Harris, R.F.; Karlen, D.L.; Mulla, D.J. A conceptual framework for assessment and management of soil quality and health. In *Methods for Accessing Soil Quality*; Doran, J.W., Jones, A.J., Eds.; Soil Science Society of American Special Publication: Madison, WI, USA, 1996; pp. 61–82.
18. Wang, S.H.; Cheng, C.H.; Chang, L.C.; Lee, L.S.; Lin, B.H.; Li, M.Y. *Investigation and Research of Organic Fertilizers and Existing Compost Resources in Taiwan (Project Final Report)*; Science and Technology Advisory Group of the Executive Yuan: Taipei, Taiwan, 1977.
19. Tsai, C.C.; Chen, Z.S.; Hseu, Z.Y.; Guo, H.Y. Representative soils selected from arable and slope soils in Taiwan and their database establishment. *Soil Environ.* **1998**, *1*, 73–88.
20. Thomas, G.W. Soil pH and soil acidity. In *Methods of Soil Analysis, Part 3. Chemical Methods*; Bigham, J.M., Ed.; Agronomy Society of America and Soil Science Society of America: Madison, WI, USA, 1986; pp. 475–489.
21. Rhoades, J.D. Salinity: Electrical conductivity and total dissolved solids. In *Methods of Soil Analysis, Part 3. Chemical Methods*; Bigham, J.M., Ed.; Agronomy Society of America and Soil Science Society of America: Madison, WI, USA, 1986; pp. 417–435.
22. Gee, G.W.; Bauder, J.W. Particle-size analysis. In *Methods of Soil Analysis, Part 1*, 2nd ed.; Klute, A., Ed.; Agronomy Society of America and Soil Science Society of America: Madison, WI, USA, 1986; pp. 383–411.
23. Tabatabai, M.A.; Bremner, J.M. Automated instruments for determination of total carbon, nitrogen, and sulfur in soils by combustion techniques. In *Soil Analysis: Modern Instrumental Techniques*; Smith, K.A., Ed.; Marcel Dekker: New York, NY, USA, 1991; pp. 261–289.
24. Rhoades, J.D. Cation exchange capacity. In *Methods of Soil Analysis, Part 2*, 2nd ed.; Page, A.L., Miller, R.H., Keeney, D.R., Eds.; Agronomy Society of America and Soil Science Society of America: Madison, WI, USA, 1982; pp. 149–157.
25. Mehlich, A. Mehlich-3 soil test extractant: A modification of Mehlich-2 extractant. *Commun. Soil Sci. Plant Anal.* **1984**, *15*, 1409–1416. [CrossRef]
26. Hwang, G.S.; Ho, C.L.; Yu, H.Y.; Su, Y.C. Bamboo vinegar collected during charcoal making with an earth kiln and its basic properties. *Taiwan J. For. Sci.* **2006**, *21*, 547–557.
27. Lin, Y.J.; Ho, C.L.; Yu, H.Y.; Hwang, G.S. Study on charcoal production with branches and tops wood of *Cryptomeria japonica* using an earth kiln. *Q. J. Chin. For.* **2008**, *41*, 549–558.

28. Tsai, C.C.; Chang, Y.F.; Hwang, G.S.; Hseu, Z.Y. Impact of wood biochar addition on nutrient leaching and fertility in a rural Ultisols of Taiwan. *Taiwan. J. Agric. Chem. Food Sci.* **2013**, *51*, 80–93.
29. Tsai, C.C.; Chang, Y.F. Viability of biochar on reducing C mineralization and improving nutrients status in a compost-treated Oxisols. *Taiwan. J. Agric. Chem. Food Sci.* **2016**, *54*, 74–89.
30. Ribeiro, H.M.; Fangueiro, D.; Alves, F.; Vasconcelos, E.; Coutinho, J.; Bol, R. Carbon-mineralization kinetics in an organically managed Cambic Arenosol amended with organic fertilizers. *J. Plant Nutr. Soil Sci.* **2010**, *173*, 39–45. [CrossRef]
31. Zibilske, L.M. Carbon Mineralization. In *Methods of Soil Analysis, Part 2, Microbiological and Biochemical Properties*; Weaver, R.W., Angle, J.S., Bottomly, P., Eds.; Soil Science of America Inc.: Madison, WI, USA, 1994; pp. 835–863.
32. Díez, J.A.; Polo, A.; Guerrero, F. Effect of sewage sludge on nitrogen availability in peat. *Biol. Fertil. Soils* **1992**, *13*, 248–251. [CrossRef]
33. Méndez, A.; Gómez, A.; Paz-Ferreiro, J.; Gascó, G. Effects of sewage sludge biochar on plant metal availability after application to a Mediterranean soil. *Chemosphere* **2012**, *89*, 1354–1359. [CrossRef] [PubMed]
34. Yun, S.I.; Ro, H.M. Natural ^{15}N abundance of plant and soil inorganic-N as evidence for over-fertilization with compost. *Soil Biol. Biochem.* **2009**, *41*, 1541–1547. [CrossRef]
35. Lehmann, J.; Pereira da Silva, J.; Steiner, C.; Nehls, T.; Zech, W.; Glaser, B. Nutrient availability and leaching in an archaeological anthrosol and a ferralsol of the central Amazon basin: Fertilizer, manure and charcoal amendments. *Plant Soil* **2003**, *249*, 343–357. [CrossRef]
36. Spokas, K.A. Review of the stability of biochar in soils: Predictability of O:C molar ratios. *Carbon Manag.* **2010**, *1*, 289–303. [CrossRef]
37. Liang, B.; Lehmann, J.; Solomon, D.; Sohi, S.; Thies, J.E.; Skjemstad, J.O.; Luizão, F.J.; Engelhard, M.H.; Neves, E.G.; Wirick, S. Stability of biomass-derived black carbon in soils. *Geochim. Cosmochim. Acta* **2008**, *72*, 1598–1610. [CrossRef]
38. Streubel, J.D.; Collins, H.P.; Garcia-Perez, M.; Tarara, J.; Granatstein, D.; Kruger, C.E. Influence of contrasting biochar types on five soils at increasing rates of application. *Soil Sci. Soc. Am. J.* **2011**, *75*, 1402–1413. [CrossRef]
39. Luo, X.X.; Wang, L.Y.; Liu, G.C.; Wang, X.; Wang, Z.Y.; Zheng, H. Effects of biochar on carbon mineralization of coastal wetland soils in the Yellow River Delta, China. *Ecol. Eng.* **2016**, *94*, 329–336. [CrossRef]
40. Mukome, F.N.D.; Six, J.; Parikh, S.J. The effects of walnut shell and wood feedstock biochar amendments on greenhouse gas emissions from a fertile soil. *Geoderma* **2013**, *200–201*, 90–98. [CrossRef]
41. Hamer, U.; Marschner, B.; Brodowski, S.; Amelung, W. Interactive priming of black carbon and glucose mineralisation. *Org. Geochem.* **2004**, *35*, 823–830. [CrossRef]
42. Kuzyakov, Y.; Subbotina, I.; Chen, H.; Bogomolova, I.; Xu, X. Black carbon decomposition and incorporation into soil microbial biomass estimated by ^{14}C labeling. *Soil Biol. Biochem.* **2009**, *41*, 210–219. [CrossRef]
43. Liang, B.; Lehmann, J.; Sohi, S.P.; Thies, J.E.; O'Neill, B.; Trujillo, L.; Gaunt, J.; Solomon, D.; Grossman, J.; Neves, E.G.; et al. Black carbon affects the cycling of non-black carbon in soil. *Org. Geochem.* **2010**, *41*, 206–213. [CrossRef]
44. Novak, J.M.; Busscher, W.J.; Watts, D.W.; Laird, D.A.; Ahmedna, M.A.; Niandou, M.A.S. Short-term CO_2 mineralization after additions of biochar and switchgrass to a Typic Kandiudult. *Geoderma* **2010**, *154*, 281–288. [CrossRef]
45. Agegnehu, G.; Bass, A.M.; Nelson, P.N.; Muirhead, B.; Wright, G.; Bird, M.I. Biochar and biochar-compost as soil amendments: Effects on peanut yield soil properties and greenhouse gas emissions in tropical North Queensland, Australia. *Agric. Ecosyst. Environ.* **2015**, *213*, 72–85. [CrossRef]
46. Agegnehu, G.; Bird, M.; Nelson, P.; Bass, A. The ameliorating effects of biochar and compost on soil quality and plant growth on a Ferralsol. *Soil Res.* **2015**, *53*, 1–12. [CrossRef]
47. Agegnehu, G.; Bass, A.M.; Nelson, P.N.; Bird, M.I. Benefits of biochar, compost and biochar–compost for soil quality, maize yield and greenhouse gas emissions in a tropical agricultural soil. *Sci. Total Environ.* **2016**, *543*, 295–306. [CrossRef] [PubMed]
48. Schulz, H.; Glaser, B. Effects of biochar compared to organic and inorganic fertilizers on soil quality and plant growth in a greenhouse experiment. *J. Plant Nutr. Soil Sci.* **2012**, *175*, 410–422. [CrossRef]

49. Naeem, M.A.; Khalid, M.; Aon, M.; Abbas, G.; Amjad, M.; Murtaza, B.; Khan, W.D.; Ahmad, N. Combined application of biochar with compost and fertilizer improves soil properties and grain yield of maize. *J. Plant Nutr.* **2017**, *41*, 112–122. [CrossRef]
50. Speratti, A.B.; Johnson, M.S.; Martins Sousa, H.; Nunes Torres, G.; Guimarães Couto, E. Impact of different agricultural waste biochars on maize biomass and soil water content in a Brazilian Cerrado Arenosol. *Agronomy* **2017**, *7*, 49. [CrossRef]
51. Speratti, A.B.; Johnson, M.S.; Martins Sousa, H.; Dalmagro, H.J.; Guimarães Couto, E. Biochar feedstock and pyrolysis temperature effects on leachate: DOC characteristics and nitrate losses from a Brazilian Cerrado Arenosol mixed with agricultural waste biochars. *J. Environ. Manag.* **2018**, *211*, 256–268. [CrossRef] [PubMed]
52. Fang, Y.; Singh, B.; Singh, B.P.; Krull, E. Biochar carbon stability in four contrasting soils. *Eur. J. Soil Sci.* **2014**, *65*, 60–71. [CrossRef]
53. Berek, A.K.; Hue, N.V.; Radovich, T.J.K.; Ahmad, A.A. Biochars improve nutrient phyto-availability of Hawai'i's highly weathered soils. *J. Agron.* **2018**, *8*, 203. [CrossRef]

© 2019 by the authors. Licensee MDPI, Basel, Switzerland. This article is an open access article distributed under the terms and conditions of the Creative Commons Attribution (CC BY) license (http://creativecommons.org/licenses/by/4.0/).

Article

Valorization of Vineyard By-Products to Obtain Composted Digestate and Biochar Suitable for Nursery Grapevine (*Vitis vinifera* L.) Production

Domenico Ronga [1,2,*], Enrico Francia [1], Giulio Allesina [3], Simone Pedrazzi [3], Massimo Zaccardelli [4], Catello Pane [4], Aldo Tava [2] and Cristina Bignami [1]

1. Department of Life Science, Centre BIOGEST-SITEIA, University of Modena and Reggio Emilia, Via Amendola, n. 2, 42122 Reggio Emilia (RE), Italy
2. Council for Agricultural Research and Economics-Research Centre for Animal Production and Aquaculture, Viale Piacenza, 29, 26900 Lodi, Italy
3. Department of Engineering 'Enzo Ferrari', University of Modena and Reggio Emilia, Via Vivarelli 101, 41125 Modena, Italy
4. Council for Agricultural Research and Economics-Research Centre for Vegetable and Ornamental Crops, Via Cavalleggeri, 25, 84098 Pontecagnano Faiano (SA), Italy
* Correspondence: domenico.ronga@unimore.it; Tel.: +39-0522-522064

Received: 27 May 2019; Accepted: 22 July 2019; Published: 1 August 2019

Abstract: Although compost and biochar received high attention as growing media, little information is available on the potential of vineyard by-products for the production and use of composted solid digestate (CSD) and biochar (BC). In the present study, two experiments are reported on CSD and BC mixed with commercial peat (CP) for grapevine planting material production. Four doses (0, 10%, 20%, 40% vol.) of CSD and BC were assessed in the first and second experiment, respectively. CSD mixed at a dose of 10% recorded the highest values of shoot dry weight (SDW) and a fraction of total dry biomass allocated to shoot (FTS), both cropping bench-graft and bare-rooted vine. On the other hand, CSD mixed at a dose of 40% displayed the highest values of SDW and FTS, cropping two-year-old vine. BC used at a dose of 10% improved SDW, root dry weight, total dry weight, FTS, shoot diameter, and height on bare-rooted vine. The present study shows that CSD and BC, coming from the valorization of vineyard by-products, can be used in the production of innovative growing media suitable for nursery grapevine production. Further studies are needed to assess the combined applications of CSD and BC in the same growing media.

Keywords: vineyard by-products; composted solid digestate; biochar; grapevine planting material

1. Introduction

The growing demand for grapevine (*Vitis vinifera* L.) planting materials, due to the increasing worldwide viticulture, is promoting research studies to obtain useful guidelines for improving vineyard sustainability [1]. In addition, the losses caused by the failures, such as poor establishment or vigor of the vines and infection of trunk disease pathogens, are a significant but often unrecognized burden for nurseries, farmers, and wineries [2–4].

Among several agronomic practices, soil and vineyard canopy management received particular attention. In fact, soil and canopy management play a fundamental role on the vegetative and reproductive development of the vines [5,6]. Soil management practices include the fertilizer administration able to improve and/or maintain soil fertility, satisfying grapevine nutrient requests [7].

Field production and planting in the vineyard of dormant bare-rooted vines may be negatively affected by some biotic and abiotic stresses, such as root dehydration, contamination by soil borne pathogens, and frost damage [8]. Thus, to overcome these problems, the production in greenhouses of forced bench-grafts or dormant bench-grafts grown in containers is a successful alternative to open-field propagation, allowing the transplanting of high-quality plants at any time of the year.

In greenhouse nursery, peat is the most common growing medium used by growers; however, peat is an expensive and non-renewable material [9]. Hence, the reduction of exploitation of peatlands received high attention by researchers, and currently different materials are investigated as substitute growing media [10].

Growing media originated from the valorization of agro-industrial by-products might be an alternative to the consumption of peat. Among them, compost and biochar can be an interesting ingredient for alternative and sustainable substrates [11,12]. On the other hand, one critical point when compost and biochar are used at high doses is ascribed to the alkalinization of growing media, causing a reduced nutrient availability for plants [12].

Although several studies performed in a greenhouse nursery showed that compost and biochar, used at low doses, are able to improve plant growth [9,13], few studies were performed on grapevine planting material productions [14–16]. On the other hand, working in the open field, compost and biochar were used in large amounts, up to 30 and 44 t ha^{-1}, respectively [14,17]. Moreover, to expand the range of agro-industrial by-products used as alternative substrates, the investigation of new feedstock materials should be performed. In this framework, little attention was paid to vineyard by-products.

Vineyard biomass is characterized by underutilized by-products (such as vine pruning residues and winery outstream, such as grape stalks) that can be valorized [18] by processing them in compost and other products, thus reducing the vineyard carbon footprint [19,20]. Vineyard winter pruning wood is usually destroyed by infield burning or crushing onto the soil, and the same occurs for grape stalks. On the other hand, a better valorization of vineyard pruning and winery out stream as feedstock for biogas plants [21–23] and pyrolysis [24] might increase vineyard sustainability providing green energy and promising fertilizers, such as digestate and biochar [25]. Considering that little information is reported on the use of compost and biochar coming from the valorization of vineyard by-products, the present study aimed to close the waste cycle in vineyards via the reuse of by-products.

In a vineyard, a high correlation between the development of the belowground and aboveground organs were demonstrated [26,27]. Thus, our objective was to assess the potential benefits of composted solid digestate (CSD) and biochar (BC) coming from vineyard by-products on different grapevine planting materials grown in the greenhouse. Two independent experiments were carried out. In the first experiment, CSD was assessed on greenhouse forced bench-grafts, bare-rooted vines, and two-year-old vines, while in the second experiment, BC was assessed on bare-rooted vines.

2. Materials and Methods

2.1. Composted Solid Digestate and Biochar Productions

Digestate containing grape stalks and coming from a biogas plant was composted with chips (1 cm in length) of vineyard winter pruning residues. Moreover, chips of vineyard prunings were also used in a gasifier to obtain biochar. Grape stalks and prunings derived from a vineyard of 'Lambrusco Salamino', grafted on 'Kober 5 BB' rootstock, located in the provinces of Modena and Reggio Emilia, Italy.

CSD was produced at the University of Modena and Reggio Emilia, located at Reggio Emilia, Italy, through a static pile on a farm composting for 105 days, turned weekly. Solid digestate (SD) was conferred by a local biogas plant (CAT, Correggio, Italy) and composted with vineyard prunings chips, mixed at a ratio of 15.0%–83.3% in dry weight, respectively. Feedstocks used in the anaerobic digestion were maize (*Zea mays* L.) silage (43%), triticale (X Triticosecale Wittmack) silage (22%), cow slurry (27%), and grape stalks (8%). The 1 m^3 pile was wetted through an irrigation system and manually activated on demand when the pile gravimetrically determined relative humidity (RH) was <50%. An aliquot of 1.7% dry weight of mature compost was added to the pile as a composting starter. Composting temperatures were measured by thermoresistance sensors (PT100, Gandolfi, Parma, Italy), placed in the center of the pile at 15 cm from its base. A total of 35 days of the thermophilic phase were followed by a further 2-month curing period.

The biochar used in this study was produced in a PP20 gasifier, a commercial biomass-to-power unit manufactured by the US company, ALL Power Labs (Berkeley, CA, USA). The gasification reactor consists of a single throat, fixed bed, downdraft system. The reactor is coupled with a 4 cylinders internal combustion engine capable of producing 20 kW of electrical power at 60 Hz, or 16 kW at 50 Hz.

The biochar is extracted below the reduction zone of the reactor, where temperature decreased from the 900–950 °C of the combustion zone down to 650–700 °C of the end of the endothermic gasification zone. The temperature in the extraction point is much higher than the average dew point of polycyclic aromatic hydrocarbons, thus reducing the chance to find these compounds as toxic contaminants into the char. Biomass preparation consisted in the manual chipping of the vineyard prunings using a specifically designed rotary valve (Torex RWN05, Modena, Italy).

2.2. Compost and Biochar Characterisations

CSD and BC were analyzed according to the respective procedures indicated below. The content of different elements (Ca, Mg, Na, Cr, Cu, Cd, Ni, Fe, Pb, and Zn) was determined after acid digestion with a microwave oven, according to EPA 3052 (EPA 3052, 1996), with an ICP-OES (iCAP 6000 Series, Thermo Scientific, Waltham, MA, USA) on the basis of the EPA 6010D 2014 standard. The following parameters were detected according to the respective procedures: Total organic carbon (C) (UNI EN 13137:2002); total nitrogen (N) (UNI EN 13654–1:2001 and ISO 11261:1995); P_2O_5 and K_2O contents (UNI EN 13650:2002 and UNI EN ISO 11885:2009). The water content of the CSD and BC was determined after drying at 105 °C for 72 h. pH and electrical conductivity (EC) were determined on wet material (1:5 ratio) using a CRISON pH meter basic 20 (Crison Instrument, Barcelona, Spain) and CRISON GLP 31 EC meter (Crison Instrument, Barcelona, Spain), respectively. The growing media water capacity—the amount of water content held in the growing media after excess water has drained away and the downward movement has stopped—was determined by the gravimetric method, with two replicates for each control and treatment.

2.3. Phytotoxicity Characterisations of CSD and BC

CSD was evaluated for phytotoxicity, as suggested by Zucconi et al. [28]. A total of 20 seeds of garden cress (*Lepidium satibum* L.) were incubated at 20 °C in three-replicated Petri dishes, in which 4 mL of CSD water extract (50 g L^{-1}) was poured over sterile filter paper. At 36 h after germination, the germination index percentage (GI%) was calculated on both roots and shoots. For roots, the following formula was used:

$$GI\% = 100 \times (G1/G2) \times (R1/R2) \quad (1)$$

where G1 and G2 are germinated seeds in the sample and control and R1 and R2 are the mean root length for the sample and for the control, respectively. For shoots, a similar formula adopted for roots, was used:

$$GI\% = 100 \times (G1/G2) \times (S1/S2) \quad (2)$$

where G1 and G2 are germinated seeds in the sample and control and S1 and S2 are the mean shoot length of the sample and control, respectively.

2.4. Microbiological and Suppressiveness Characterisations of CSD

The abundance of culturable filamentous fungi, total bacteria, and spore-forming bacteria in CSD were evaluated by a serial 10-fold (10^{-1} to 10^{-7}) dilution method in triplicate. Fungi were counted on the potato dextrose agar (PDA, Oxoid) pH 6, supplemented with 150 mg L^{-1} of nalidixic acid and 150 mg L^{-1} of streptomycin. Total bacteria were counted on the selective medium (glucose 1 g L^{-1}, protease peptone 3 g L^{-1}, yeast extract 1 g L^{-1}, potassium phosphate buffer 1 g L^{-1}, agar 15 g L^{-1})

supplemented with 100 mg L^{-1} of actidione. Spore-forming bacteria were counted by plating 10-fold dilutions of CSD on nutrient agar previously heated at 90 °C for 10 min. Population densities expressed as a colony forming unit (CFU) g^{-1} dry weight of CSD. Coliform, *Escherichia coli*, and *Salmonella* spp. detection were performed following the methods reported by Cekmecelioglu et al. [29].

The suppressiveness ability of CSD on two important soil-borne pathogens, such as *Rhizoctonia solani* and *Sclerotinia minor*, was assessed testing garden cress as a host plant. Potting mixes were done by amending commercial peat (CP) with CSD at a rate of 20% by vol. [30]. The bioassays were performed on 10 pots per treatment, and the pots filled up only with non-amended CP were used as the control. Pathogen inoculations and pot assessment were performed as reported by Pane et al. [31]. The bioassay was done in duplicate.

2.5. Growing Media Preparation and Characterisation

In the first experiment, four different growing media (GM) were composed by mixing ingredients as follows (% vol.): Commercial peat (CP) (Fondolinfa® Universale, Linfa Spa, RE, Italy) 100% (GM1); CP 90% + CSD 10% (GM2); CP 80% + CSD 20% (GM3); CP 60% + CSD 40% (GM4). In the second experiment, four different GM were tested mixing ingredients as follows (% vol.): CP 100%; CP 90% + biochar (BC) 10% (GM2); CP 80% + BC 20% (GM3); CP 60% + BC 40% (GM4). CP exhibited 70% of organic matter, 35% of organic carbon, and 0.6% N. Fertigation was performed once a week throughout the growth cycle, administering, in total, 6 g of N per pot. Growing media pH and EC were determined on the wet material (1:5 ratio) using a CRISON pH meter basic 20 and a CRISON GLP 31 EC meter, respectively.

2.6. Nursery Greenhouse Experiments

Potting experiments were performed in a nursery greenhouse, located at Reggio Emilia, Italy, with programmed temperature ranging from 19 to 25 °C (day/night), relative humidity ranging from 50% to 70%, and with a natural photoperiod and solar radiation. At present, different grapevine planting materials are available in commerce. For the planting of a new vineyard, the most used material is the 1-year-old dormant bare-rooted vines (hereafter called 'bare-rooted vines'). This material is obtained by bench grafting dormant one-bud cuttings of the cultivar onto dormant hardwood cuttings of the rootstock, followed by callusing in a greenhouse and subsequently transferring in the open field for the development of roots and shoots. For marketability, the materials are excavated, the roots and shoot pruned, and plants sold with the bare root. For late planting in soil with lower water content, or to replace the dead vines in a new vineyard, nursery growers suggest the use of material similar to bare-rooted vines, but obtained using a scion bench grafted on rooted rootstocks and potted for callusing and hardening in a controlled greenhouse (hereafter called 'bench-grafts'). In fact, this material, having a preformed root-ball, could be more suitable for overcoming abiotic stresses than bare-rooted vines. Finally, to replace adult plants (having two or more years, but affected by diseases), the use of 2-year-old vines (obtained cropping bare-rooted vines in pots for two years, and hereafter called '2-year-old vines') is suggested by nursery growers in the vineyard. Hence, considering the availability of different planting material, in the first experiment, the following kinds of vines were used: (1) bench-grafts; (2) bare-rooted vines; (3) 2-year-old vines. Meanwhile, in the second experiment, only the bare-rooted vine was assessed. The combination scion/rootstock was cultivar 'Lambrusco Salamino' grafted on 'Kober 5BB' in all the three types of grapevine planting materials and in both experiments.

Grapevine planting materials were transplanted manually (one plant per pot) in plastic pots (4.5 L *per* bench-graft and bare-rooted vine and 9.0 L per 2-year-old vine) filled with the GM mixtures. Pots were arranged in a completely randomized design with five replicates. Pots were irrigated every night by spinner-type sprinklers in order to maintain the substrate at water capacity. Pests were controlled according to the integrated production rules of the Emilia Romagna Region, Italy. Grapevine planting materials were pruned, leaving one shoot that was trimmed at 120 cm at 259 day of year (DOY) and at 289 DOY in the first and second experiments, respectively. During the growing season, the main phenological growth stages were recorded following the BBCH-scale (BBCH = Biologische Bundesantalt, Bundessortenamt and CHemische Industrie, Germany) [32].

In the first and second experiments, at 177 DOY and 170 DOY, respectively, the following parameters were measured: the leaf chlorophyll content (CHL), leaf flavonoid content (FLAV), nitrogen balance index (NBI) (the ratio between CHL and FLAV, NBI), the leaf anthocyanin content (ANT), the basal shoot diameter, shoot height, number of leaves and nodes, and the shoot height/shoot diameter (H/D) ratio. CHL, FLAV, NBI, and ANT were estimated on the youngest fully expanded leaf by Dualex 4 Scientific (Dx4, FORCE-A, Orsay, France), an optical leaf-clip meter for non-destructive assessment of the physiological status of the plants [33]. At 273 DOY in the first experiment and 275 DOY in the second experiment, CHL, FLAV, NBI, ANT, shoot diameter, shoot height, and H/D ratio were recorded. At 314 DOY in the first experiment and 362 DOY in the second experiment, the shoot fresh weight (SFW), the aboveground part of the rootstock fresh weight (RSFW), the root fresh weight (RTFW), total fresh weight (TFW), shoot dry weight (SDW), aboveground part of rootstock dry weight (RSDW), root dry weight (RTDW), total dry weight (TDW), fraction of total dry weight allocated to shoot (FTS), fraction of total dry weight allocated to the aboveground part of rootstock (FTRS), and fraction of total dry weight allocated to root (FTRT), were recorded.

2.7. Data Analysis

Considering that the agronomic and physiological performance of the fruit-trees could be affected by the plant age and by the nutrients accumulated in the previous cropping seasons, the experimental data were analyzed separately for each grapevine planting material and experiment, using GenStat 17th software (VSN International, Hemel Hempstead, UK) for analysis of variance (ANOVA). Significant means were separated by Duncan's test at $p < 0.05$. Experimental data were also processed for a principal component analysis (PCA) in order to evaluate the existing relationships with original variables and growing media for each grapevine planting material and experiment.

3. Results and Discussion

Grapevines are relatively easy to propagate in the nursery greenhouse. However, high skill and organization are required to produce planting materials with the high standard quality requested every year by growers for new planting or replanting uneconomic vineyards or to replace plants affected by trunk disease pathogens [8].

Chemical and physical growing media characteristics can affect cutting establishment [34]. Growing media should be friable, free of weeds and pathogens, with good water capacity and drainage [35,36]. Among them, peat is the most commonly used, due to its positive hydrological, physiochemical, and agronomic characteristics [9,37]. However, to improve agricultural sustainability, researchers are called to study alternatives to peat utilization as innovative growing media. In the present study, vineyard winter prunings and grape stalks were valorized to obtain CSD and BC. CSD and BC were tested in two experiments as innovative ingredients for growing media preparation, used for the production in nursery greenhouses of grapevine planting material.

3.1. Compost, Biochar, and Growing Media Characteristics

During the composting process, CSD exceeded 60 °C for 16 days, achieving sanitation [38]. The main chemical characteristics of CSD and BC are reported in Table 1.

CSD and BC showed pH values of 7.9 and 10.0 with an EC of 1.5 and 2.5 dS m^{-1}, respectively. Total N in CSD was more than six times compared to CP (3.2% vs. 0.6%), while BC showed a similar value to CP. The C/N ratio was 13.2 and 80.9 for CSD and BC, respectively, suggesting that the CSD was a suitable ingredient for growing media, since a C/N ratio less than 20.0 is required [39]. On the other hand, a direct use of BC is not indicated, suggesting dilution with other growing media. For organic C and total N, CSD and BC were within the framework previously reported by other studies [7,40]. CSD and BC showed levels of heavy-metal content below the limit established for the commercialization of amendment in the European Community, according to the European Regulation CE 2003/2003. According to chemical parameters, CSD and BC appeared to be suitable for growing media preparation, although the high pH of CSD and BC and the high C/N ratio of BC encourages their dilution in a peat-based mixture, as previously suggested by other authors [7,41]. For this reason, in the present study, different growing media were composed by replacing CP from 10% to 40% vol. with CSD and BC.

Table 1. Main chemical characteristics of composted solid digestate (CSD) and biochar (BC).

Parameter	CSD Value	BC Value
pH	7.93	10.00
EC (dS m^{-1})	1.51	2.45
TOC (%)	41.79	56.60
Total N (%)	3.17	0.70
P_2O_5 (%)	18.88	0.11
K_2O (%)	27.47	19.09
Ca (%)	21.26	0.12
Mg (%)	9.81	0.31
Na (%)	0.88	0.22
C/N (–)	13.18	80.85
Cd (mg kg^{-1})	<0.50	<0.50
Ni (mg kg^{-1})	7.42	10.51
Cr (mg kg^{-1})	14.00	4.58
Zn (mg kg^{-1})	152.00	82.26
Cu (mg kg^{-1})	35.40	59.50
Hg (mg kg^{-1})	<0.20	<0.20
Pb (mg kg^{-1})	13.80	1.81
Fe (mg kg^{-1})	3.75	64.20
H_2O (%)	77.30	71.37

Total organic carbon (TOC). All values as dry weight.

The germination assays indicated no phytotoxicity problems, both for the CSD and BC tested in the present study. In fact, the germination index displayed values higher than 50% for both CSD and BC, which is considered the threshold value for phytotoxicity [28]. In particular, the germination index showed higher values for the root and shoot of the sensitive reference species 'garden cress' when treated with a water extract of CSD (Table 2).

Table 2. Phytotoxicity assessment of CSD and BC and microbiological and suppressiveness characterization of CSD.

Index	Value
CSD-GI root (%)	138.00 *
CSD-GI shoot (%)	113.00 *
BC-GI root (%)	68.00 *
BC-GI shoot (%)	59.00 *
Fungi (CFU g^{-1})	9.63E + 06
Bacteria (CFU g^{-1})	3.61E + 02
Coliform bacteria (CFU g^{-1})	0.40E + 01
Escherichia coli (CFU g^{-1})	Absent
Salmonella spp. (CFU g^{-1})	Absent
Clostridia spp. (CFU g^{-1})	Absent
Rhizoctonia solani damping-off (%)	98.47
Sclerotinia minor damping-off (%)	46.56 *

Composted solid digestate (CSD), biochar (BC), germination index (GI). * = statistically different compared to the control.

The microbial community within the compost was shown to be one of the major factors involved in the biological control of soilborne disease through different antagonistic mechanisms linked to the relationship between microbes [42]. Moreover, fungal populations were suggested as the main contributors of biological control in organic matrices [43]. Levels of microbial populations in CSD are shown in Table 2. Populations of total fungi, total bacteria, and *Coliform* bacteria were 9.6E + 06, 3.6E + 02, and 0.4E + 01 (CFU g^{-1}), respectively (Table 2). In addition, a total absence was recorded for *Escherichia coli*, *Clostridia* spp., and *Salmonella* spp., in line with the ecolabel criteria established by Decision 2001/688/CE.

The ability of compost to suppress soil-borne disease is also an important added value when compost is used as a component of growing media [44]. In the present study, suppressive bioassays displayed that garden cress damping-off caused by *S. minor* was significantly reduced by CSD, whereas no effects were reported on the control of *R. solani* (Table 2).

Growing media displayed pH and EC ranging between 7.6 and 8.5 and between 0.2 and 1.5 dS m^{-1}, respectively (Table 3); values which are suitable for the production of several horticultural crops [12,45].

Table 3. Growing media assayed on grapevine planting materials.

Substrates	Formulation	pH		EC (dS m^{-1})		GMWC (%)	
		CSD	BC	CSD	BC	CSD	BC
GM1	CP	7.6	7.6	0.21	0.21	85.0	85.0
GM2	CP 90% + CSD or BC 10%	7.6	7.8	0.22	0.46	83.2	81.3
GM3	CP 80% + CSD or BC 20%	7.7	8.1	0.39	0.87	84.6	79.1
GM4	CP 70% + CSD or BC 40%	7.8	8.5	0.44	1.48	84.8	77.0

Growing media (GM), commercial peat (CP), composted solid digestate (CSD), biochar (BC), electrical conductivity (EC), growing media water capacity (GMWC).

3.2. Grapevine Planting Material Productions

In the first experiment, morpho-physiological and agronomic parameters of bench-grafts, bare-rooted vines, and two-year-old vines were affected by CSD applications that, in general, induced a plant growth comparable to plant crops using only CP (Tables 4–6). In particular, bench-grafts and bare-rooted vines, when grown in amended pots with CSD at 10% (GM2), showed similar or higher values of FLAV, NBI, shoot diameter, shoot height, number of leaves and nodes, and H/D ratio, compared to CP at 177 DOY, corresponding to the BBCH-scale 19 "9 or more unfolded leaves". In addition, the same growing media (GM2) recorded the highest values of RSFW, RTFW, TFW, SDW, and FTS at 314 DOY, corresponding to the BBCH-scale 97 "end of leaf-fall", both on the bench-grafts and bare-rooted vines (Tables 4 and 5).

For two-year-old vines, plants grown in amended pots with CSD at 40% (GM4) highlighted similar or higher values of the CHL, FLAV, ANT, NBI, shoot diameter, shoot height, number of leaves and nodes, and H/D ratio, compared to CP at 177 DOY. Moreover, GM4 recorded the highest values of SDW, RTDW, TDW, FTS, and FTRT at 314 DOY (Table 6).

Table 4. Parameters assessed on greenhouse bench-grafts grown on growing media containing composted solid digestate.

(A) Morphological and physiological parameters recorded at 177 day of year (DOY)

Treatment	CHL (-)		FLAV (-)	NBI (-)	ANT (-)		DIAMETER (mm)	H (cm)	LEAVES (no.)	NODE (no.)	H/D (-)
GM1	22.34	a	1.25	18.15	0.15	b	4.10	140.00	30.20	26.20	357.00a
GM2	17.74	b	1.06	17.65	0.21	a	3.74	135.00	30.20	24.80	370.40a
GM3	17.12	b	1.06	17.19	0.20	ab	4.32	107.00	29.60	23.80	231.90b
GM4	16.60	b	1.10	15.70	0.24	a	4.12	115.00	30.80	24.20	280.1ab

(B) Morphological and physiological parameters recorded at 273 DOY

Treatment	CHL (-)		FLAV (-)		NBI (-)		ANT (-)		DIAMETER (mm)		H/D (-)	
GM1	12.98	a	3.38	a	3.70	b	0.85	a	5.15	a	234.50	b
GM2	10.98	b	3.06	b	3.59	b	0.68	b	4.22	b	283.10	a
GM3	9.45	c	2.58	d	3.67	b	0.46	c	4.15	b	292.90	a
GM4	11.26	b	2.82	c	4.22	a	0.34	d	4.61	ab	261.70	ab

(C) Morphological and physiological parameters recorded at 314 DOY

Treatment	SFW (g plant^{-1})		RSFW (g plant^{-1})		RTFW (g plant^{-1})		TFW (g plant^{-1})		SDW (g plant^{-1})		RSDW (g plant^{-1})		RTDW (g plant^{-1})		TDW (g plant^{-1})	FTS (-)		FTRS (-)		FTRT (-)	
GM1	20.88	a	24.38	d	113.30	c	158.60	c	7.08	b	23.80	a	21.58	d	47.80	0.16	b	0.52	a	0.48	c
GM2	17.62	b	35.17	a	117.90	b	169.90	a	8.61	a	19.32	d	23.94	c	45.00	0.20	a	0.45	b	0.55	b
GM3	13.59	d	30.15	b	105.40	d	149.20	d	5.11	d	22.80	b	27.40	b	46.33	0.10	d	0.45	b	0.55	b
GM4	15.64	c	26.77	c	123.50	a	165.90	b	6.50	c	20.26	c	32.21	a	55.00	0.12	c	0.39	c	0.61	a

GM1 = commercial peat (CP) 100%; GM2 = CP 90% + composted solid digestate (CSD) 10%; GM3 = CP 80% + CSD 20%; GM4 = CP 60% + CSD 40%. Leaf chlorophyll content (CHL); leaf flavonoid content (FLAV); nitrogen balance index (NBI); leaf anthocyanins (ANT); shoot diameter (DIAMETER); plant height (H); number of leaves (LEAVES); number of nodes (NODE); height diameter ratio (H/D); shoot fresh weight (SFW); rootstock fresh weight (RSFW); root fresh weight (RTFW); total fresh weight (TFW); shoot dry weight (SDW); rootstock dry weight (RSDW); root dry weight (RTDW); total dry weight (TDW); fraction to shoot (FTS); fraction to rootstock (FTRS); fraction to root (FTRT); parameter without unit of measure (-). Means followed by the same letter do not significantly differ at $p < 0.05$.

Table 5. Parameters recorded on bare-rooted vines grown on growing media containing composted solid digestate.

(A) Morphological and physiological parameters recorded at 177 day of year (DOY)

Treatment	CHL (-)	FLAV (-)	NBI (-)	ANT (-)	DIAMETER (mm)	H (cm)	LEAVES (no.)	NODE (no.)	H/D (-)
GM1	28.30 b	0.76	39.28	0.10 a	3.94	136.40 ab	27.20	24.20	350.30
GM2	35.92 a	0.96	38.11	0.06 b	3.78	124.80 ab	30.80	25.80	353.10
GM3	29.30 b	0.77	42.54	0.10 a	3.66	114.00 b	32.40	26.00	311.20
GM4	30.50 b	0.73	42.34	0.10 a	4.24	140.80 a	32.00	25.00	340.00

(B) Morphological and physiological parameters recorded at 273 DOY

Treatment	CHL (-)	FLAV (-)	NBI (-)	ANT (-)	DIAMETER (mm)	H/D (-)
GM1	32.09 a	0.78 b	41.80 a	0.06 a	5.01 ab	240.90
GM2	35.28 a	1.05 a	33.69 b	0.03 b	5.12 ab	235.50
GM3	28.37 b	0.69 bc	41.49 a	0.05 ab	5.82 a	210.50
GM4	25.48 b	0.60 c	42.72 a	0.06 a	4.90 b	247.90

(C) Agronomic parameters recorded at 314 DOY

Treatment	SFW (g plant^{-1})	RSFW (g plant^{-1})	RTFW (g plant^{-1})	TFW (g plant^{-1})	SDW (g plant^{-1})	RSDW (g plant^{-1})	RTDW (g plant^{-1})	TDW (g plant^{-1})	FTS (-)	FTRS (-)	FTRT (-)
GM1	24.05 ab	18.99 a	102.40 b	145.40 b	9.55 b	12.34 a	32.60 a	53.91 a	0.18 b	0.23 a	0.61 b
GM2	24.61 a	19.42 a	109.77 a	154.20 a	11.53 a	10.97 c	31.32 b	53.70 a	0.21 a	0.20 b	0.58 b
GM3	23.43 b	16.27 b	71.02 d	120.90 d	6.27 d	9.68 d	30.40 c	46.76 c	0.13 c	0.21 b	0.65 a
GM4	22.23 c	15.56 b	83.92 c	133.30 c	8.48 c	11.43 b	29.36 d	49.01 b	0.17 b	0.23 a	0.60 b

GM1 = commercial peat (CP) 100%; GM2 = CP 90% + composted solid digestate (CSD) 10%; GM3 = CP 80% + CSD 20%; GM4 = CP 60% + CSD 40%; Leaf chlorophyll content (CHL); leaf flavonoid content (FLAV); nitrogen balance index (NBI); leaf anthocyanins (ANT); shoot diameter (DIAMETER); plant height (H); number of leaves (LEAVES); number of nodes (NODE); height diameter ratio (H/D); shoot fresh weight (SFW); rootstock fresh weight (RSFW); root fresh weight (RTFW); total fresh weight (TFW); shoot dry weight (SDW); rootstock dry weight (RSDW); root dry weight (RTDW); total dry weight (TDW); fraction to shoot (FTS); fraction to rootstock (FTRS); fraction to root (FTRT); parameter without unit of measure (-). Means followed by the same letter do not significantly differ at $p < 0.05$.

Table 6. Parameters recorded on two-year-old vines grown on growing media containing composted solid digestate.

(A) Morphological and physiological parameters recorded at 177 day of year (DOY)

Treatment	CHL (-)	FLAV (-)	NBI (-)	ANT (-)	DIAMETER (mm)	H (cm)	LEAVES (no.)	NODE (no.)	H/D (-)
GM1	19.34	2.50	7.95	0.18	5.40	83.40 c	21.80 ab	13.60 b	154.70
GM2	18.64	2.21	8.56	0.14	5.50	119.60 b	21.80 ab	16.00 a	220.50b
GM3	19.16	2.66	7.36	0.21	5.20	131.40 ab	24.40 a	16.20 a	264.70
GM4	21.98	2.31	9.52	0.16	5.50	147.40 a	17.00 b	13.40 b	282.60

(B) Morphological and physiological parameters recorded at 273 DOY

Treatment	CHL (-)	FLAV (-)	NBI (-)	ANT (-)	DIAMETER (mm)	H/D (-)
GM1	15.12 a	3.45 a	4.47 a	0.72 a	6.23 a	193.80 b
GM2	12.08 c	3.23 b	3.74 b	0.56 b	5.63 b	213.20 a
GM3	13.62 b	3.14 b	4.30 a	0.33 c	6.31 a	190.30 b
GM4	9.38 d	2.75 c	3.43 c	0.16 d	6.52 a	184.50 b

(C) Agronomic parameters recorded at 314 DOY

Treatment	SFW (g plant^{-1})	RSFW (g plant^{-1})	RTFW (g plant^{-1})	TFW (g plant^{-1})	SDW (g plant^{-1})	RSDW (g plant^{-1})	RTDW (g plant^{-1})	TDW (g plant^{-1})	FTS (-)	FTRS (-)	FTRT (-)
GM1	145.40 b	42.29 d	182.40 a	243.40 b	8.44 c	27.40 c	47.58 d	83.42 d	0.10 ab	0.32 a	0.58 b
GM2	154.20 a	54.13 b	156.60 d	224.20 d	7.48 d	29.38 b	51.90 c	88.76 c	0.08 c	0.33 a	0.59 b
GM3	120.90 d	62.71 a	174.70 b	258.60 a	9.74 b	33.46 a	57.94 b	101.14 b	0.10 b	0.34 a	0.56 b
GM4	133.30 c	45.65 c	167.10 c	238.00 c	11.36 a	25.58 d	71.04 a	107.98 a	0.11 a	0.25 b	0.64 a

GM1 = commercial peat (CP) 100%; GM2 = CP 90% + composted solid digestate (CSD) 10%; GM3 = CP 80% + CSD 20%; GM4 = CP 60% + CSD 40%; Leaf chlorophyll content (CHL); leaf flavonoid content (FLAV); nitrogen balance index (NBI); leaf anthocyanins (ANT); shoot diameter (DIAMETER); plant height (H); number of leaves (LEAVES); number of nodes (NODE); height diameter ratio (H/D); shoot fresh weight (SFW); rootstock fresh weight (RSFW); root fresh weight (RTFW); total fresh weight (TFW); shoot dry weight (SDW); rootstock dry weight (RSDW); root dry weight (RTDW); total dry weight (TDW); fraction to shoot (FTS); fraction to rootstock (FTRS); fraction to root (FTRT); parameter without unit of measure (-). Means followed by the same letter do not significantly differ at $p < 0.05$.

The most likely explanation for the different behavior shown by bench-grafts and bare-rooted vines vs. two-year-old vines could be ascribed to the different plant age and is probably linked to the different root expansion (higher in the two-year-old vine) as already suggested by Bozzolo et al. [7]. In addition, as reported by Ericsson et al. [46] the growth dynamics of the fruit-tree depend on the changes in the source-sink balance. The same authors reported that a competition for intensive shoot growth can affect the root development, and a similar trend was displayed in the present study. In fact, a higher percentage of biomass was allocated to the shoot (and lower to the root) in bench-grafts and bare-rooted vines, grown on GM2. On the other hand, for the two-year-old vines, the highest values of biomass allocation, both to the shoot and root, were recorded using GM4. These results support one of the nursery grower requests, which is a planting material with a high dry matter content in the shoot.

Results about improved biomass production (using CSD) are in agreement with those of Raviv et al. [47] and Ronga et al. [12], who reported that composts obtained from separated cow manure with wheat straw, grape marc, orange peels, and spent coffee grounds positively affected potted plants grown in greenhouses. Similarly, vine root development in the vineyard was stimulated by the addition of compost from vine pruning waste supplied in the under-row of cultivar 'Cabernet Sauvignon' [48]. The improved biomass was ascribed to the presence of some substances working as growth promoters in the CSD as suggested by Bernal–Vicente et al. [49]. However, further research is needed to validate this hypothesis. In addition, the similar or highest dry weight displayed by growing media containing CSD is fundamental for successful transplanting as it is linked with an increase resistance to environmental stresses [9].

In the second experiment, BC was assessed on bare-rooted vines. BC applied at 10% (GM2) positively affected the ANT, shoot diameter, shoot height, number of leaves, and nodes at 170 DOY, corresponding to the BBCH-scale 19 (Table 7). Moreover, GM2 also increased the FLAV, ANT, and shoot diameter at 275 DOY, corresponding to the BBCH-scale 95 (50% of leaves fallen) and SFW, RTFW, TFW, SDW, RTDW, TDW, FTS, and FTRT at 362 DOY, corresponding to the BBCH-scale 97 (end of leaf-fall). Our results are in accordance with those of other authors, who reported increments in plants aboveground and belowground biomass, applying biochar at low doses to different species [15,50,51]. In grapevine, a significant increase of fine root biomass has been observed following the application of biochar as an amendment to the vineyard soil [52].

Table 7. Parameters recorded on bare-rooted vines grown on growing media containing biochar.

(A) Morphological and physiological parameters recorded at 170 day of year (DOY)

Treatment	CHL (-)		FLAV (-)		NBI (-)		ANT (-)		DIAMETER (mm)		H (cm)		LEAVES (no.)		NODE (no.)		H/D (-)
GM1	18.80	b	3.84	a	4.93	b	0.27	b	4.90	c	85.00	c	19.20	b	14.80	b	113.90a
GM2	15.40	c	3.80	a	4.05	c	0.34	a	5.90	a	121.40	a	24.20	a	19.40	a	94.20b
GM3	17.18	b	3.50	a	4.89	b	0.30	b	5.41	b	107.80	b	18.20	b	12.20	b	74.70c
GM4	21.78	a	2.99	b	6.55	a	0.26	b	3.54	d	67.20	d	8.80	c	6.00	c	39.80d

(B) Morphological and physiological parameters recorded at 275 DOY

Treatment	CHL (-)		FLAV (-)		NBI (-)		ANT (-)		DIAMETER (mm)		H (cm)		H/D (-)	
GM1	19.92	a	3.14	b	6.43	a	1.18	b	5.01	b	115.60	c	231.00	bc
GM2	12.76	b	3.66	a	3.49	b	2.09	a	5.89	a	126.70	b	215.30	c
GM3	13.78	b	3.25	b	4.24	b	1.87	a	5.48	ab	134.50	a	246.00	ab
GM4	12.94	b	3.06	b	4.29	b	1.26	b	3.86	c	102.20	d	261.60	a

(C) Agronomic parameters recorded at 362 DOY

Treatment	SFW (g plant^{-1})		ESFW (g plant^{-1})		RTFW (g plant^{-1})		TFW (g plant^{-1})		SDW (g plant^{-1})		RSDW (g plant^{-1})		RTDW (g plant^{-1})		TDW (g plant^{-1})		FTS (-)		FTRS (-)		FTRT (-)	
GM1	7.02	b	34.30	c	58.40	b	100.20	b	4.19	b	20.90	b	18.88	b	43.97	b	0.09	b	0.48	b	0.43	b
GM2	13.48	a	35.30	bc	81.80	a	131.20	a	6.84	a	19.83	b	22.55	a	49.23	a	0.14	a	0.40	c	0.49	a
GM3	14.80	a	48.30	a	49.80	b	112.60	b	7.36	a	28.44	a	15.85	c	51.66	a	0.14	a	0.55	a	0.36	c
GM4	4.30	c	40.40	b	59.60	b	104.30	b	2.62	c	22.82	b	17.75	b	43.19	b	0.06	c	0.53	a	0.41	b

GM1 = commercial peat (CP) 100%; GM2 = CP 90% + biochar (BC) 10%; GM3 = CP 80% + BC 20%; GM4 = CP 60% + BC 40%; Leaf chlorophyll content (CHL); leaf flavonoid content (FLAV); nitrogen balance index (NBI); leaf anthocyanins (ANT); shoot diameter (DIAMETER); plant height (H); number of leaves (LEAVES); number of nodes (NODE); height diameter ratio (H/D); shoot fresh weight (SFW); rootstock fresh weight (RSFW); root fresh weight (RTFW); total fresh weight (TFW); shoot dry weight (SDW); lklrootstock dry weight (RSDW); root dry weight (RTDW); total dry weight (TLW); fraction to shoot (FTS); fraction to rootstock (FTRS); fraction to root (FTRT); parameter without unit of measure (-). Means followed by the same letter do not significantly differ at $p < 0$.

Summarizing, the results of the two experiments are broadly in agreement with those of previous works, which reported that an increase of plant biomass is one of the macroscopic effects induced by the partial substitution of peat in the growing media with agro-industrial by-products, such as compost, digestate, and biochar [12,22,23]. The ability of agro-industrial by-products to increase plant biomass is probably due to an improved nutrient availability and uptake by the plant and/or to the presence of some microorganisms and compounds able to increase the plant growth [49]. In fact, when these substances are added in the growing media, they are able to improve the biochemical activity of the plants, similar to plant hormone-like promoters [53].

From a physiological point of view, the addition of CSD and BC reduced the CHL values, apart from bare-rooted vine grown on GM2 and GM4 containing CSD and BC, respectively, which recorded higher values compared to GM1. However, the lower values of CHL did not affect the biomass production and its distribution to shoot in each grapevine planting material. These results are in accordance with those reported by Bozzolo et al. [7].

Chlorophyll, flavonoid, and anthocyanin leaf contents are indices of leaf photosynthetic capacity and plant vigor status, which are linked to the N uptake [33]. In particular, high levels of anthocyanins in the leaf might allow the plant to increase resistance to abiotic and biotic stresses [54]. Thus, our results suggest that CSD and BC applied at 10% may increase plant resistance to stress on bare-rooted vines cropped in a pot in a greenhouse.

CSD and BC used at doses higher than 10% negatively affected the biomass production of bench-grafts and bare-rooted vines. These results were due to the high dosage of CSD and BC that induced alkaline pH to the relative growing media. Bozzolo et al. [7] also reported the same trend on grapevine and on other horticultural crops, such as tomato, basil, and lettuce [9,12].

One interesting aspect of our results is that the FTS was stimulated by CSD and BC applications used up to 10%, compared to CP. Moreover, BC applied at 10% (GM2) also showed the highest FTRT. On the other hand, working with the two-year-old vines, the highest values of FTS and FTRT were recorded applying CSD at a dose of 40% (GM4). The highest allocation of dry matter to the aboveground part of the vine might suggest the highest N uptake [55], confirming the role of CSD and BC in improving the nutrient plant uptake [7,50]. In addition, these results suggested that shoot growth response to CSD and BC was not only due to a higher nutrient uptake by plants, but that also other factors might be involved, such as the presence of humic substances and an improved porosity in the growing media, respectively [56,57].

3.3. Relationships between Recorded Parameters and Growing Media

The correlations between data of growing media variables measured on grapevine planting materials were studied by PCA analysis. Figures 1–3 report ordination biplots of the PCA output of the first experiment on bench-grafts, bare-rooted vines, and two-year-old vines, respectively, assessing CSD. Figure 4 reports ordination biplot of the PCA output of the second experiment on bare-rooted vine, assessing BC. In the first experiment, for the bench-grafts, the PC1 accounted for 53.99% of the variance, while PC2 accounted for 32.87%. For the bare-rooted vines dataset, the two principal components, 1 and 2, accounted for 51.64% and 26.86% of the variation, respectively. Meanwhile, for two-year-old vines, the PC1 accounted for 49.14% of the variance, while PC2 accounted for 31.52%.

Growing media containing CSD at 10% (GM2) clustered along the positive side of PC1 both on the bench-grafts (Figure 1) and bare-rooted vines (Figure 2). GM2 performed as well as CP (GM1) and was associated with the most important parameters, such as CHL, FLAV, SDW, and FTS (Figures 1 and 2).

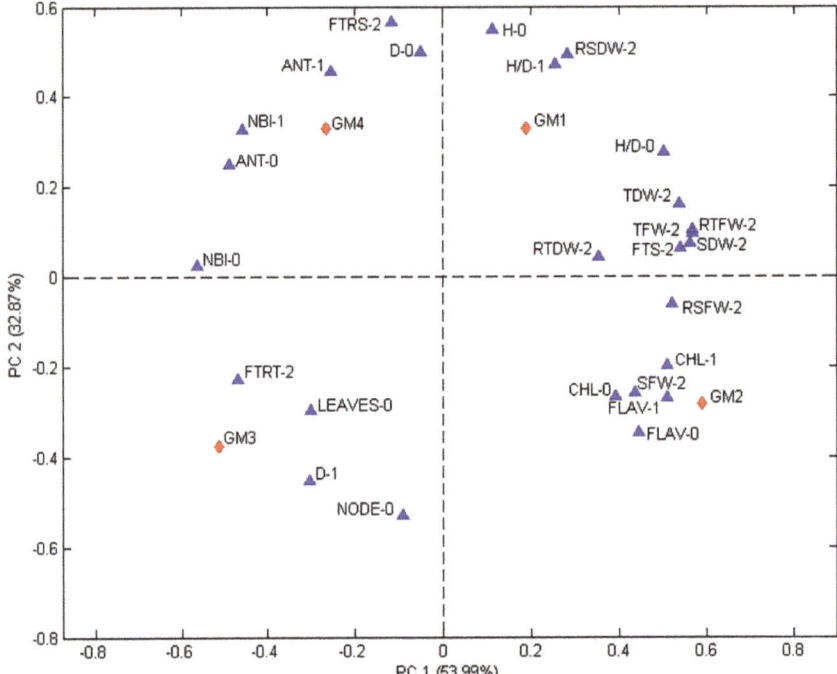

Figure 1. Ordination biplots of principal component analysis outputs of the data from bench-grafts grown on growing media (GM), containing composted solid digestate (CSD) at different rates. GM1 = commercial peat (CP) 100%; GM2 = CP 90% + CSD 10%; GM3 = CP 80% + CSD 20%; GM4 = CP 60% + CSD 40%. Labels in the triangles indicate the investigated parameters: Leaf chlorophyll content (CHL); leaf flavonoid content (FLAV); nitrogen balance index (NBI); leaf anthocyanins (ANT); plant height (H); height diameter ratio (H/D); shoot fresh weight (SFW); rootstock fresh weight (RSFW); root fresh weight (RTFW); total fresh weight (TFW); shoot dry weight (SDW); rootstock dry weight (RSDW); root dry weight (RTDW); total dry weight (TDW); fraction to shoot (FTS); fraction to rootstock (FTRS); fraction to root (FTRT). 0 = parameter recorded at 170 day of year (DOY); 1 = parameter recorded at 275 DOY; 2 = parameter recorded at 362 DOY.

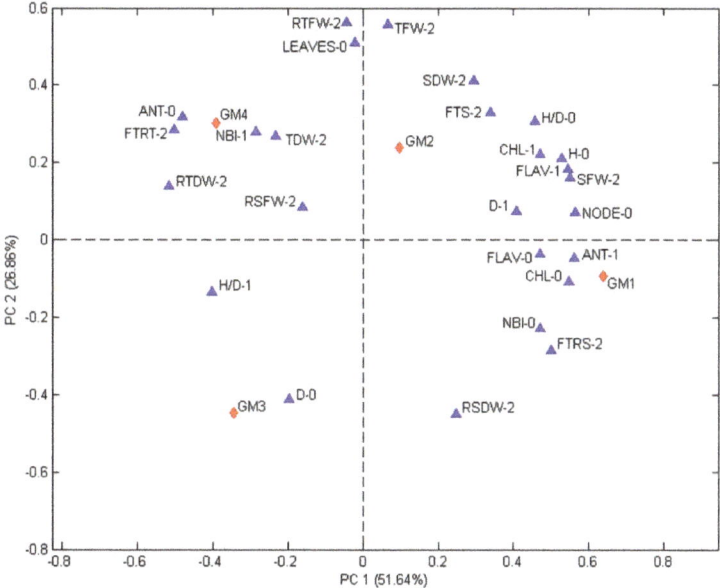

Figure 2. Ordination biplots of principal component analysis outputs of the data from bare-rooted vines grown on growing media (GM) containing composted solid digestate (CSD) at different rates. GM1 = commercial peat (CP) 100%; GM2 = CP 90% + CSD 10%; GM3 = CP 80% + CSD 20%; GM4 = CP 60% + CSD 40%. Labels in the triangles indicate the investigated parameters: Leaf chlorophyll content (CHL); leaf flavonoid content (FLAV); nitrogen balance index (NBI); leaf anthocyanins (ANT); plant height (H); height diameter ratio (H/D); shoot fresh weight (SFW); rootstock fresh weight (RSFW); root fresh weight (RTFW); total fresh weight (TFW); shoot dry weight (SDW); rootstock dry weight (RSDW); root dry weight (RTDW); total dry weight (TDW); fraction to shoot (FTS); fraction to rootstock (FTRS); fraction to root (FTRT). 0 = parameter recorded at 170 day of year (DOY); 1 = parameter recorded at 275 DOY; 2 = parameter recorded at 362 DOY.

For two-year-old vines, CSD applied at 40% (GM4) recorded the best agronomic performance and was positively linked with several parameters, such as SDW, FTS, TDW, RTDW, and the shoot diameter (Figure 3).

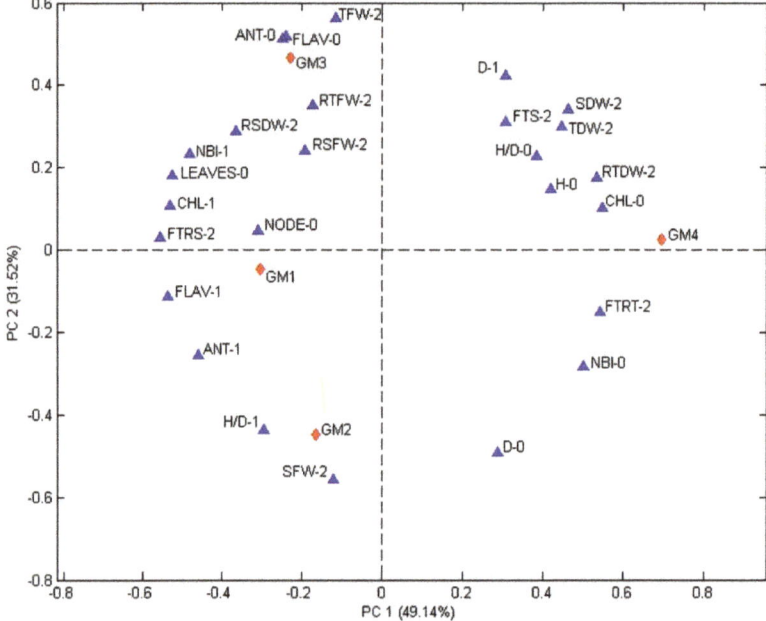

Figure 3. Ordination biplots of principal component analysis outputs of the data from two-year-old vines grown on growing media (GM) containing composted solid digestate (CSD) at different rates. GM1 = commercial peat (CP) 100%; GM2 = CP 90% + CSD 10%; GM3 = CP 80% + CSD 20%; GM4 = CP 60% + CSD 40%. Labels in the triangles indicate the investigated parameters: Leaf chlorophyll content (CHL); leaf flavonoid content (FLAV); nitrogen balance index (NBI); leaf anthocyanins (ANT); plant height (H); height diameter ratio (H/D); shoot fresh weight (SFW); rootstock fresh weight (RSFW); root fresh weight (RTFW); total fresh weight (TFW); shoot dry weight (SDW); rootstock dry weight (RSDW); root dry weight (RTDW); total dry weight (TDW); fraction to shoot (FTS); fraction to rootstock (FTRS); fraction to root (FTRT). 0 = parameter recorded at 170 day of year (DOY); 1 = parameter recorded at 275 DOY; 2 = parameter recorded at 362 DOY.

In the second experiment, using BC for the bare-rooted vines, the PC1 accounted for 62.55% of the variance, while PC2 accounted for 23.42%. Growing media containing BC at 10% (GM2) clustered along the positive side of PC1, performing as well as growing media containing BC at 20% (GM3), and were associated with the most important parameters, such as CHL, FLAV, the number of leaves and nodes, the shoot height, SDW, TDW, and FTS (Figure 4).

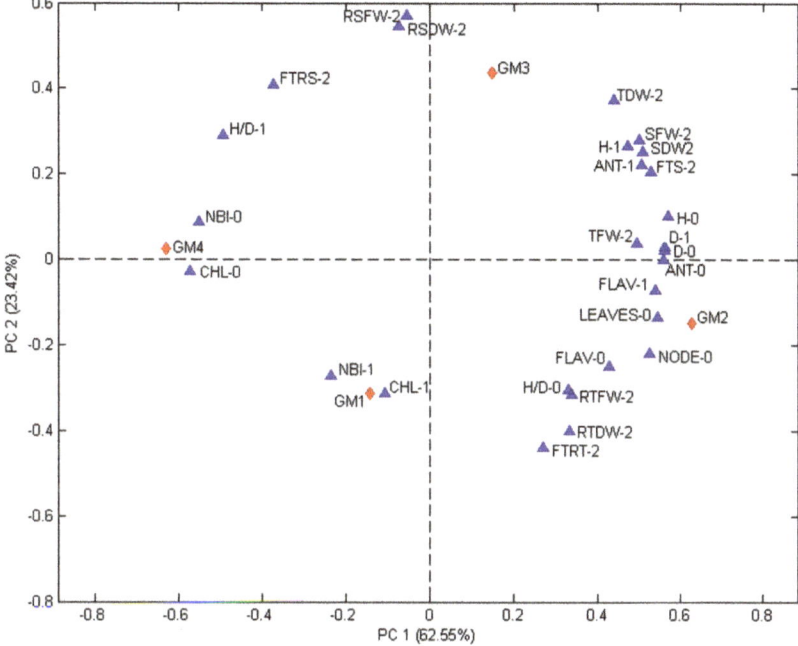

Figure 4. Ordination biplots of principal component analysis outputs of the data from bare-rooted vines grown on growing media (GM) containing biochar (BC) at different rates. GM1 = commercial peat (CP) 100%; GM2 = CP 90% + BC 10%; GM3 = CP 80% + BC 20%; GM4 = CP 60% + BC 40%. Labels in the triangles indicate the investigated parameters: Leaf chlorophyll content (CHL); leaf flavonoid content (FLAV); nitrogen balance index (NBI); leaf anthocyanins (ANT); plant height (H); height diameter ratio (H/D); shoot fresh weight (SFW); rootstock fresh weight (RSFW); root fresh weight (RTFW); total fresh weight (TFW); shoot dry weight (SDW); rootstock dry weight (RSDW); root dry weight (RTDW); total dry weight (TDW); fraction to shoot (FTS); fraction to rootstock (FTRS); fraction to root (FTRT). 0 = parameter recorded at 170 day of year (DOY); 1 = parameter recorded at 275 DOY; 2 = parameter recorded at 362 DOY.

These results indicated that CSD and BC applied at 10% are able to increase the growth of potted bench-grafts and bare-rooted vines, reducing the consumption of CP. Moreover, working with plants with a preformed and well-developed root apparatus, such as in a two-year-old vine, CSD might be applied up to 40%. Finally, CSD and BC applied in low doses did not show any signs of abiotic and biotic stresses, in accordance with several previous reports [7,58,59].

4. Conclusions

Results obtained in the present study suggest that CSD and BC could be a suitable ingredient for alternative growing media, recycling, and valorizing by-products coming from vineyards, such as winter prunings and grape stalks. CSD and BC showed good agronomic performances compatible with the development of grapevine planting materials. To our knowledge, this is the first study that investigated the suitability of CSD and BC valorizing vineyard prunings and grape stalks as an innovative ingredient for alternative growing media able to replace peat aliquots, for the production of different grapevine planting materials. Bench-grafts and bare-rooted vines grown in mixtures containing CSD at a dose of 10% displayed agronomic parameters higher than those recorded using only CP, and the same was displayed using BC on bare-rooted vines. On the other hand, CSD at a dose of 40% showed the highest agronomic performance on two-year-old vines. Agronomic measurements suggest that the CSD or BC induce an enhanced plant-growth response that is particularly expressed at shoot level. The replacing of CP aliquots with CSD or BC could be an excellent sustainable practice to reduce the consumption of CP, recycling, and valorizing the underutilized vineyard prunings and grape stalks, thus supporting the current tendency to enhance the soilless propagation of grapevine. However, further research is needed to assess the combined use of CSD and BC as an ingredient of growing media for grapevine planting material productions.

Author Contributions: Conceptualization, D.R. and C.B.; methodology, D.R., M.Z. and C.P.; investigation, D.R., M.Z., G.A., S.P., C.P., A.T. and C.B.; resources, C.B.; data curation, D.R. and C.B.; writing—original draft preparation, D.R.; writing—review and editing, D.R., M.Z., G.A., S.P., C.P., A.T., E.F. and C.B.; supervision, E.F. and C.B.; project administration, C.B.; funding acquisition, C.B.

Funding: This research was funded by the Fondi di Ateneo per la Ricerca di UNIMORE, FAR 2015 "Valorizzazione dei digestati da impianti di biogas e dei sarmenti del vigneto: preparati organici innovativi per la qualità vivaistica e la gestione agronomica della vite (VADI.SAVI.)", FAR 2017 "La valorizzazione degli scarti agroindustriali tra diritto e scienza: processi innovativi dalla sperimentazione all'industrializzazione nel contesto legale" and by EU-ERFD, Emilia-Romagna Regional Operational Programme, project SOSTINNOVI (PG/2015/737442).

Acknowledgments: The authors acknowledge and are grateful to Guido Bezzi (CIB), Leonardo Setti, and Federica Caradonia (UniMORE) for their helpful assistance in this study. The authors want to thank Marco Veroni (Veroni's farm) for providing the vineyard prunings, Massimo Zaghi (CAT) for providing digestate and dr. Giovanni Grazzi (Foliae) for providing the greenhouse space and some of the materials used in this study.

Conflicts of Interest: The authors declare no conflict of interest.

References

1. Reynolds, A.G. *Managing Wine Quality: Viticulture and Wine Quality*; Woodhead Publishing: Cambridge, UK, 2010.
2. Borsellino, V.; Galati, A.; Schimmenti, E. Survey on the innovation in the Sicilian grapevine nurseries. *J. Wine Res.* **2012**, *23*, 1–13. [CrossRef]
3. Waite, H.; May, P.; BossingEr, G. Variations in phytosanitary and other management practices in Australian grapevine nurseries. *Phytopathol. Mediterr.* **2013**, 369–379.
4. Whitelaw-Weckert, M.A.; Rahman, L.; Appleby, L.M.; Hall, A.; Clark, A.C.; Waite, H.; Hardie, W.J. Co-infection by B otryosphaeriaceae and I lyonectria spp. fungi during propagation causes decline of young grafted grapevines. *Plant Pathol.* **2013**, *62*, 1226–1237. [CrossRef]
5. Tesic, D.; Keller, M.; Hutton, R.J. Influence of vineyard floor management practices on grapevine vegetative growth, yield, and fruit composition. *Am. J. Enol. Vitic.* **2007**, *58*, 1–11.
6. Salomé, C.; Coll, P.; Lardo, E.; Metay, A.; Villenave, C.; Marsden, C.; Blanchart, E.; Hinsinger, P.; Le Cadre, E. The soil quality concept as a framework to assess management practices in vulnerable agroecosystems: A case study in Mediterranean vineyards. *Ecol. Indic.* **2016**, *61*, 456–465. [CrossRef]
7. Bozzolo, A.; Pizzeghello, D.; Cardinali, A.; Francioso, O.; Nardi, S. Effects of moderate and high rates of biochar and compost on grapevine growth in a greenhouse experiment. *AIMS Agric. Food* **2017**, *2*, 113–128. [CrossRef]
8. Waite, H.; Whitelaw-Weckert, M.; Torley, P. Grapevine propagation: Principles and methods for the production of high-quality grapevine planting material. *N. Z. J. Crop Hortic. Sci.* **2015**, *43*, 144–161. [CrossRef]
9. Herrera, F.; Castillo, J.E.; Chica, A.F.; López Bellido, L. Use of municipal solid waste compost (MSWC) as a growing medium in the nursery production of tomato plants. *Bioresour. Technol.* **2008**, *99*, 287–296. [CrossRef]
10. Vaughn, S.F.; Deppe, N.A.; Palmquist, D.E.; Berhow, M.A. Extracted sweet corn tassels as a renewable alternative to peat in greenhouse substrates. *Indic. Crops Prod.* **2011**, *33*, 514–517. [CrossRef]
11. Steiner, C.; Harttung, T. Biochar as growing media additive and peat substitute. *Solid Earth* **2014**, *5*, 995–999. [CrossRef]
12. Ronga, D.; Pane, C.; Zaccardelli, M.; Pecchioni, N. Use of spent coffee ground compost in peat-based growing media for the production of basil and tomato potting plants. *Commun. Soil Sci. Plant* **2016**, *47*, 356–368. [CrossRef]
13. Schulz, H.; Glaser, B. Effects of biochar compared to organic and inorganic fertilizers on soil quality and plant growth in a greenhouse experiment. *J. Plant. Nutr. Soil Sci.* **2012**, *175*, 410–422. [CrossRef]

14. Baronti, S.; Vaccari, F.P.; Miglietta, F.; Calzolari, C.; Lugato, E.; Orlandinie, S.; Pini, R.; Zulian, C.; Genesio, L. Impact of biochar application on plant water relations in *Vitis vinifera* (L.). *Eur. J. Agron.* **2014**, *53*, 38–44. [CrossRef]
15. Schmidt, H.P.; Kammann, C.; Niggli, C.; Evangelou, M.W.; Mackie, K.A.; Abiven, S. Biochar and biochar-compost as soil amendments to a vineyard soil: Influences on plant growth, nutrient uptake, plant health and grape quality. *Agric. Ecosyst. Environ.* **2014**, *191*, 117–123. [CrossRef]
16. Genesio, L.; Miglietta, F.; Baronti, S.; Vaccari, F.P. Biochar increases vineyard productivity without affecting grape quality: Results from a four years field experiment in Tuscany. *Agric. Ecosyst. Environ.* **2015**, *201*, 20–25. [CrossRef]
17. Ronga, D.; Villecco, D.; Zaccardelli, M. Effects of compost and defatted oilseed meals as sustainable organic fertilizer on Cardoon (*Cynara Cardunculus* L.) production in the Mediterranean basin. *J. Hortic. Sci. Biotechnol.* **2019**, *94*, 1–12. [CrossRef]
18. Puglisi, R.; Severgnini, A.; Tava, A.; Montedoro, M. In Vitro Assessment of the Antioxidant Properties of Aqueous Byproduct Extracts of Vitis vinifera. *Food Technol. Biotechnol.* **2019**, *57*, 119–125. [CrossRef]
19. Rubio, R.; Pérez-Murcia, M.D.; Agulló, E.; Bustamante, M.A.; Sánchez, C.; Paredes, C.; Moral, R. Recycling of Agro-food Wastes into Vineyards by composting: Agronomic Validation in Field Conditions. *Commun. Soil Sci. Plant* **2013**, *44*, 502–516. [CrossRef]
20. Beres, C.; Costa, G.N.S.; Cabezudo, I.; da Silva-James, N.K.; Teles, A.S.C.; Cruz, A.P.G.; Mellinger-Silva, C.; Tonon, R.V. Towards integral utilization of grape pomace from winemaking process: A review. *Waste Manag.* **2017**, *68*, 581–594. [CrossRef]
21. Pulvirenti, A.; Ronga, D.; Zaghi, M.; Tomasselli, A.R.; Mannella, L.; Pecchioni, N. Pelleting is a successful method to eliminate the presence of Clostridium spp. from the digestate of biogas plants. *Biomass Bioenerg.* **2015**, *81*, 479–482. [CrossRef]
22. Ronga, D.; Pellati, F.; Brighenti, V.; Laudicella, K.; Laviano, L.; Fedailaine, M.; Benvenuti, S.; Pecchioni, N.; Francia, E. Testing the influence of digestate from biogas on growth and volatile compounds of basil (*Ocimum basilicum* L.) and peppermint (*Mentha x piperita* L.) in hydroponics. *J. Appl. Res. Med. Aromat. Plants* **2018**, *11*, 18–26. [CrossRef]
23. Ronga, D.; Caradonia, F.; Setti, L.; Hagassou, D.; Giaretta Azevedo, C.V.; Milc, J.; Pedrazzi, S.; Allesina, G.; Arru, L.; Francia, E. Effects of innovative biofertilizers on yield of processing tomato cultivated in organic cropping systems in northern Italy. *Acta Hortic.* **2019**, *1233*, 129–136. [CrossRef]
24. Allesina, G.; Pedrazzi, S.; Puglia, M.; Morselli, N.; Allegretti, F.; Tartarini, P. Gasification and wine industry: Report on the use vine pruning as fuel in small-scale gasifiers. In *Proceedings of the European Biomass Conference and Exhibition*; ETA: Bristol, England, 2018; Renewable Energies; pp. 722–725.
25. Ronga, D.; Setti, L.; Salvarani, C.; De Leo, R.; Bedin, E.; Pulvirenti, A.; Milc, J.; Francia, E. Effects of solid and liquid digestate for hydroponic baby leaf lettuce (*Lactuca sativa* L.) cultivation. *Sci. Hortic.* **2019**, *244*, 172–181. [CrossRef]
26. Morlat, R.; Jacquet, A. The soil effects on the grapevine root system in several vineyards of the Loire Valley (France). *Vitis* **1993**, *32*, 35–42.
27. Vršič, S.; Kocsis, L.; Pulko, B. Influence of substrate pH on root growth, biomass and leaf mineral contents of grapevine rootstocks grown in pots. *J. Agric. Sci. Technol.* **2016**, *18*, 483–490.
28. Zucconi, F.; Pera, A.; Forte, M.; De Bertoldi, M. Evaluating toxicity of immature compost. *BioCycle* **1981**, *22*, 54–57.
29. Cekmecelioglu, D.; Demirci, A.; Graves, R.E.; Davitt, N.H. Applicability of optimised in vessel food waste composting for windrow systems. *Biosyst. Eng.* **2005**, *91*, 479–486. [CrossRef]
30. Pane, C.; Piccolo, A.; Spaccini, R.; Celano, G.; Villecco, D.; Zaccardelli, M. Agricultural waste-based composts exhibiting suppressivity to diseases caused by the phytopathogenic soil-borne fungi *Rhizoctonia solani* and *Sclerotinia minor*. *Appl. Soil Ecol.* **2013**, *65*, 43–51. [CrossRef]
31. Pane, C.; Spaccini, R.; Piccolo, A.; Scala, F.; Bonanomi, G. Compost amendments enhance peat suppressiveness to *Pythium ultimum*, *Rhizoctonia solani* and *Sclerotinia minor*. *Biol. Control* **2011**, *56*, 115–124. [CrossRef]
32. Lorenz, D.H.; Eichhorn, K.W.; Blei-Holder, H.; Klose, R.; Meier, U.; Weber, E. Phenological growth stages and BBCH—identification keys of grapevine (*Vitis vinifera* L. ssp. vinifera). In *Growth Stages of Mono and Dicotyledonous Plants—BBCH Monograph*, 2nd ed.; Meier, U., Ed.; Federal Biological Research Centre for Agriculture and Forestry: Berlin, Germany, 2001; p. 158.

33. Cerovic, Z.G.; Masdoumier, G.; Ghozlen, N.B.; Latouche, G. A new optical leaf-clip meter for simultaneous non-destructive assessment of leaf chlorophyll and epidermal flavonoids. *Physiol Plant.* **2012**, *146*, 251–260. [CrossRef]
34. Baker, K.F.; Chandler, P.A. *UC System for Producing Healthy Container-Grown Plants*; University of California Experimental Station Service: Riverside, CA, USA, 1957; pp. 23–331.
35. Daughtrey, M.L.; Benson, D.M. Principles of plant health management for ornamental plants. *Annu. Rev. Phytopathol.* **2005**, *43*, 141–169. [CrossRef]
36. Hartmann, H.T.; Kester, D.E.; Davies, F.T. *Plant Propagation: Principles and Practices*, 5th ed.; Prentice-Hall: Englewood Cliffs, NJ, USA, 1990; p. 647.
37. Schmilewski, G. The role of peat in assuring the quality of growing media. *Mires Peat* **2008**, *3*, 2.
38. Vinnerås, B.; Björklund, A.; önsson, H. Thermal composting of faecal matter as treatment and possible disinfection method—laboratory-scale and pilot-scale studies. *Bioresour. Technol.* **2003**, *88*, 47–54. [CrossRef]
39. Golueke, C.G. Principles of biological resources recovery. *Biocycle* **1981**, *22*, 36–40.
40. Hachicha, R.; Rekik, O.; Hachicha, S.; Ferchichi, M.; Woodward, S.; Moncef, N.; Cegarra, J.; Mechichi, T. Cocomposting of spent coffee ground with olive mill wastewater sludge and poultry manure and effect of Trametes versicolor inoculation on the compost maturity. *Chemosphere* **2012**, *88*, 677–682. [CrossRef]
41. Raviv, M. Horticultural uses of composted material. *Acta Hortic.* **1998**, *469*, 225–234. [CrossRef]
42. Hadar, Y. Suppressive compost: When plant pathology met microbial ecology. *Phytoparasitic* **2011**, *39*, 311–314. [CrossRef]
43. Hardy, G.E.; Sivasithampram, K. Antagonism of fungi and actinomycetes isolated from composted eucalyptus bark to Phytophthora drechsleri in a steamed and non-steamed composted medium. *Soil Biol. Biochem.* **1995**, *27*, 243–246. [CrossRef]
44. Avilés, M.; Borrero, C.; Trillas, M.I. Review on compost as an inducer of disease suppression in plant grown in soilless culture. *Dyn. Soil Dyn. Plant* **2001**, *5*, 1–11.
45. Abad, M.; Patricia, N.; Burés, S. National inventory of organic wastes for use as growing media for ornamental potted plant production: Case study in Spain. *Bioresour. Technol.* **2001**, *77*, 197–200. [CrossRef]
46. Ericsson, T.; Rytter, L.; Vapaavuori, E. Physiology of carbon allocation in trees. *Biomass Bioenerg.* **1996**, *11*, 115–127. [CrossRef]
47. Raviv, M.; Oka, Y.; Katan, J.; Hadar, Y.; Yogev, A.; Medina, S.; Krasnovskya, A.; Ziadna, H. High-nitrogen compost as a medium for organic container-grown crops. *Bioresour. Technol.* **2005**, *96*, 419–427. [CrossRef]
48. Gaiotti, F.; Marcuzzo, P.; Belfiore, N.; Lovat, L.; Fornasier, F.; Tomasi, D. Influence of compost addition on soil properties, root growth and vine performances of *Vitis vinifera* cv Cabernet sauvignon. *Sci. Hortic.* **2017**, *225*, 88–95. [CrossRef]
49. Bernal-Vicente, A.; Ros, M.; Tittarelli, F.; Intrigliolo, F.; Pascual, J.A. Citrus compost and its water extract for cultivation of melon plants in greenhouse nurseries. Evaluation of nutriactive and biocontrol effects. *Bioresour. Technol.* **2008**, *99*, 8722–8728. [CrossRef]
50. Vaccari, F.P.; Maienza, A.; Miglietta, F.; Baronti, S.; Di Lonardo, S.; Guagnoni, L.; Lagomarsino, A.; Pozzi, A.; Pusceddu, E.; Ranieri, R.; et al. Biochar stimulates plant growth but not fruit yield of processing tomato in a fertile soil. *Agric. Ecosyst. Environ.* **2015**, *207*, 163–170 [CrossRef]
51. Sigua, G.C.; Novak, J.M.; Watts, D.W.; Johnson, M.G.; Spokas, K. Efficacies of designer biochars in improving biomass and nutrient uptake of winter wheat grown in a hard setting subsoil layer. *Chemosphere* **2016**, *142*, 176–183. [CrossRef]
52. Amendola, C.; Montagnoli, A.; Terzaghi, M.; Trupiano, D.; Oliva, F.; Baronti, S.; Miglietta, F.; Chiatante, D.; Scippa, G.S. Short-term effects of biochar on grapevine fine root dynamics and arbuscular mychorrhizae production. *Agric. Ecosyst. Environ.* **2017**, *239*, 236–245. [CrossRef]
53. Jindo, K.; Martim, S.A.; Navarro, E.C.; Pérez-Alfocea, F.; Hernandez, T.; Garcia, C.; Aguiar, N.O.; Canellas, L.P. Root growth promotion by humic acids from composted and non composted urban organic wastes. *Plant Soil* **2012**, *353*, 209–220. [CrossRef]
54. Chalker-Scott, L. Environmental significance of anthocyanins in plant stress responses. *Photochem. Photobiol.* **1999**, *70*, 1–9. [CrossRef]
55. Poorter, H.; Fiorani, F.; Stitt, M.; Schurr, U.; Finck, A.; Gibon, Y.; Usadel, B.; Munns, R.; Atkin, O.K.; Tardieu, F.; et al. The art of growing plants for experimental purposes: A practical guide for the plant biologist. *Funct. Plant Biol.* **2012**, *39*, 821–838. [CrossRef]

56. Graber, E.R.; Tsechansky, L.; Mayzlish-Gati, E.; Shema, R.; Koltai, H. A humic substances product extracted from biochar reduces Arabidopsis root hair density and length under P-sufficient and P-starvation conditions. *Plant Soil* **2015**, *395*, 21–30. [CrossRef]
57. Vaccaro, S.; Muscolo, A.; Pizzeghello, D.; Spaccini, R.; Piccolo, A.; Nardi, S. Effect of a compost and its water-soluble fractions on key enzymes of nitrogen metabolism in maize seedlings. *J. Agric. Food Chem.* **2009**, *57*, 11267–11276. [CrossRef]
58. García-Gómez, A.; Bernal, M.P.; Roig, A. Growth of ornamental plants in two composts prepared from agroindustrial wastes. *Bioresour. Technol.* **2002**, *83*, 81–87. [CrossRef]
59. Kostov, O.; Tzvetkov, Y.; Kaloianova, N.; VanCleemput, O. Production of tomato seedlings on composts of vine branches and grape prunings, husks, and seeds. *Compos. Sci. Util.* **1996**, *4*, 55–61. [CrossRef]

© 2019 by the authors. Licensee MDPI, Basel, Switzerland. This article is an open access article distributed under the terms and conditions of the Creative Commons Attribution (CC BY) license (http://creativecommons.org/licenses/by/4.0/).

Article

Charcoal Fine Residues Effects on Soil Organic Matter Humic Substances, Composition, and Biodegradability

Otávio dos Anjos Leal [1,*], Deborah Pinheiro Dick [2], José María de la Rosa [3], Daniela Piaz Barbosa Leal [2], José A. González-Pérez [3], Gabriel Soares Campos [4] and Heike Knicker [3]

1. Catarinense Federal Institute of Technology, Science and Education (IFC), Rua das Rosas s/n, Santa Rosa do Sul 88965-000, Brazil
2. Departamento de Físico-Química (DFQ), Federal University of Rio Grande do Sul (UFRGS), Avda. Bento Gonçalves, 7712, Porto Alegre 91540-000, Brazil
3. Instituto de Recursos Naturales y Agrobiología de Sevilla (IRNAS), Avda. Reina Mercedes, 10, 41012 Seville, Spain
4. Department Animal and Dairy Science, University of Georgia, River Road 425, Athens, GA 30605, USA
* Correspondence: otavioleal@hotmail.com; Tel.: +55-53-981260239

Received: 16 June 2019; Accepted: 12 July 2019; Published: 16 July 2019

Abstract: Biochar has been shown as a potential mean to enhance carbon sequestration in the soil. In Brazil, approximately 15% of the produced charcoal is discarded as charcoal fines, which are chemically similar to biochar. Therefore, we aimed to test charcoal fines as a strategy to increase soil carbon sequestration. Charcoal fines of hardwood *Mimosa scabrella* were incorporated into a Cambisol down to 10 cm (T1 = 0 and T4 = 40 Mg ha^{-1}) in Southern Brazil. Soil samples were collected (0–30 cm) 20 months after charcoal amendment. Soil organic matter (SOM) acid extract, humic acid, fulvic acid, and humin fractions were separated. Solid-state ^{13}C nuclear magnetic resonance (NMR) spectra from charcoal and SOM in T1 and T4 were obtained before and after 165 days of incubation under controlled conditions. Charcoal increased soil carbon as fulvic (10–20 cm) and humic acids (10–30 cm) and, especially, as humin (0–5 cm), which probably occurred due to the hydrophobic character of the charcoal. The ^{13}C NMR spectra and mean residence times (MRT) measured from incubation essays indicated that the charred material decomposed relatively fast and MRT of T1 and T4 samples were similar. It follows that the charcoal fines underwent similar decomposition as SOM, despite the high charcoal dose applied to the soil and the high aryl C contribution (78%) to the total ^{13}C intensity of the charcoal NMR spectra.

Keywords: field experiment; incubation; ^{13}C NMR; mean residence time; slow pool

1. Introduction

In recent years, the literature has reported the potential of biochar as a strategy to mitigate global warming. Basically, biochar can enhance carbon (C) sequestration directly when buried into the soil or indirectly by improving soil quality and crop production, thus, enhancing CO_2 capture from the atmosphere [1,2].

Conceptually, biochar is a C-enriched material intentionally produced via pyrolysis of biomass to be applied to the soil as a means to improve C sequestration, soil quality, and crop yield [1,3]. Therefore, biochar is distinguished from charcoal since energy generation, industrial use, and domestic cooking are the main purposes of charcoal production. Nevertheless, the thermochemical conversion of biomass in pyrogenic C (PyC) is a common process for both biochar and charcoal production [3].

In this way, several authors have reported positive effects of charcoal on soil fertility, crop production, and C sequestration [4–7].

Brazil is one of the world's greatest agricultural producers and consequently generates great amounts of biomass residues [8,9]. Thus, numerous organic residues, such as sugarcane straw and filtercake, rice husk, poultry manure, sawdust, and sewage sludge, have been converted in biochar and applied to agricultural soils, originating promising field and greenhouse experiments [10–14].

In parallel to this scenario, Brazil is the world's greatest charcoal producer with an annual production of 10 million tons of charcoal, supplying mainly the steel industry. However, approximately 15% of this production is lost as charcoal fines, since they are not suitable for industrial use [6,15]. Despite biochar and charcoal own similar composition, minor importance has been directed to the charcoal fine residues in Brazil as a potential soil conditioner and, especially, as a means to sequester C and mitigate global warming. In addition, the utilization of charcoal fines for such purposes concomitantly contributes in reducing environmental liabilities originated by the improper discard of this residue and also in reducing biochar production from organic residues, which can be used for other purposes.

The greater stability of PyC in the soil compared to that of soil organic matter (SOM) has been attributed to the high aromaticity and hydrophobicity of the pyrogenic materials, characteristics supposed to confer high biochemical recalcitrance to this matrix [16,17]. Nevertheless, PyC can undergo transformation and biodegradation in the soil at certain rates, depending on the material source, temperature, and duration of heating, and on oxygen supply during the pyrolysis process [18,19]. Hydrophobicity and oxygenation of the pyrogenic materials in the soil may affect the C distribution in humic fractions. Increments of C as humin (HU) can indicate preservation of PyC in the soil since charcoal is essentially hydrophobic, while increments in humic acid (HA) and fulvic acid (FA) suggest oxidation of aromatic structures, thus, affecting organic matter functions and persistence in the soil [20].

The mean residence time (MRT) of the PyC has been reported to be generally one or two orders of magnitude greater than those of their fresh organic precursors [21–24]. Knicker et al. [25] estimated an MRT of the slow SOM pool to be 3 to 4 times (40 years) longer in soil with charcoal produced via wildfire than in the fire unaffected soil. On the other hand, Schneider et al. [7] did not observe changes in the chemical composition and content of charcoal derived from burned vegetation even after exposure to intense weathering during 100 years in a tropical Humic Nitosol. These findings are in line with Vasilyeva et al. [26]. The authors evaluated total C (TC) and PyC contents of a fire-affected Chernozem under fallow for 55 years. A smaller loss of PyC stock (6%) in comparison to soil organic C (33%) was observed, and changes in the aromatic condensation degree of PyC were not detected. In general, these findings support the hypothesis that the high aromaticity of the charcoal can lead to longer SOM MRT.

In a previous study, we observed favorable effects of the charcoal fines on soil fertility and consistent increase of SOM thermostability after charcoal incorporation into a subtropical Cambisol at a rate of 40 Mg ha^{-1} in Southern Brazil [27]. These findings are in line with above-mentioned literature and support the hypothesis that this residue deserves to be tested not only as a soil conditioner but, moreover, as a means to efficiently promote C sequestration in the soil. However, the utilization of charcoal fines as a strategy to sequester C needs elucidation. Additionally, there is a lack of information and ambiguous results in the literature concerning changes in the SOM after charcoal amendments to the soil [28,29].

In this context and based on the high aromaticity of the charcoal fines observed in our previous study, it is hypothesized that such pyrogenic material may alter SOM humic fractions distribution and slow down SOM biodegradability. In order to test this hypothesis, soil samples were collected at field experimental plots with or without charcoal amendment. These samples were incubated under controlled conditions, and SOM composition and degradation were investigated.

2. Materials and Methods

2.1. Site Description, Experimental Design, and Soil Sampling

The experimental area is located in Irati, Center-South of Paraná State, Brazil, at approximately 855 m above the sea level. The climate is humid subtropical mesothermic-Cfb (Köppen), with frequent and severe frosts during winter (June–September). The annual mean temperature is 17.2 °C, the average rainfall is 194 mm month^{-1}, and the relative humidity is 79.6%. The study was carried out in the *Campus* Irati of the State University of Centro-Oeste (25°27′56″ S 50°37′51″ W). The relief is undulated and strongly undulated, and the soil is classified as Haplic Cambisol [30]. Sand, silt, and clay contents of the soil at 0–5 cm depth are: 441, 167, and 392 g kg^{-1}, respectively; at 5–10 cm depth: 492, 170, and 338 g kg^{-1}, respectively; at 10–20 cm depth: 467, 162, and 371 g kg^{-1}, respectively; at 20–30 cm depth: 437, 182, and 381 g kg^{-1}, respectively. The experimental area has agricultural use history (soybean - *Glycine max*, five years), but for the last three years, before the beginning of the experiment, the soil was under fallow.

The experiment was established in February 2010 when the area was manually and mechanically mowed. About 2.5 Mg ha^{-1} of dolomitic limestone (85% of the relative power of total neutralization) was applied on the soil surface and subsequently incorporated at 10 cm depth using a light disk harrow.

The charcoal used in this experiment was originated from the pyrolysis of hardwood native Brazilian species, mainly *Mimosa scabrella* Bentham. The charcoal fine residues (<6.3 mm) were acquired from a local producer that supplies the Brazilian steel industry. About 45% of the used charcoal particles were smaller than 2 mm. The pyrolysis of the hardwood was performed under artisanal conditions, which might have contributed to its low fixed C (7.6%) and high volatile matter (84%) contents, as discussed in Leal et al. [27]. These data characterize this material as low condensed charcoal produced at low temperatures [31]. The C (46.6%) and the N (1.0%) contents of the charcoal were determined by dry combustion, and its ash content (8.2%) was determined after heating at 750 °C during 4 h. A more exhaustive physicochemical characterization of the charcoal is available in Leal et al. [27].

The study was conducted in a complete randomized block design with four treatments arranged in four blocks: T1 = 0 Mg ha^{-1} (without charcoal-control); T2 = 10 Mg ha^{-1}; T3 = 20 Mg ha^{-1}; T4 = 40 Mg ha^{-1}. Each field replicate (each block) was composed of three subsamples collected within a 144 m^2 plot. The treatments were implemented just after soil liming (February 2010). Firstly, the charcoal was applied on the soil surface and, thereafter, it was incorporated up to 10 cm using a light disk harrow. In March 2010, seedlings of *Eucalyptus benthamii* were planted in the experimental plots, and plant height and diameter were monitored up to 210 days after planting. A clear effect of charcoal doses on Eucalyptus growth was not observed [32]. Considering that the most outstanding differences regarding the effects of charcoal doses on soil characteristics and C contents were observed between T1 and T4 [27], these treatments were selected for the present study.

The soil samples were collected in September 2012 (20 months after charcoal incorporation in the soil) at 0–5; 5–10; 10–20; 20–30 cm depths. Before analysis, the soil samples were air dried and passed through a 2 mm sieve.

2.2. Total C content and SOM Fractionation

The TC content of the soil samples was determined by dry combustion (975 °C) (Thermo Fisher Scientific-Flash EA1112, Waltham, MA, USA, detection limit = 0.01%, $n = 4$).

The SOM chemical fractionation was performed according to Swift [33], adapted by Dick et al. [34]. In a centrifuge tube, one gram of soil was shaken with 60 mL of 0.5 M HCl for 2 h. The acid extract was separated by centrifugation. This procedure was repeated three times, and the final extraction volume was measured. Hereafter, the material remaining in the tube was shaken with 60 mL of 0.5 M NaOH for 3 h to extract the soluble humic substances (SHS), namely HA and FA. This procedure was repeated until the supernatant became colorless, and the final SHS volume was measured. Aliquots

(5 mL) of the acid and of the SHS extracts were collected for C content determination. The pH of the material remaining in the tube was lowered to 2 with 4 M HCl solution, and the suspension was allowed to settle overnight. Centrifugation was used to separate HA fraction (precipitated) and FA fraction (supernatant). The FA extract volume was measured, and an aliquot (5 mL) was collected for C content determination. The residue of the extraction contained the HU fraction. The C content in the acid (C_{HCl}), SHS (C_{SHS}), and FA (C_{FA}) extracts was quantified by measuring the absorbance at 580 nm (Shimadzu-UV-160A, Kyoto, Japan) after C oxidation with K dichromate in acidic medium at 60 °C during 4 h. For discussion purposes, the concentration of the three fractions (acid extract, SHS, and of FA) was based on the C content allocated in each fraction. Likewise, the C concentration in the HA fraction (C_{HA}) was determined as follows: $C_{HA} = C_{SHS} - C_{FA}$, and the C concentration in the HU fraction (C_{HU}) was determined by the difference: $C_{HU} = TC - C_{HCl} - C_{SHS}$.

2.3. Incubation of Soil Samples and Charcoal

Before incubation, soil samples (10 g) were inoculated with 1 mL of a microbial suspension, which was extracted from a gardening soil after manual shaking with deionized water and subsequent filtering (5 µm pore size). Soil samples (four field replicates) were placed into individual closed incubation vessels (250 mL), and their water content was adjusted to ca. 60% of the maximum soil water holding capacity. In addition to the soil samples, four replicates of charcoal alone (10 g, milled to pass a 2 mm sieve) were also incubated following the same methodology. Considering that part of the charcoal is usually hydrophobic and that it can affect the water distribution within the sample and, therefore, charcoal mineralization rates, the wetted samples were carefully homogenized before incubation.

Soil and charcoal samples were incubated for 165 days at 20 °C under aerobic conditions in a Respicond Apparatus IV (Nordgren Innovations, Alnarp, Sweden). The respiration was measured every three hours by determining changes in the electrical conductivity induced by absorption of CO_2 in a KOH solution (10 mL, 0.6 M), which was allocated inside the incubation vessel [35]. The cumulative C loss was calculated by normalizing the CO_2 production to the C content of each sample and by employing a calibration constant value informed by the producer of the equipment, which takes into account the temperature used during the incubation period. The proportion of remaining C at a given time (A(t)) was calculated by subtracting the accumulated C loss from 100%. At the end of the incubation period, data were fitted to a double exponential decay model using Sigmaplot 11.0 according to equation 1. This model separates the decomposition curve in two different compartments, corresponding to the fast and the slow turnover pools, both following the first order kinetics model.

$$A(t) = A_1 \times e^{-k_1 t} + A_2 \times e^{-k_2 t} \qquad (1)$$

where $A(t)$ = remaining C (% of TC); A_1 = amount of C relatively labile against mineralization (% of TC); A_2 = amount of C more stable against mineralization (% of TC); t = incubation time; k_1 and k_2 = apparent first order mineralization rate constants for the labile and stable pool (y^{-1}), respectively. The mean residence times of the first-order reactions were $MRT_1 = 1/k_1$ and $MRT_2 = 1/k_2$, whereas the half-life time of A_2 was calculated as $t_{1/2long} = 0.693/k_2$.

Besides the soil and the charcoal samples, one blank (without soil or charcoal) was prepared in order to monitor variations due to temperature changes or background noise. The blank sample showed that no background CO_2 production occurred.

2.4. Solid-State ^{13}C Nuclear Magnetic Resonance (NMR) Cross Polarization Magic-Angle Spinning CPMAS Spectroscopy

Since differences between T1 and T4 regarding TC contents were observed only at 0–5 and 10–20 cm depths (Table 1), soil samples from these depths were selected for the ^{13}C NMR analysis. Prior to the analysis, composite samples (formed by the four field replicates) were treated with 10%

hydrofluoric acid (HF) solution to concentrate the SOM and to remove paramagnetic materials [36]. Thereafter, samples were washed five times with deionized water and freeze-dried before analysis.

Prior to the ^{13}C NMR analysis, charcoal composite samples (3 g) were placed into centrifuge tubes and treated with 75 mL of 2 M HCl solution for removing soluble ashes and paramagnetic compounds. After 2 h of mechanical shaking and subsequent centrifugation at 2000 g for 10 min, the supernatant was removed and discarded. The material remaining into the containers was washed four times with deionized water and freeze-dried. Aiming to elucidate chemical alterations on soil and charcoal samples resulting from degradation during incubation, these samples were subjected to ^{13}C NMR analysis before and after incubation.

The solid-state ^{13}C NMR spectra were obtained with a Bruker Avance III 600 MHz spectrometer, Billerica, MA, USA from General Services and facilities of the University of Seville (CITIUS, Seville, Spain) operating at a resonance frequency of 150.91 MHz, and CPMAS approach [37] was applied with a spinning speed of 15 kHz. A ramped ^{1}H pulse was used during the contact time of 1 ms to circumvent spin modulation during the Hartmann-Hahn contact [38,39]. For each sample, about 300–400 mg of finely crushed and homogenized material were packed into 4 mm zirconium rotors. Subsequently, depending on the C content of the samples, about 13,000 to 28,000 scans were accumulated for soil samples (T1 and T4), and 3000 to 8000 scans for charcoal samples with a pulse delay of 300 ms and line broadenings between 50 and 100 Hz. The ^{13}C chemical shifts were calibrated relative to tetramethylsilane (0 ppm) with glycine (COOH at 176.08 ppm). The contributions of the various C groups to the total ^{13}C intensity were calculated using the MestreNova 8.1 software. Firstly, the integration of the signal intensity of all C groups was performed. Subsequently, the area of each chemical shift region was divided by the total area, and this result was multiplied by 100. The calculation was carried out by taking into account the spinning sideband disturbance [40]. The chemical shift assignments of the CPMAS ^{13}C NMR spectra were performed as follows: 0–45 ppm, alkyl C; 45–60 ppm, N-alkyl C; 60–110 ppm, O-alkyl C; 110–160 ppm, aryl C; 160–220 ppm, carboxyl C [25].

In order to calculate the loss of each C group resulting from the microbial activity during the incubation experiment, the intensities of each chemical region in the spectra of samples, after incubation, were multiplied by the percentage of remaining C (% of the initial TC) in the sample at the end of the incubation [25]. Therefore, for samples after incubation, the difference between 100% and the sum of ^{13}C intensities corresponds to the C loss occurring during incubation. Finally, these values were subtracted from the intensities obtained for the spectra of the samples before the incubation.

2.5. Statistical Analysis

Paired *t*-tests were used to compare the means of T1 and T4 variables within soil depth. All statistical analyses were performed using the software R [41].

3. Results and Discussion

3.1. Total C Content and Humic Substances

The application of 40 Mg ha^{-1} of charcoal (T4) increased TC contents at 0–5 and 10–20 cm depths by 38 and 23%, respectively, in comparison to T1. At 0–5 cm depth, the greater concentration of C found in T4 could be attributed to the considerable amount of charcoal particles remaining near the soil surface (even after charcoal incorporation up to 10 cm depth), as evidenced, in a previous study, by scanning electron microscopy (SEM) and by the higher C content in the particulate SOM fraction [27]. The higher TC content observed in T4 at 10–20 cm depth was probably associated with the vertical transport of smaller charcoal particles.

In general, the charcoal application did not affect C_{HCl} contents, which varied from 1.0 to 2.2 g kg^{-1} (Table 1). The contribution of C_{HCl} to the TC content ranged from 2 to 9% (Figure 1). The C_{HCl} fraction is mainly composed of organic compounds that are originated from the microbial activity and from root exudations. Such chemical structures are smaller and more labile than that of the HA and FA, and due to their fast turnover, C_{HCl} is usually found in low proportions in subtropical soils [42].

Table 1. Total carbon content (TC), carbon content in the acid extract (C_{HCl}), fulvic acid (C_{FA}), humic acid (C_{HA}), and humin (C_{HU}), and C_{HA}/C_{FA} and ($C_{FA} + C_{HA})/C_{HU}$ ratios of a Cambisol without charcoal (T1) and with charcoal −40 Mg ha^{-1} (T4) application.

Treatments	TC	C_{HCl}	C_{FA}	C_{HA}	C_{HU}	C_{HA}/C_{FA}	$(C_{FA} + C_{HA})/C_{HU}$
			——————g kg^{-1}——————				
				0–5 cm			
T1	41.1 ***	1.4 ns	7.6 ns	9.2 ns	22.8 **	1.3 ns	0.8 *
T4	56.6	1.4	8.1	11.5	35.6	1.4	0.6
				5–10 cm			
T1	36.2 ns	1.2 ns	5.2 ns	8.8 ns	20.9 ns	1.8 ns	0.7 ns
T4	40.8	1.0	6.8	8.1	24.9	1.2	0.6
				10–20 cm			
T1	29.3 **	1.8 ns	4.3	5.4 **	17.8 ns	1.2 ns	0.6 *
T4	36.0	1.6	6.7 *	8.8	18.9	1.3	0.8
				20–30 cm			
T1	24.8 ns	2.2 ns	3.2 ns	5.0 **	14.5 ns	1.7 ns	0.6 *
T4	28.8	1.7	4.3	7.9	14.9	1.9	0.8

Values represent means of four replicates and ns, *, **, and *** indicate t-test results of $p > 0.10$, $p < 0.10$, $p < 0.05$, and $p < 0.01$, respectively.

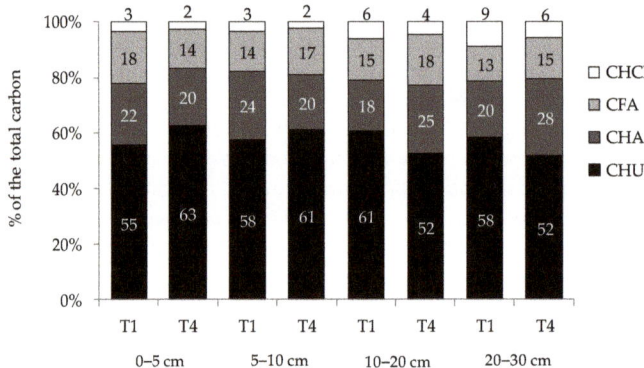

Figure 1. Contribution of C content in the acid extract (C_{HCl}), fulvic acid (C_{FA}), humic acid (C_{HA}), and humin (C_{HU}) fractions to the total carbon content of a Cambisol without charcoal (T1) and with charcoal −40 Mg ha^{-1} (T4) application.

Different from C_{HCl} contents, the C distribution in the other SOM chemical fractions was considerably affected by the charcoal application. The C_{FA} contents varied from 3.2 to 8.1 g kg^{-1} (Table 1) and represented 13 to 18% of TC (Figure 1). Compared to T1, C_{FA} content in T4 was 55% higher at 10–20 cm depth (Table 1). Significant differences were not observed in other depths.

The C_{HA} contribution to TC ranged from 18 to 28% (Figure 1). Interestingly, the increments in C_{HA} contents in response to the charcoal application were observed not near to soil surface, where charcoal particles were concentrated, but at 10–20 and 20–30 cm depths, where C_{HA} contents in T4 were 64% and 59% higher than in T1, respectively. According to the $\delta^{13}C$ isotopic ratio data of the soil samples presented in a previous study [27], after incorporation, most of the charcoal particles remained at the 0–5 cm depth and a lower proportion of them was moved to 10–20 cm depth. The presence at 10–20 cm depth of humified compounds derived from the charcoal material, might explain the higher

C_{FA} and C_{HA} contents noticed at this depth. The mild conditions of the charcoal production along with the weathering of the charcoal during the time span of the field experiment until soil sampling (20 months) might have contributed to its partial oxidation, originating carboxylic groups directly linked to aromatic structures [43–45]. Possibly, these organic compounds have migrated downward from upper depths with a preferential accumulation in the alkaline extractable fraction (SHS). In fact, literature has reported that aging of charcoal can start very quickly after entering the soil, enhancing the number of functional groups (mostly carboxyl) in the charcoal structure, leading to an increase of polar sorptive sites and, thus, facilitating the movement of charcoal compounds downward the soil profile [46–48]. The leaching of charcoal particles in soils has been also evidenced by the increase of SOM aromaticity degree in deeper depths, as a result of functionalization of aromatic charcoal structures [49–51]. At 20–30 cm depth, the greater content of C_{HA} and C_{FA} in T4 compared to T1 might be associated to an indirect effect of the charcoal on the endogenous SOM dynamics of the upper depths, promoting its functionalization and, thus, an increase of the SHS fraction, since $\delta^{13}C$ isotopic ratio data of the soil samples did not indicate charcoal presence at 20–30 cm depth [27].

The C_{HU} contents varied from 14.5 to 35.6 g kg^{-1} (Table 1), and regardless of treatment and soil depth, the contribution of C_{HU} to TC was higher than 50% (Figure 1). Similarly, to C_{FA} and C_{HA}, charcoal incremented C_{HU} contents in the soil. At 0–5 cm depth, C_{HU} content in T4 was 56% higher than in T1. This result is probably related to the concentration of charcoal particles in the soil surface even after its incorporation and to the hydrophobic character and particulate size of the charcoal, leading this material to accumulate in the non-alkaline extractable fraction, HU [20].

The relative enrichment of C as HU at 0–5 cm depth after charcoal incorporation was evidenced by the lower (C_{FA} + C_{HA})/C_{HU} ratio in T4 compared to T1 (Table 1). At 10–20 and 20–30 cm depths, where charcoal did not affect C_{HU} contents, higher (C_{FA} + C_{HA})/C_{HU} ratios were observed in T4 in comparison to T1 due to the increment of C_{HA} and/or C_{FA} content. The intensification of the humification process, particularly HA formation, in soils with PyC (e.g., "Terra Preta" in Amazon) has been reported in the literature [52–54] and can be attributed to the increase of N, P, and Ca contents in the soil due to ashes addition and/or to the increase of the effective cation exchange capacity (ECEC) of the soil as a consequence of PyC amendments. In this sense, higher P and Ca contents, as well as ECEC in T4 in comparison to T1, especially at 0–5 cm depth, were observed in our previous study [27] and might support such interpretation.

3.2. SOM and Charcoal Biodegradability

Figure 2 shows the curves of remaining C (% of the initial C) versus the incubation time (165 days expressed in hours) for T1 and T4 samples at each soil depth and for charcoal alone. In order to facilitate graphs visualization, only every tenth measured data is presented. All the coefficients of determination (R^2) were greater than 0.96 (Table 2).

During the incubation experiment, the data acquisition was interrupted for a few days (gaps between symbols in the graphs, Figure 2) due to electric power cuts and consequent computer instabilities. However, such interruption did not interfere in the amount of CO_2 accumulated in the KOH solution and, thus, the cumulative C loss measurement was not affected.

The C loss during the incubation of the soil samples ranged from 3.8 to 6.5% of the initial C and tended to decrease with soil depth regardless of treatment (Table 2). These findings suggest that SOM at the soil surface was biochemically more labile than that at deeper depths. Stabilization of SOM at deeper depths occurs mainly via organo-mineral interactions, thus, hindering its biodegradability. In the present study, the chemical SOM composition seemed to have a minor role in the SOM stabilization, since the more easily degradable functional group, i.e., O-alkyl C, decreased with depth, neither in T1 nor in T4 (Table 3).

Differences between T1 and T4 regarding C loss during incubation were not observed (Table 2) regardless of the soil depth. Despite the accumulation of charcoal particles and the substantially higher

C_{HU} content in T4 compared to T1 at 0–5 cm depth, the SOM biodegradability remained unaffected in terms of organic matter conversion to CO_2.

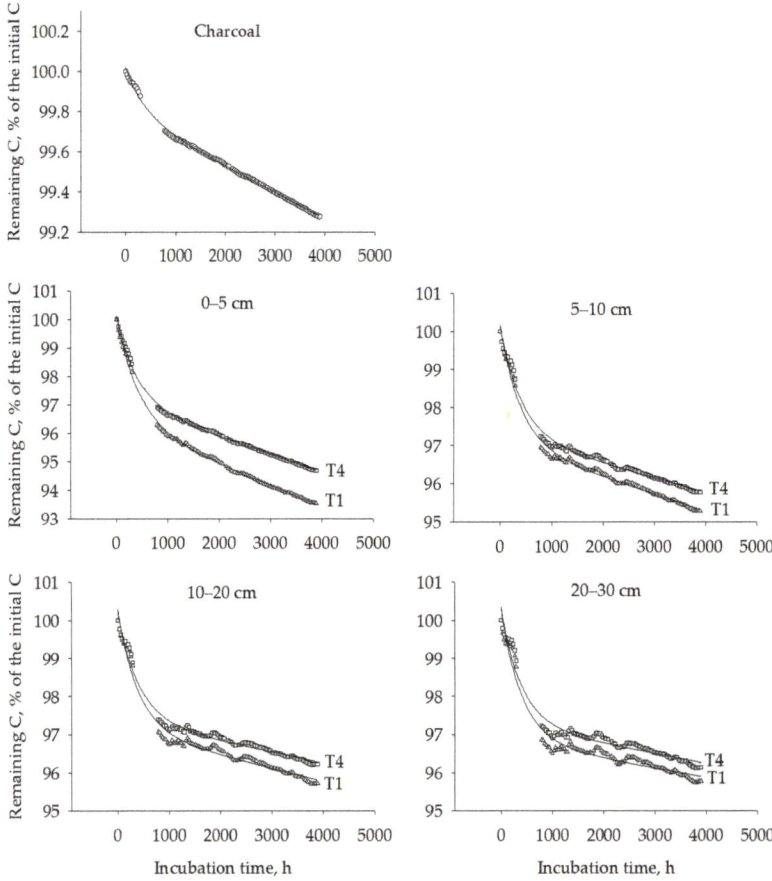

Figure 2. Remaining carbon (C) as a function of incubation time for a Cambisol without charcoal (T1) and with charcoal −40 Mg ha^{-1} (T4) and for charcoal alone (not mixed with soil).

Mineralization of charcoal (alone) observed in the present study (0.73% of the initial C) was in the range of values reported by Hamer et al. [55] (from 0.3 to 1.2%) for charred maize and rye straw (thermally altered at 350 °C) and wood (thermally altered at 800 °C) incubated for 60 days. Slower mineralization (<0.25%) for biochars, derived from wheat and eucalyptus shoot incubated for 74 days, was reported by Farrel et al. [56], and faster mineralization of hardwood derived charcoal pyrolyzed at 500 °C (2.8%) after 84 days of incubation was reported by Zavalloni et al. [57]. These comparisons suggest that the potential of the charcoal fine residues for C sequestration is comparable to that presented by biochars, including biochars produced at high temperatures and derived from hardwood, which are usually considered as highly resistant against degradation. However, it is important to highlight that when charcoal was mixed with soil and subjected to weathering under field conditions (20 months) and posterior incubation, it did not alter SOM biodegradability. The TC contents of the Cambisol (without charcoal) were considerable, 24.8 up to 41.1 g kg^{-1} (Table 1), and mainly composed of easily decomposable organic compounds (as O-alkyl C) (Table 3), which might suggest that availability of labile organic compounds possibly stimulated microorganisms to degrade assumed less labile compounds, like that from charcoal. Acceleration of PyC (charred maize and rye residues and oak wood) mineralization due to the presence of labile organic material (glucose) in a 60-day incubation experiment (air temperature of 20 °C) was already reported in the literature [55].

Table 2. Cumulative carbon (C) loss after 165 days incubation experiment, proportion of the fast (A_1) and slow (A_2) organic matter pools, their respective constant mineralization rates (k_1 and k_2) and mean residence times (MRT_1 and MRT_2), and half-life time of the slow organic matter pool ($t_{1/2\ long}$) of the charcoal alone (not mixed with soil) and of a Cambisol without charcoal (T1) and with charcoal -40 Mg ha^{-1} (T4) application.

Treatments	C loss % of TC	A_1 % of TC	k_1 year^{-1}	MRT_1 years	A_2 % of TC	k_2 year^{-1}	MRT_2 years	$t_{1/2\ long}$ years	R^2 *
				0–5 cm					
T1	6.5 ns	3.7 ns	19.5 ns	0.05 ns	96.3 ns	0.066 ns	15.7 ns	10.9 ns	0.998
T4	5.2	2.9	22.7	0.04	97.3	0.058	17.4	12.1	0.998
				5–10 cm					
T1	4.7 ns	3.1 ns	22.6 ns	0.05 ns	97.1 ns	0.041 ns	25.2 ns	17.5 ns	0.991
T4	4.3	3.0	21.2	0.05	97.2	0.034	29.7	20.6	0.990
				10–20 cm					
T1	4.2 ns	3.1 ns	23.3 ns	0.04 ns	97.2 ns	0.030 ns	34.3 ns	23.8 ns	0.979
T4	3.8	2.8	22.1	0.05	97.4	0.027	37.8	26.2	0.982
				20–30 cm					
T1	4.3 ns	3.6 ns	21.5 ns	0.05 ns	96.9 ns	0.025 ns	42.1 ns	29.2 ns	0.965
T4	3.9	3.1	20.6	0.05	97.3	0.024	41.9	29.1	0.967
				Charcoal					
	0.73 (0.09)[1]	0.23 (0.04)	21.9 (0.00)	0.05 (0.00)	99.8 (0.05)	0.012 (0.00)	87.0 (8.19)	60.3 (5.68)	0.998

Values represent means of four replicates, and ns indicates t-test results of $p > 0.10$. [1] Values within parentheses refer to the standard deviation of the charcoal data. * Coefficient of determination of the fit correlating the cumulative C loss versus time according to a double exponential decay model. TC: total carbon.

The lower mineralization of the charcoal (0.73%) in comparison to that of soil samples (from 3.9 to 6.5%) could be assigned to the higher aryl C and lower O-alkyl and N-alkyl C proportions in the charcoal compared to the SOM (Table 3) assuming that aryl C compounds are more resistant to biodegradation than O-alkyl and N-alkyl C [58–60].

The proportion of the fast SOM pool (A_1) to the TC ranged from 3.1 to 3.7% in T1 and from 2.8 to 3.1% in T4 (Table 2), and charcoal did not affect this SOM compartment. The charcoal is mainly composed of less labile organic compounds, as aryl C compounds, which accounted for 78% of total ^{13}C intensity in the charcoal NMR spectrum (before incubation) (Table 3). Labile compounds are less representative in charcoal materials, and possibly these compounds were degraded within the 20 months of field experiment and exposition of charcoal to weathering. In fact, relatively rapid oxidation of charcoal after entering the soil has been reported in the literature, especially for those materials produced under mild temperatures [25,43,47], as in the case of the charcoal used in this study [27].

The mineralization rate of the fast SOM pool (k_1) for T1 and T4 ranged from 19.5 to 23.3 year^{-1} and for charcoal alone was 21.9 year^{-1} (Table 2). Differences between T1 and T4 k_1 values were not observed regardless of the soil depth (Table 2). In this way, T1 and T4 did not differ with respect to the mean residence time of the fast SOM pool (MRT_1). MRT_1 for soil samples varied from 0.04 to 0.05 years and were comparable to the MRT_1 of the charcoal alone (0.05 years), reinforcing that labile charcoal-derived compounds undergo similar decomposition as labile SOM compounds. Similar MRT_1 values (0.03 to 0.05 years) were reported by authors evaluating the biodegradability of the SOM in fire-affected and unaffected soils from Southern Spain [25].

In charcoal samples, the amount of C more stable against mineralization (A_2) accounted for 99.8% of the TC (Table 2). In the soil samples, regardless of the treatment, values were high as well, varying between 96 and 97% (Table 2). Similar A_2 values (91 to 96%) were reported by other researchers after 206 days incubation of fire-affected and unaffected soil samples [25].

The k_2 observed for pure charcoal (0.012 year^{-1}) was about 2 to 5-fold lower than that observed for soil samples (from 0.024 to 0.066 year^{-1}), indicating that for this SOM pool, the chemical composition of the C source is relevant. However, when applied to the soil, even at a rate of 40 Mg ha^{-1}, which is a considerably high amount of charcoal application to the soil, the charcoal did not promote higher A_2 and k_2 values in T4 compared to T1 (Table 2). More remarkable effect in A_2 values due to charcoal application was expected due to a coupled effect resulting from: i) charcoal particles concentration at 0–5 cm [27] and ii) the aromatic character of the charcoal (78% aryl C contribution, Table 3), which is supposed to confer slow biodegradability to this material. Mean residence time of the slow pool (MRT$_2$) for pure charcoal was of 87 years, about 2 to 6-fold higher than those observed for T1 and T4, from 15.7 to 42.1 years (Table 2). However, charcoal addition did not increase MRT$_2$ in T4 compared to T1 despite the high aryl C intensity in the charcoal ^{13}C NMR spectra (Table 2). Yet, based on the MRT$_2$ of the charcoal alone, a more relevant increase of MRT$_2$ for T4 samples was expected. As discussed before, labile organic compounds from SOM possibly stimulated soil microorganisms to degrade charcoal [31,55]. Although not statistically significant, MRT$_2$ at 0–5, 5–10, and 10–20 cm depths tended to be higher in T4 than in T1, suggesting the influence of the added charcoal on the greater MRT$_2$ values. At 20–30 cm depth, MRT$_2$ values observed for T1 and T4 were similar. In fact, charcoal particles were mainly concentrated within 0–20 cm depth and not evidenced at 20–30 cm depth [27], justifying why such tendency was restricted to the first 20 cm depth. One alternative to increase SOM MRT$_2$ more consistently aiming C sequestration into the soil through charcoal fines addition would be to increase charcoal doses applied to the soil, i.e., higher than 40 Mg ha^{-1} in this case. However, it is important to consider that applying higher doses of charred material to the soil, on the other side, could be prejudicial to soil biota, plant growth, and food chain, a part of the uncertain effects of these materials to the environment in the long-term. Also, improper deposition of such materials could occur as a result of water erosion in charcoal or biochar-amended soils [61–64].

The $t_{1/2\ long}$ of the charcoal alone (60.3 years) was 2 to 6-fold higher than those of soil samples, which ranged from 10.9 to 29.2 years (Table 2). Regardless of the soil depth, $t_{1/2\ long}$ values did not differ between T1 and T4 (Table 2). Similar to MRT$_2$, $t_{1/2\ long}$ values tended to be greater in T4 within 0–20 cm depth, most probably due to the concentration of charcoal particles up to 20 cm depth, as discussed previously. At 20–30 cm, $t_{1/2\ long}$ values were quite similar in T1 and T4, 29.2 and 29.1 years, respectively, corroborating suppositions that large charcoal fragments, which are less prone to fast degradation and leaching [65], were not relevantly transported to depths below 20 cm [27].

3.3. Alteration of SOM and Charcoal Chemical Composition after Incubation

The ^{13}C NMR spectrum of the charcoal alone before the incubation experiment was dominated by the ^{13}C intensity band corresponding to the aryl C region, which accounted for 78% of the total ^{13}C intensity (Table 3). The other signals found in the charcoal ^{13}C NMR spectrum, and their contribution to the total ^{13}C intensity were: O-alkyl C –8%, carboxyl C –6%, alkyl C –5%, which can be assigned to short alkyl chains from lignin that remained in the charcoal after the mild charring process [40], and N-alkyl C –3% (Table 3). The low signal in the 100–60 ppm region (O-alkyl C) can be assigned to aromatic C compounds, which may have caused resonance lines mainly in the 110–90 ppm region or to the few easily decomposable structures, as carbohydrates from *Mimosa scabrella* trees, which may have resisted to the charring [40,59]. After 165 days of incubation, the ^{13}C NMR spectrum of the charcoal exhibited the same pattern as prior to incubation (Figure 3). However, the contributions of the chemical groups to the total ^{13}C intensity were altered, despite the low mineralization of the charcoal during the incubation experiment (0.73% of C loss of the initial TC). The aryl C contribution after the incubation period decreased to 67%, suggesting that some microorganisms were able to use the aromatic compounds as a C source. This event might have been magnified when charcoal was mixed with soil (in the field), at least partially explaining why charcoal did not increment SOM resistance against biodegradability, especially MRT$_2$. Consequently, after incubation, the relative contribution of the alkyl C increased from 5% to 11% and that of O-alkyl C from 8% to 12% (Table 3).

The ^{13}C NMR spectra of T1 and T4 samples, before and after incubation experiment, presented the same pattern and differed markedly from that of charcoal (Figure 3). Soil samples spectra were dominated by the ^{13}C intensity in the O-alkyl C chemical shift region (30 to 38% of the total ^{13}C intensity), which is assigned mainly to carbohydrates from the microbial biomass and plant residues [59,66]. Due to the aromaticity of the charcoal and its concentration near to the soil surface, higher aryl C ^{13}C intensity in T4 compared to T1 was expected, but difference of aryl C contribution to the total ^{13}C intensity in T4 and T1 was only 2.5% (Table 3), which can more likely be attributed to instrumental variations and samples heterogeneity. These results suggest that despite the aromaticity and thermostability of the charcoal fines [27], once they were applied to the Cambisol (40 Mg ha^{-1}), they were not able to increment aryl C contribution to the SOM composition, probably due to the susceptibility of this material to biodegradation. At 10–20 cm depth, relevant differences in ^{13}C NMR spectra between treatments, before incubation experiment, were not evidenced as well. Decrease of aryl C contribution to the total ^{13}C intensity from 46% to 23%, when PyC amended A-horizon samples were collected four weeks and 7 years after forest fire in Southern Spain, was recently reported in the literature and attributed to the microbial decomposition, leaching, or erosion of such material [67]. These findings reinforce that PyC can suffer rapid microbial decomposition into the soil, and moreover, that the assumption of PyC as a stable C compartment in the soil should be taken carefully.

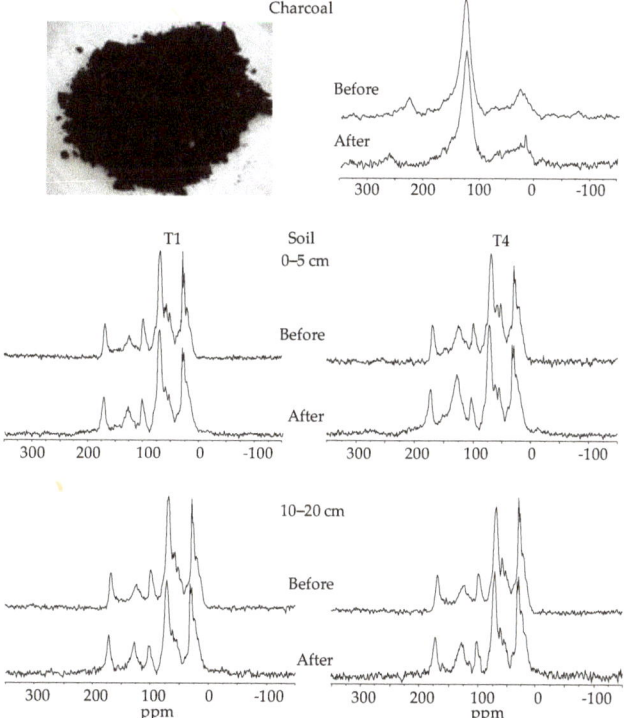

Figure 3. Charcoal illustration and Solid-state ^{13}C NMR spectra of the HCl-treated charcoal alone (not mixed with soil) and of HFtreated samples of a Cambisol without charcoal (T1) and with charcoal −40 Mg ha^{-1} (T4) application, before and after 165 incubation days, at 0–5 and 10–20 cm depth.

After 165 days of incubation, the ^{13}C NMR spectra of T1 and T4 samples were not altered compared to the spectra before incubation, regardless of the soil depth, and ^{13}C intensity variations of up to 4% within chemical shift regions (Table 3) were attributed to instrumental aspects. In general, these results indicate that preferential preservation of charcoal derived-aromatic structures against degradation did not occur as could be expected when assuming charcoal as a highly recalcitrant material. Moreover, aryl C ^{13}C intensity enrichment in T4 after incubation was not observed, supporting why charcoal did not increment MRT$_2$, as discussed before.

Table 3. Comparison of the relative intensity distribution (%) of Solid-State ^{13}C NMR spectra, before and after 165 days of incubation, of the charcoal and of the Cambisol samples collected at 0–5 and 10–20 cm depth after charcoal application.

Treatments [1]	Incubation	Carboxyl C 220–160	Aryl C 160–110	O-Alkyl C 110–60	N-Alkyl C 60–45	Alkyl C 45–0
				%		
				0–5 cm		
T1	Before	6.5	12.3	38.3	12.9	30.0
	After	8.4	12.7	35.2	9.0	28.3
	Before-after	−2	0	3	4	2
T4	Before	8.7	22.9	33.1	8.7	24.8
	After	8.8	25.4	31.6	8.3	24.0
	Before-after	0	−2.5	1.5	0	1
				10–20 cm		
T1	Before	6.5	13.4	37.0	12.4	30.7
	After	9.5	13.4	33.8	8.7	30.5
	Before-after	−3	0	3	4	0
T4	Before	7.5	15.4	34.8	11.3	30.9
	After	9.2	15.5	32.0	8.4	31.1
	Before-after	−2	0	3	3	0
				Charcoal		
	Before	6.1	78.3	8.0	2.7	4.9
	After	6.7	66.6	11.6	3.8	10.5
	Before-after	−1	12	−4	−1	−6

[1] T1 = 0 Mg ha^{-1} (control); T4 = 40 Mg ha^{-1}.

4. Conclusions

The charcoal fragments located at 0–5 cm depth were preferentially accumulated in the humin fraction, most probably due to the charcoal hydrophobic character. The greater content of humic acids and fulvic acids in the charcoal amended soil was related to charcoal oxidation and to the effect of the charcoal on endogenous SOM humification dynamics.

Despite the concentration of charcoal particles at 0–5 cm depth, leading to carbon content increase (as humin) and the aromaticity of the charcoal verified by ^{13}C NMR, charcoal did not increment the slow soil organic matter pool. Apparently, the availability of labile organic compounds from SOM stimulated the biodegradation of charcoal aromatics when charcoal was applied to the soil and suffered weathering. Preferential preservation of aromatic structures in the charcoal-amended soil after incubation was not evident, supporting such interpretations.

Overall, our findings suggest that incorporation of charcoal fine residues at a rate of 40 Mg ha^{-1} to a subtropical Cambisol was not an efficient strategy to promote carbon sequestration in the soil and that this material could be preferentially used as a means to improve soil chemical agronomical attributes.

Author Contributions: Conceptualization, O.d.A.L. and D.P.D.; Formal analysis, O.d.A.L., G.S.C., and H.K.; Investigation, O.d.A.L., D.P.D., D.P.B.L., J.A.G.-P., and H.K.; Methodology, O.d.A.L., D.P.D., J.M.D.l.R., and H.K.; Project administration, O.d.A.L. and D.P.D.; Resources, D.P.D., J.A.G.-P., and H.K.; Supervision, D.P.D., J.M.D.l.R., J.A.G.-P., and H.K.; Visualization, O.d.A.L., D.P.D., and H.K.; Writing–original draft, O.d.A.L., D.P.D., J.M.D.l.R., D.P.B.L., and H.K.; Writing–review & editing, O.d.A.L., D.P.D., J.M.D.l.R., D.P.B.L., G.S.C., and H.K.

Funding: This research received no external funding

Acknowledgments: The authors are grateful to Kátia Cylene Lombardi (Unicentro), Federal University of Rio Grande do Sul, Institute of Natural Resources and Agrobiology of Seville, National Council for Scientific and Technological Development (CNPq), Coordination for Improvement of Higher Education Personnel (CAPES) and Spanish MINECO INTERCARBON project (CGL2016-78937-R) for supporting this work. Leal thanks the International Humic Substances Society and the Alexander von Humboldt Foundation who encouraged the development of this work.

Conflicts of Interest: The authors declare no conflict of interest.

References

1. Lehman, J.; Gaunt, J.; Rondin, M. Bio-char sequestration in terrestrial Ecosystems—A review. *Mitig. Adapt. Strateg. Glob.* **2006**, *11*, 403–427. [CrossRef]
2. Qian, K.; Kumar, A.; Zhang, H.; Bellmer, D.; Huhnke, R. Recent advances in utilization of biochar. *Renew Sustain. Energy Rev.* **2015**, *42*, 1055–1064. [CrossRef]
3. Lehman, J.; Joseph, S. *Biochar for Environmental Management*, 2nd ed.; Routledge: London, UK, 2015; pp. 1–14.
4. Glaser, B.; Lehman, J.; Zech, W. Ameliorating physical and chemical properties of highly weathered soils in the tropics with charcoal—A review. *Biol. Fertil. Soils* **2002**, *35*, 219–230. [CrossRef]
5. Steiner, C.; Teixeira, W.G.; Lehman, J.; Nehls, T.; Macêdo, J.L.V.; Blum, W.E.H.; Zech, W. Long term effects of manure, charcoal and mineral fertilization on crop production and fertility on a highly weathered Central Amazonian upland soil. *Plant Soil* **2007**, *291*, 275–290. [CrossRef]
6. Maia, C.M.B.F.; Madari, B.E.; Novotny, E.H. Advances in biochar research in Brazil. *Dyn. Soil Dyn. Plant* **2011**, *5*, 53–58.
7. Schneider, M.P.W.; Lehmann, J.; Schmidt, M.W.I. Charcoal quality does not change over a century in a tropical agro-ecosystem. *Soil Biol. Biochem.* **2011**, *43*, 1992–1994. [CrossRef]
8. Bayer, C.; Zschornack, T.; Pedroso, G.M.; Da Rosa, C.M.; Camargo, E.S.; Boeni, M.; Marcolin, E.; Reis, C.E.S.; Santos, D.C. A seven-year study on the effects of fall soil tillage on yield-scaled greenhouse gas emission from flood irrigated rice in a humid subtropical climate. *Soil Till. Res.* **2015**, *145*, 118–125. [CrossRef]
9. Godoi, E.G.; Neufeld, A.D.H.; Ibarr, M.A.; Ferreto, D.O.C.; Bayer, C.; Lorentz, L.H.; Vieira, F.C.B. The conversion of grassland to acacia forest as an effective option for net reduction in greenhouse gas emissions. *J. Environ. Manag.* **2016**, *15*, 91–102. [CrossRef] [PubMed]
10. Melo, L.C.A.; Coscione, A.R.; Abreu, C.A.; Puga, A.P.; Camargo, O.A. Influence of pyrolisis temperature on Cadmiun and Zinc sorption capacity of sugar cane straw-derived biochar. *Bioresources* **2013**, *8*, 4992–5004. [CrossRef]
11. Eykelbosh, A.J.; Johnson, M.S.; Queiroz, E.S.; Dalmagro, H.J.; Couto, E.G. Biochar from sugarcane filtercake reduces soil CO_2 emissions relative to raw residue and improves water retention and nutrient availability in a highly- weathered tropical soil. *PLoS ONE* **2014**, *9*, 1–9. [CrossRef] [PubMed]
12. Andrade, C.A.; Bibar, M.P.S.; Coscione, A.R.; Pires, A.M.M.; Soares, A.G. Mineralização e efeitos de biocarvão de cama de frango sobre a capacidade de troca catiônica do solo. *Pesq. Agropec. Bras.* **2015**, *50*, 407–416. [CrossRef]
13. Conz, R.F. Caracterização de matérias-primas e biochars para aplicação na agricultura. Master's Thesis, Universidade de São Paulo, São Paulo, Brazil, 2015.
14. Sousa, A.A.T.C.; Figueiredo, C.C. Sewage sludge biochar: Effects on soil fertility and growth of radish. *Biol. Agric. Hortic.* **2015**, *32*, 1–12. [CrossRef]
15. Angelo, L.C.; Mangrich, A.S.; Mantovani, K.M.; Santos, S.S. Loading of VO^{2+} and Cu^{2+} to partially oxidized charcoal fines rejected from Brazilian metallurgical industry. *J. Soils Sediments* **2014**, *14*, 353–359. [CrossRef]
16. Spokas, K.A. Review of the stability of biochar in soils: Predictability of O:C molar ratios. *Carbon Manag.* **2010**, *1*, 289–303. [CrossRef]
17. Verheijen, F.G.A.; Montanarella, L.; Bastos, A.C. Sustainability, certification, and regulation of biochar. *Pesq. Agropec. Bras.* **2012**, *47*, 649–653. [CrossRef]
18. Combrie, K.; Masek, O.; Sohi, S.P.; Brownsort, P.; Cross, A. The effect of pyrolysis conditions on biochar stability as determined by three methods. *GCB Bioenergy* **2013**, *5*, 122–131. [CrossRef]
19. Novotny, E.H.; Maia, C.M.B.F.; Carvalho, M.T.M.; Madari, B.E. Biochar: Pyrogenic carbon for agricultural use—A critical review. *Rev. Bras. Cienc. Solo* **2015**, *39*, 321–344. [CrossRef]

20. Novotny, E.H.; Hayes, M.H.B.; Madari, B.E.; Bonagamba, T.J.; Azevedo, E.R.; Souza, A.A.; Song, G.; Nogueira, C.M.; Mangrich, A.S. Lessons from the *Terra Preta de Índios* of the Amazon region for the utilization of charcoal for soil amendment. *J. Braz. Chem. Soc.* **2009**, *20*, 1003–1010. [CrossRef]
21. Knoblauch, C.; Maarifat, A.A.; Pfeiffer, E.M.; Haefele, S.M. Degradability of black carbon and its impact on trace gas fluxes and carbon turnover in paddy soils. *Soil Biol. Biochem.* **2011**, *43*, 1768–1778. [CrossRef]
22. Bruun, S.; El-Zehery, T. Biochar effect on the mineralization of soil organic matter. *Pesq. Agropec. Bras.* **2012**, *47*, 665–671. [CrossRef]
23. Santos, F.; Torn, M.S.; Bird, J.A. Biological degradation of pyrogenic organic matter in temperate forest soils. *Soil Biol. Biochem.* **2012**, *51*, 115–124. [CrossRef]
24. Maestrini, B.; Herrmann, A.M.; Nannipieri, P.; Shmidt, M.W.I.; Abiven, S. Ryegrass-derived pyrogenic organic matter changes organic matter and nitrogen mineralization in a temperate forest soil. *Soil Biol. Biochem.* **2014**, *69*, 291–301. [CrossRef]
25. Knicker, H.; González-Vila, F.J.; González-Vásquez, R. Biodegradability of organic matter in fire-affected mineral soils of Southern Spain. *Soil Biol. Biochem.* **2013**, *56*, 31–39. [CrossRef]
26. Vasilyeva, N.A.; Abiven, S.; Milanovskiy, Y.; Hilf, M.; Rizkov, O.V.; Schmidt, M.W.I. Pyrogenic carbon quantity and quality unchanged after 55 years of organic matter depletion in a Chernozem. *Soil Biol. Biochem.* **2011**, *43*, 1985–1988. [CrossRef]
27. Leal, O.A.; Dick, D.P.; Lombardi, K.C.; Maciel, V.G.; González-Pérez, J.A.; Knicker, H. Soil chemical properties and organic matter composition of a subtropical Cambisol after charcoal fine residues incorporation. *J. Soils Sediments* **2015**, *15*, 805–815. [CrossRef]
28. Kuzyakov, Y.; Subbotina, I.; Chen, H.; Bogomolova, I.; Xu, X. Black carbon decomposition and incorporation into soil microbial biomass estimated by ^{14}C labeling. *Soil Biol. Biochem.* **2009**, *41*, 210–219. [CrossRef]
29. De la Rosa, J.M.; Rosado, M.; Paneque, M.; Miller, A.Z.; Knicker, H. Effects of aging under field conditions on biochar structure and composition: Implications for biochar stability in soils. *Sci. Total Environ.* **2018**, *613–614*, 969–976. [CrossRef]
30. FAO. *World Reference Base for Soils Resources 2006*; Food and Agriculture Organization of the United Nations: Rome, Italy, 2006; p. 128.
31. Zimmerman, A.R.; Gao, B.; Ahn, M.Y. Positive and negative carbon mineralization priming effects among a variety of biochar-amended soils. *Soil Biol. Biochem.* **2011**, *43*, 1169–1179. [CrossRef]
32. Woiciechowski, T. Evaluation of the Attributes of a Cambisol and Early Growth of *Eucalyptus benthamii* after Application of Biochar in Irati City, Paraná State, Brazil. Master's Thesis, Universidade Estadual do Centro-Oeste, Paraná, Brazil, 2011.
33. SWIFT, R.S. Organic matter characterization. In *Methods of soil Analysis. Part 3. Chemical Methods*, 1st ed.; Sparks, D.L., Ed.; SSSA: Madison, WI, USA, 1996; Volume 1, pp. 1001–1069.
34. Dick, D.P.; Gomes, J.; Rosinha, P.B. Caracterização de substâncias húmicas extraídas de solos e de lodo orgânico. *Rev. Bras. Cienc. Solo* **1998**, *22*, 603–611. [CrossRef]
35. Nordgren, A. Apparatus for the continuous, long-term monitoring of soil respiration rate in large numbers of samples. *Soil Biol. Biochem.* **1988**, *20*, 955–957. [CrossRef]
36. Gonçalves, C.N.; Dalmolin, R.S.D.; Dick, D.P.; Knicker, H.; Klamt, E.; Kögel-Knaber, I. The effect of 10% HF treatment on the resolution of CPMAS ^{13}C NMR spectra and on the quality of organic matter in Ferrasols. *Geoderma* **2003**, *116*, 373–392. [CrossRef]
37. Schaefer, J.; Stejskal, E.O. C-13 nuclear magnetic-resonance of polymers spinning at magic angle. *J. Am. Chem. Soc.* **1976**, *98*, 1031–1032. [CrossRef]
38. Peersen, O.B.; Wu, X.L.; Kustanovich, I.; Smith, S.O. Variable-amplitude crosspolarization MAS NMR. *J. Magn. Reson. Ser. A* **1993**, *104*, 334–339. [CrossRef]
39. Cook, R.L.; Langford, C. A modified crosspolarization magic angle spinning C-13 NMR procedure for the study of humic materials. *Anal. Chem.* **1996**, *68*, 3979–3986. [CrossRef]
40. Knicker, H.; Totsche, K.U.; Almendros, G.; González-Vila, F.J. Condensation degree of burnt peat and plant residues and the reliability of solid-state VACP MAS ^{13}C NMR spectra obtained from pyrogenic humic material. *Org. Geochem.* **2005**, *36*, 1359–1377. [CrossRef]
41. R Core Team. The R Project for Statistical Computing. 2013. Available online: http://www.R-project.org (accessed on 4 July 2019).

42. Santana, G.S.; Dick, D.P.; Jacques, A.V.A.; Chitarra, G.S. Substâncias húmicas e suas interações com Fe e Al em Latossolo subtropical sob diferentes sistemas de manejo de pastagem. *Rev. Bras. Cienc. Solo* **2011**, *35*, 461–472. [CrossRef]
43. De la Rosa, J.M.; Knicker, H. Bioavailability of N released from N-rich pyrogenic organic matter: An incubation study. *Soil Biol. Biochem.* **2011**, *43*, 2368–2373. [CrossRef]
44. De la Rosa, J.M.; Miller, A.Z.; Knicker, H. Soil-borne fungi challenge the concept of long-term biochemical recalcitrance of pyrochar. *Sci. Rep.* **2018**, *8*, 1–9. [CrossRef] [PubMed]
45. Zimmerman, A. Abiotic and microbial oxidation of laboratory-produced black carbon (biochar). *Environ. Sci. Tech.* **2010**, *44*, 1295–1301. [CrossRef] [PubMed]
46. Abiven, S.; Hengartner, P.; Schneider, M.P.W.; Singh, N.; Schmidt, M.W.I. Pyrogenic carbon soluble fraction is larger and more aromatic in aged charcoal than in fresh charcoal. *Soil Biol. Biochem.* **2011**, *43*, 1615–1617. [CrossRef]
47. Hilscher, A.; Knicker, H. Degradation of grass-derived pyrogenic organic material, transport of the residues within a soil column and distribution in soil organic matter fractions during a 28 month microcosm experiment. *Org. Geochem.* **2011**, *42*, 42–54. [CrossRef]
48. Jimenez-Gonzalez, M.A.; De la Rosa, J.M.; Jimenez-Morillo, N.T.; Almendros, G.; Gonzalez-Perez, J.A.; Knicker, H. Post-fire recovery of soil organic matter in a Cambisol from typical Mediterranean forest in Southwestern Spain. *Sci. Total Environ.* **2016**, *572*, 1414–1421. [CrossRef] [PubMed]
49. Dieckow, J.; Mielniczuk, J.; Knicker, H.; Bayer, C.; Dick, D.P.; Kögel-Knaber, I. Composition of organic matter in a subtropical Acrisol as influenced by land use cropping and N fertilization, assessed by CPMAS ^{13}C NMR spectroscopy. *Eur. J. Soil Sci.* **2005**, *56*, 705–715. [CrossRef]
50. Knicker, H.; Almendros, G.; González-Vila, F.J.; González-Pérez, J.A.; Povillo, O. Characteristic alterations of quantity and quality of soil organic matter caused by forest fires in continental Mediterranean ecosystems: A solid-state ^{13}C NMR study. *Eur. J. Soil Sci.* **2006**, *57*, 558–569. [CrossRef]
51. Knicker, H.; Nikolova, R.; Dick, D.P.; Dalmolin, R.S.D. Alteration of quality and stability of organic matter in grassland soils of Southern Brazil highlands after ceasing biannual burning. *Geoderma* **2012**, *181–182*, 11–21. [CrossRef]
52. Lima, H.N.; Schaefer, C.E.R.; Mello, J.W.V.; Gilkes, R.J.; Ker, J.C. Pedogenesis and pre-Colombian land use of "Terra Preta Anthrosols" ("Indian black earth") of Western Amazonia. *Geoderma* **2002**, *110*, 1–17. [CrossRef]
53. Cunha, T.J.F.; Madari, B.E.; Benites, V.M.; canellas, L.P.; Novotny, E.H.; Moutta, R.O.; Trompowsky, P.M.; Santos, G.A. Fracionamento químico da matéria orgânica e características de ácidos húmicos de solos com horizonte a antrópico da amazônia (Terra Preta). *Acta Amazon.* **2007**, *37*, 91–98. [CrossRef]
54. Cunha, T.J.F.; Novotny, E.H.; Madari, B.E.; martin-Neto, L.; Rezende, M.O.O.; Canellas, L.P.; Benites, V.M. Spectroscopy characterization of humic acids isolated from Amazonian dark earth soils (Terra preta de índio). In *Amazonian Dark Earths: Wim Sombroek's Vision*, 1st ed.; Woods, W.I., Teixeira, W.G., Lehmann, J., Steiner, C., WinklerPrins, A., Rebellato, L., Eds.; Springer: Dordrecht, The Netherlands, 2009; Volume 1, pp. 363–372.
55. Hamer, U.; Marschner, B.; Brodowski, S.; Amelung, W. Interactive priming of black carbon and glucose mineralisation. *Org. Geochem.* **2004**, *35*, 823–830. [CrossRef]
56. Farrel, M.; Kuhn, T.K.; Macdonald, L.M.; Maddern, T.M.; Murphy, D.V.; Hall, P.A.; Singh, B.P.; Baumann, K.; Krull, E.S.; Baldock, J.A. Microbial utilisation of biochar-derived carbon. *Sci. Total Environ.* **2013**, *465*, 288–297. [CrossRef]
57. Zavalloni, C.; Alberti, G.; Biasiol, S.; Vedove, G.D.; Fornasier, F.; Liu, J.; Peressotti, A. Microbial mineralization of biochar and wheat straw mixture in soil: A short-term study. *Appl. Soil Eco.* **2011**, *50*, 45–51. [CrossRef]
58. De La Rosa, J.M.; Knicker, H.; López-Capel, E.; Maning, D.A.C.; González-Pérez, J.A.; González-Vila, F.J. Direct detection of black carbon in soils by Py-GC/MS, Carbon-13 NMR spectroscopy and thermogravimetric techniques. *Soil Sci. Soc. Am. J.* **2008**, *72*, 258–267. [CrossRef]
59. Knicker, H. How does fire affect the nature and stability of soil organic nitrogen and carbon? A review. *Biogeochem.* **2007**, *85*, 91–118. [CrossRef]
60. Knicker, H.; Hilscher, A.; González-Vila, F.J.; Almendros, G. A new conceptual model for the structural properties of char produced during vegetation fires. *Org. Geochem.* **2008**, *39*, 935–939. [CrossRef]
61. Major, J.; Lehmann, J.; Rondon, M.; Goodale, C. Fate of soil-applied black carbon: Downward migration, leaching and soil respiration. *Glob. Chang. Biol.* **2010**, *16*, 1366–1379. [CrossRef]

62. MacDonald, L.M.; Farrel, M.; Zwieten, L.V.; Krull, E.S. Plant growth responses to biochar addition: An Australian soils perspective. *Biol. Fertil. Soils.* **2014**. [CrossRef]
63. Lahori, A.H.; Zhanyu, G.; Zengqiang, Z.; Ronghua, L.; Mahar, A.; Awasthi, M.; Feng, S.; Sial, T.A.; Kumbhar, F.; Ping, W.; et al. Use of biochar as an amendment for remediation of heavy metal-contaminated soils: Prospects and challenges. *Pedosphere* **2017**, *6*, 991–1014. [CrossRef]
64. Vista, S.P.; Khadka, A. Determining appropriate dose of biochar for vegetables. *J. Pharm. Phytochem.* **2017**, *SP1*, 673–677.
65. Spokas, K.A.; Novak, J.M.; Masiello, C.A.; Johnson, M.G.; Colosky, E.C.; Ippolito, J.A.; Trigo, C. Physical disintegration of biochar: An overlooked process. *Environ. Sci. Tech.* **2014**, *1*, 326–332. [CrossRef]
66. Baldock, J.A.; Oades, J.M.; Waters, A.G.; Peng, X.; Vassalo, A.M.; Wilson, M.A. Aspects of the chemical structure of soil organic materials as revealed by solid-state ^{13}C NMR spectroscopy. *Biogeochemistry* **1992**, *16*, 1–42. [CrossRef]
67. López-Martins, M.; González-Vila, F.J.; Knicker, H. Distribution of black carbon and black nitrogen in physical fractions from soils seven years after an intense forest fire and their role as C sink. *Sci. Total Environ.* **2018**, *637–638*, 1187–1196. [CrossRef] [PubMed]

© 2019 by the authors. Licensee MDPI, Basel, Switzerland. This article is an open access article distributed under the terms and conditions of the Creative Commons Attribution (CC BY) license (http://creativecommons.org/licenses/by/4.0/).

Article

Hydrothermal Carbonization and Pyrolysis of Sewage Sludge: Effects on *Lolium perenne* Germination and Growth

Marina Paneque [1,*], Heike Knicker [1], Jürgen Kern [2] and José María De la Rosa [1]

[1] Instituto de Recursos Naturales y Agrobiología de Sevilla, Consejo Superior de Investigaciones Científicas (IRNAS-CSIC), Reina Mercedes, Av. 10, 41012 Seville, Spain
[2] Leibniz-Institut für Agrartechnik und Bioökonomie, Max-Eyth-Allee 100, 14469 Postdam, Germany
* Correspondence: mpaneque@irnas.csic.es; Tel.: +34-9546-24711

Received: 13 May 2019; Accepted: 5 July 2019; Published: 9 July 2019

Abstract: The pyrolysis and hydrothermal carbonization (HTC) of sewage sludge (SS) resulted in products free of pathogens, with the potential for being used as soil amendment. With this work, we evaluated the impact of dry pyrolysis-treated (600 °C, 1 h) and HTC-treated (200 °C, 260 °C; 0.5 h, 3 h) SS on the germination, survival, and growth of *Lolium perenne* during an 80 day greenhouse experiment. Therefore, the hydrochars and pyrochars were amended to a Calcic Cambisol at doses of 5 and 25 t ha^{-1}. The addition of sludge pyrochars to the Cambisol did not affect *Lolium* germination, survival rates or plant yields. However, the use 25 t ha^{-1} of wood biochar reduced germination and survival rates, which may be related to the low N availability of this sample. In comparison to the control, higher or equal plant biomass was produced in the hydrochar-amended pots, even though some hydrochars decreased plant germination and survival rates. Among all the evaluated char properties, only the organic and inorganic N contents of the chars, along with their organic C values, positively correlated with total and shoot biomass production. Our work demonstrates the N fertilization potential of the hydrochar produced at low temperature, whereas the hydrochar produced at 260 °C and the pyrochars were less efficient with respect to plant yields.

Keywords: hydrochar; pyrochar; nitrogen; fertilizer; greenhouse experiment; biosolids

---

1. Introduction

Sewage sludge (SS) is a nutrient-rich organic waste, which is produced in increasing amounts. More than 10 million tons (dry weight) are produced annually in Europe [1]. Recycling of SS for agriculture can return N, P, other plant nutrients and organic matter to the soil and may help to reduce the dependency on fossil fuel-consuming synthetic N fertilizer and non-renewable P sources. However, its application poses some environmental risks such as nutrient leaching, reduced soil biodiversity, increased greenhouse gas emissions [1], and health risks if not pre-treated properly. One possibility for hygienization of SS before its application to soil represents composting. However, this process consumes not only space and time but also releases greenhouse gases such as CO_2 and volatile N. An alternative may be a thermal treatment of SS through pyrolysis or hydrothermal carbonization (HTC). These technologies allow efficient hygienization while concomitantly stabilizing organic C and N within a relative short process time. Of course, during thermal treatment greenhouse gases are also released, but this emission may be compensated by recycling the produced thermal energy for other energy requiring purposes. Both pyrolysis and HTC carbonize biomass in low oxygen environments. Temperatures between 300 and 700 °C are typically used during dry pyrolysis. Hydrochars are typically produced at temperatures between 180 and 250 °C in the presence of water, which creates autogenous pressure. The advantage of transforming SS into hydrochar rather than pyrochar lies in lower energy costs due to the lower process temperatures and the fact that pre-process drying of the feedstock is not necessary. On the other hand, they are considered to be biochemically less stable than biochar [2,3].

Commonly, SS is characterized by a high ash content yielding in pyrolyzed products with organic matter contents which are too low to meet the requirements of the International Biochar Initiative or the European Biochar Certificate [4,5] to be called biochars [6]. Therefore, we will refer to pyrolyzed SS as pyrochar.

Despite the potential of pyrochars and hydrochars to increase the amount of stable carbon stored in soils, its use in agriculture will only be economically feasible if they provide additional benefits such as increasing crop production. Whereas pyrolysis of green waste and wood commonly results in products with a high porosity which may improve some physical properties of the soil [6,7], the pyrolysis of SS turns into carbonized residues with fertilizing potential [8–11].

Solid-state ^{15}N nuclear magnetic resonance (NMR) spectroscopic studies confirmed that most of the organic N (N_{org}) in pyro- and hydrochars from SS occurs as heterocyclic N [10]. Bearing in mind that this so-called black nitrogen (BN) [12] is less bioavailable than inorganic N (N_i), a big advantage of applying such fertilizers lies in the fact that N losses due to the fact of leaching can be reduced. On the other hand, despite pyrolysis and HTC of SS decreasing P mobility [13], pyrolyzed-SS was able to increase P contents in plant tissues [11], which demonstrates that at least part of the P was bioavailable. These observations point towards the potential of HTC-treated and pyrolyzed SS to act both as P and N sources for plants once applied to the soil.

These promising results are counteracted by the fact that phytotoxic compounds may be formed during thermal treatments. Indeed, negative impacts on germination and seedling growth have been observed in other studies [14–17]. They may be eliminated by well-designed pyrolysis conditions [18] and washing treatments [16]. However, the knowledge about the most appropriate conditions for converting SS into thermally treated products suitable for agriculture is still scarce.

Bearing this in mind, the goal of the present work was to fill those knowledge gaps by complementing a former investigation on the chemical transformation of organic C and N forms during HTC and pyrolysis of two different SS [10] with an 80 day greenhouse experiment. The focus of those experiments was to obtain insights on the impact of the application of the respective hydrochars and pyrochars on germination, survival, and biomass production of *Lolium perenne*. In addition, the char properties were related to the growth of *Lolium*.

2. Materials and Methods

2.1. Characteristics of the Sample Material

Two different SS were collected at the Experimental Wastewater Treatment plant (CENTA), located in Carrion de los Céspedes, near Seville, southern Spain. The first sample, further called "A_SS", was a primary sludge produced by the settlement of suspended organic matter in a pond. The second sample, assigned "T_SS", was a secondary sludge accumulated in an extended aeration treatment system and later stored in a thickener, in order to reduce its water content. In addition, "W" pyrochar was yielded from wood chips. Our previous work [10] showed that the material collected from the pond (A_SS) was more humified, thus biochemically stabilized, than that derived from the thickener (T_SS). The total N (N_T) contents of A_SS and T_SS were 19 and 32 g kg^{-1}, respectively (Table 1), and occurred mainly as peptides and in amino sugars [10]. Regarding heavy metal content, only Zinc (Zn) for A_SS and Cadmium (Cd) for both A_SS and T_SS slightly exceeded the thresholds established in the Working document on sludge [19] (Table 1).

Table 1. The pH, organic carbon (C_{org}), total nitrogen (N_T), organic nitrogen (N_{org}), inorganic nitrogen (N_i) of N_T, C_{org}/N_T ratio, P_T, K_T, Cd, Cu, and Zn contents of primary (A_SS) and secondary (T_SS) sewage sludges and the respective hydrochars produced at 200 °C (_HTC_200) and 260 °C (_HTC_260) for 0.5 and 3 h (_0.5, _3, respectively), as well as the pyrochars (_Py) [10].

	pH	C_{org}	N_T	N_{org}	$N_i{}^a$ of N_T	C_{org}/N_T	P_T	K_T	Cd	Cu	Zn
		g kg^{-1}	g kg^{-1}	g kg^{-1}	%	(w/w)	g kg^{-1}	g kg^{-1}	mg kg^{-1}	mg kg^{-1}	mg kg^{-1}
A_SS	7.4	228	19.2	18.9	0.5	11.9	12.3	6.8	3.2	401	1329
A_HTC_200_0.5	6.5	224	16.0	15.6	2.6	14.0	13.9	7.3	3.5	423	1399
A_HTC_200_3	6.5	214	13.9	13.7	2.5	15.4	14.5	7.6	3.7	446	1459
A_HTC_260_0.5	6.4	213	11.2	10.8	1.5	19.1	15.4	8.0	4.0	477	1585
A_HTC_260_3	6.6	221	12.1	11.8	1.9	18.2	14.9	7.9	4.0	464	1554
A_Py	9.3	134	8.6	9.0	0.0	15.5	17.0	9.5	4.6	533	1766
T_SS	7.5	245	32.3	31.9	0.4	7.6	16.7	6.1	3.5	418	1821
T_HTC_200_0.5	6.7	233	25.0	24.5	2.2	9.3	20.0	6.8	4.2	493	1983
T_HTC_200_3	6.2	233	23.9	23.7	1.4	9.7	20.8	7.1	4.3	506	2017
T_HTC_260_0.5	6.3	225	19.7	19.8	1.2	11.4	21.6	7.3	4.6	528	2128
T_HTC_260_3	6.4	224	19.0	18.7	1.4	11.8	21.6	7.6	4.8	530	2162
T_Py	10.0	168	16.3	16.0	0.0	10.3	25.2	9.2	5.3	618	2432
W_Py	9.3	829	1.80	n.a.[b]	n.a.	922	0.7	4.5	n.a.	n.a.	n.a.
Soil	8.4	10	1	n.a.	n.a.	7	0.4	1.7	n.a.	n.a.	n.a.

[a] N_i: inorganic nitrogen (sum of NH_4-N, NO_2-N, and NO_3-N in %); [b] n.a.: not available.

Hydrochars were produced from the primary (A) and secondary (T) SS at 200 °C (_HTC_200) and 260 °C (_HTC_260) for 0.5 and 3 h (_0.5, _3, respectively) as described in Reference [10] in a 1 L stirred pressure reactor (Parr reactor series 4520, IL, USA). The reaction water was removed through a filter and the material was rinsed with distilled water until the rinsing water turned clear.

To obtain the A and T pyrochars, further called A_Py and T_Py, 200 g of dry SS were pyrolyzed in a closed steel container in a preheated muffle oven at 600 °C for 1 h. W_Py carries the European Biochar Certificate and was produced at 600 °C for 20 min by Swiss Biochar, Lausanne, Switzerland.

The heavy metals content of the chars tended to increase compared to the non-treated SS; however, only Cd and Zn exceeded the limits [19], similar to the non-treated SS (Table 1). The concentrations of selective volatile organic compounds in process liquid (5-hydroxymethylfurfural 2-furfural, phenol, catechol, cresol, and resorcinol) was measured using a modified ICS 3000 Dionex (Thermo Scientific) with a UV detector (wavelength 280 nm) and Knaur Eurosphere II (C 18) column. A 15% acetonitrile (85% deionizedwater) was used as mobile phase in the ion chromatography system. Column temperature was set at 23 °C and flow rate was 1.0 mL min^{-1}. No presence of these molecules was found in A_SS, T_SS or in their chars.

According to the International Biochar Initiative [4], the organic C (C_{org}) content of the hydro- and pyrochars (Table 1) allows their classification as class 3 biochar; however, only the pyrochars fulfill the second requirement of having an atomic H/C ratio <0.7. A detailed characterization of the organic matter composition is given in Reference [10]. Due to the high N contents of the source materials, all chars contained considerable amounts of N between 1% and 3%, most of which (>97%) occurred in an organic form (Table 1). In contrast to the HTC chars, the pyrochars did not concentrate N_i. Total phosphorus (P_T) and potassium (K_T) contents were determined after digestion with aqua regia (1:3 v/v concentration HNO_3/HCl) in a microwave oven (Microwave Laboratory Station Mileston ETHOS 900, Milestone s.r.l., Sorisole, Italy) by inductively coupled plasma-optical emission spectrometer (ICP-OES) spectrophotometer Varian ICP720-ES (Table 1). The pH of the samples was measured in distilled water (1:10, w/v) (Table 1).

2.2. Greenhouse Experiment

For the greenhouse experiment, 250 mL pots (16 cm height) were perforated and filled with 250 g of dried fine earth (<2 mm) from the Ah horizon of a sandy loamy Calcic Cambisol [20] mixed with amounts equivalent to 5 and 25 t ha^{-1} of each char (0.8 and 4% w/w, respectively) and topped with 25 certified grass seeds (*Lolium perenne*, ILURO Seeds Company, Barcelona, Spain). The soil derived from the experimental station "La Hampa" of the Instituto de Recursos Naturales y Agrobiología de Sevilla, in the Guadalquivir River Valley (SW Spain; 37°21.32′ N, 6°4.07′ W), Coria del Río, Seville. After sampling, the soil was dried at 40 °C for 48 h and sieved (<2 mm). Small branches, fresh mosses, and plant remains, as well as roots were removed manually. The soil material contained 21 g C kg^{-1}, of which 10 g kg^{-1} was attributed to C_{org} and 1 g N kg^{-1}. Its pH in water was 8.5 and its water holding capacity (WHC), according to Reference [21], was 24%.

For each treatment, four replicates were prepared (n = 4). Additionally, 6 controls without any char amendment but with plants were included (n = 6). However, of those 6, only 4 were used for the final analysis. Growing conditions were similar as previously described by Reference [7]. Briefly, soil moisture was adjusted to 60% of the maximum WHC, the samples were placed in a greenhouse at 25 ± 2 °C/17 ± 2 °C (day/night) maintaining a 14 h light day^{-1} cycle with the support of growing lights (120 µE m^{-2} s^{-1} of photosynthetically active radiation) for 80 days. Average relative humidity of the air in the greenhouse was maintained during the experiment in the range 60 ± 10%. The position of the samples was changed three times per week to assure comparable light and growing conditions. Chars were dried (40 °C), grounded, and sieved (<2 mm) prior to being applied to the Calcic Cambisol to avoid possible differences due to the contrasting textures or heterogeneity. No nutrient solution was added. Although the pots were placed on saucers to collect possible excess water, there was no leaching after watering. The same amount of water was added to each sample three times per week, which summed up to 145 L m^{-2} at the end of the experiment. This was equivalent to 662 L m^{-2} per year and is similar to the natural average annual precipitation in the region around the experimental station.

The number of living grass shoots was counted after 5, 9, 13, 18, 20, 25, 30, 60, and 80 days. The germination rate was determined from measurements after 20 days of incubation, since it represented the time with the highest number of living plants. The survival rate was calculated by using the 80 count-day data. In addition, the shoots were cut and then left to regrow after 18, 33, 47, and 61 days. The final harvest was after 80 days of incubation. The harvested shoot biomass was dried (48 h at 40 °C) and weighed in order to determine the shoot biomass production. After the experiment, the roots were manually separated from the soil, rinsed with distilled water, dried (48 h at 40 °C), and weighed to determine the root biomass.

2.3. Statistical Analysis

All measured variables were submitted to the same statistical analysis using SPSS version 17.0 (SPSS, Chicago, IL, USA). Shapiro–Wilk and Levene tests were used to test for normality and homoscedasticity of the data, respectively. Transformations were applied to meet model assumptions when necessary. A *t*-test was used to identify significant effects between the control and each treatment. The same test was used to evaluate the application dose effect within each char. In addition, an analysis of variance was performed followed by a comparison of means (Tukey's test) to test for significant differences among chars, independently of the dose applied. Effects were considered significant at $p \leq 0.05$.

The R version 3.4.1 was used to conduct the non-metric multidimensional scaling (NMDS) ordination. Samples variation was represented by an ordination using a Euclidean distance matrix. The "envfit" function in the "vegan" package [22] was used to draw vectors representing chars properties with a statistical effect on *Lolium perenne* response onto the NMDS ordination. The SigmaPlot version 13.0 was used to plot the previously obtained NMDS data.

3. Results

3.1. Germination and Survival Rates

Figure 1 shows that the addition of pyrochars to the soil did not affect the germination or survival rate of *Lolium perenne*. Only W_Py applied at 25 t ha^{-1} decreased both parameters (Figures 1 and 2). Hydrochars derived from the "T" sewage sludge showed lower germination and survival rates than the control. In contrast, most of the tested A hydrochars revealed no major impact on the germination or survival rate. In addition, no differences were found when comparing the different production conditions or application doses between each other.

3.2. Biomass Production

The biomass production was determined as the sum of the shoot and the root biomass. The use of pyrochars had no effect on biomass production of *Lolium perenne* except for T_Py applied at 25 t ha^{-1} (Figure 3). The total biomass production after hydrochar application was always higher or equal to the control. Note that the greatest biomass was obtained with the hydrochars produced at the lowest temperature for both A and T hydrochars, A_HTC_200 and T_HTC_200, respectively. Residence time seems to have a lower impact than temperature on this parameter. In addition, increasing the application rate did not significantly alter the total biomass production.

3.3. Root-to-Shoot Ratios

The pyrochars' addition significantly increased the root-to-shoot (R:S) ratio of *Lolium perenne*, whereas the presence of A and T hydrochars resulted in lower or equal values than the control (Figure 4). Within hydrochars, production conditions did not affect this parameter; however, the R:S ratio tended to decrease with increasing rates of hydrochar application. These changes were mainly due to the higher root biomass in the case of pyrochars and a higher shoot biomass in the case of hydrochars (data not shown).

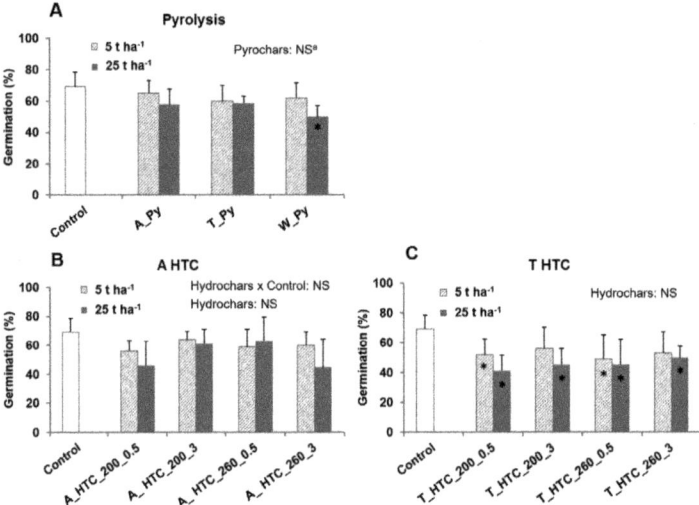

Figure 1. Germination rate (%) of *Lolium perenne* in: (**A**) Control and pyrochars-amended pots, (**B**) Control and pots amended with hydrochars derived from A_SS and (**C**) Control and pots amended with hydrochars derived from T_SS. Hydrochars were produced from primary (A_) and secondary (T_) sewage sludges at 200 °C (_HTC_200) and 260 °C (_HTC_260) for 0.5 and 3 h (_0.5, _3, respectively). Pyrochars were produced from primary (A_) and secondary (T_) sewage sludges (_Py). NS: no significance. Asterisks inside the bars show significant differences to the control according to the *t*-test.

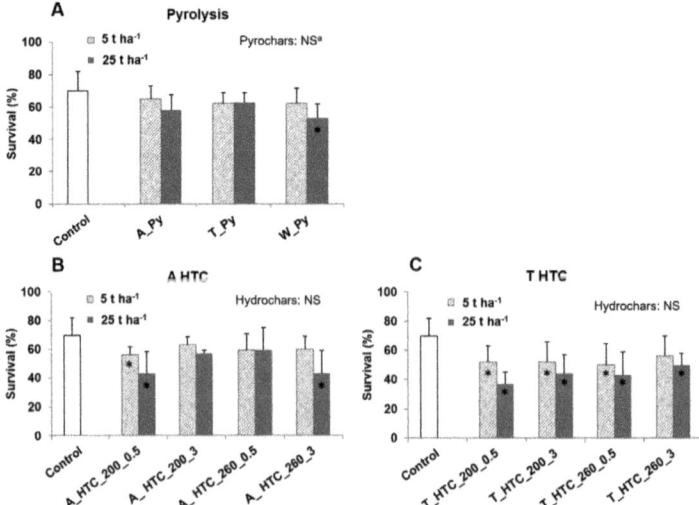

Figure 2. Survival rate (%) of *Lolium perenne* in: (**A**) Control and pyrochars-amended pots, (**B**) Control and pots amended with hydrochars derived from A_SS and (**C**) Control and pots amended with hydrochars derived from T_SS. Hydrochars were produced from primary (A_) and secondary (T_) sewage sludges at 200 °C (_HTC_200) and 260 °C (_HTC_260) for 0.5 and 3 h (_0.5, _3, respectively). Pyrochars were produced from primary (A_) and secondary (T_) sewage sludges (_Py). NS: no significance. Asterisks inside the bars show significant differences to the control according to the *t*-test.

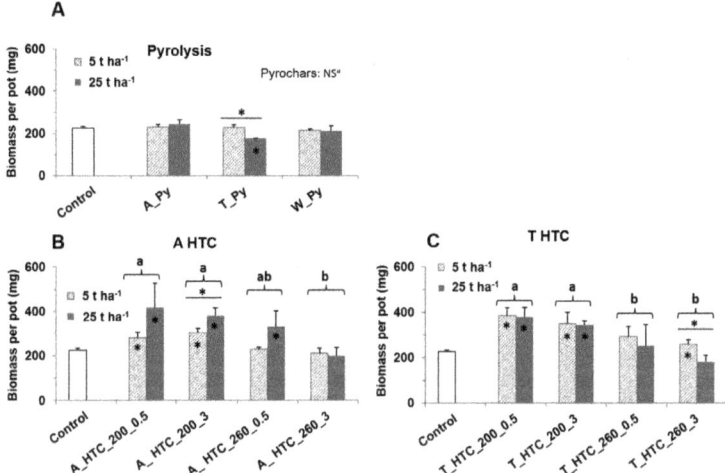

Figure 3. Biomass production per pot (mg) of *Lolium perenne* in: (**A**) Control and pyrochars-amended pots, (**B**) Control and pots amended with hydrochars derived from A_SS and (**C**) Control and pots amended with hydrochars derived from T_SS. Hydrochars were produced from primary (A_) and secondary (T_) sewage sludges at 200 °C (_HTC_200) and 260 °C (_HTC_260) for 0.5 and 3 h (_0.5, _3, respectively). Pyrochars were produced from primary (A_) and secondary (T_) sewage sludges (_Py). NS: no significance. Asterisks inside the bars show significant differences to the control according to the *t*-test. The above-line asterisks indicate significant differences among doses according to the *t*-test. The letters show significant differences among treatments according to Tukey's test.

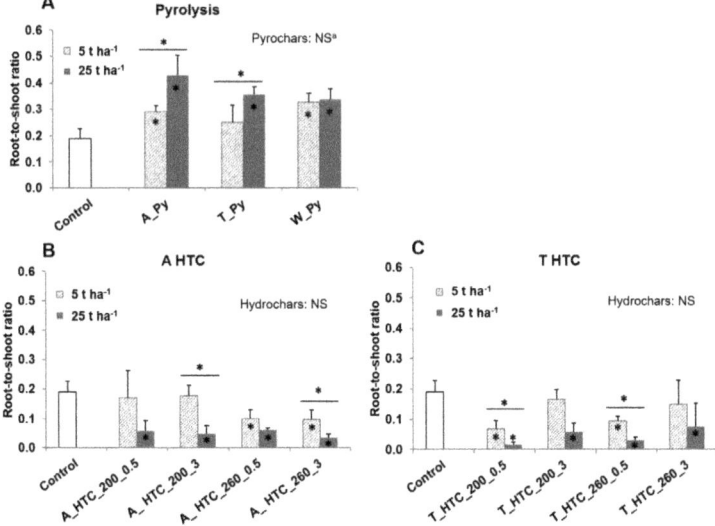

Figure 4. Root-to-shoot ratio of *Lolium perenne* in: (**A**) Control and pyrochars-amended pots, (**B**) Control and pots amended with hydrochars derived from A_SS and (**C**) Control and pots amended with hydrochars derived from T_SS. Hydrochars were produced from primary (A_) and secondary (T_) sewage sludges at 200 °C (_HTC_200) and 260 °C (_HTC_260) for 0.5 and 3 h (_0.5, _3, respectively). Pyrochars were produced from primary (A_) and secondary (T_) sewage sludges (_Py). NS: no significance. Asterisks inside the bars show significant differences with the control according to the *t*-test. The above-line asterisks indicate significant differences among doses according to the *t*-test.

3.4. Relationship between Plant Response and Chars' Properties

The NMDS graph showed that carbonization type had an overall effect on all parameters since pyrochars were separated from hydrochars (Figure 5). Pyrochars were characterized by higher R:S ratios together with higher root biomass, and to a lower extent, by higher germination and survival rates than hydrochars. In contrast, hydrochars were characterized by larger shoot and total biomass per plant. Regarding the feedstock type, T hydrochars showed larger shoots and total plant biomass than A hydrochars. In addition, total biomass per plant and shoot biomass significantly correlated with increased N_i and N_{org} contents of the chars. To a lower extent, total biomass per plant and shoot biomass also correlated with increased C_{org} levels of the chars. The p-values of the abovementioned correlations were less or equal to 0.01. The C_{org}/N ratio, P_T, and K_T parameters showed p-values higher than 0.01, and hence, were not considered significant variables.

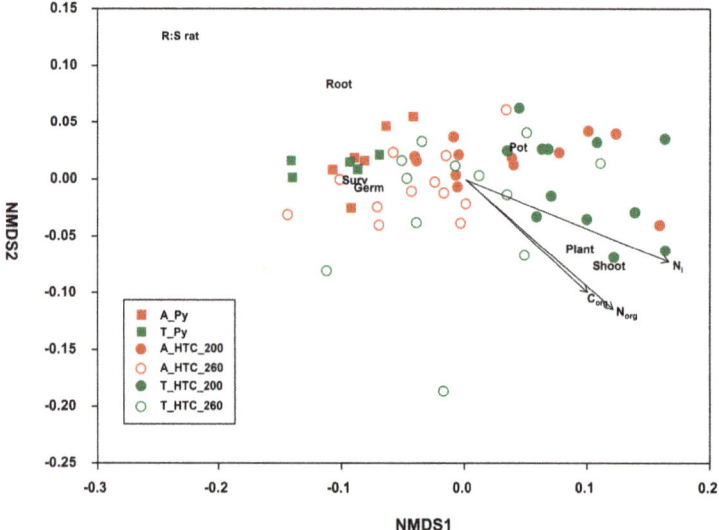

Figure 5. Non-metric multidimensional scaling (NMDS) ordination indicates how the different thermally treated sewage sludges (SS) impact on *Lolium*-measured parameters. R:S rat: root-to shoot ratio; Roots: roots biomass per plant; Surv: survival rate; Germ: germination rate; Pot: total biomass per pot; Plant: total biomass per plant; Shoot: shoot biomass per plant. Char properties significantly fitting onto the NMDS are shown as vectors ($p < 0.01$). N_i: inorganic nitrogen; N_{org}: organic nitrogen; C_{org}: organic carbon. Goodness of fit was 0.05.

4. Discussion

4.1. Germination and Survival Rates

Our data did not allow an unbiased assignment of factors causing the observed differences in germination and survival rates between hydrochars and pyrochars. The germination rates of *Lolium perenne* were similar to those previously obtained using biochars produced from different vegetation residues as an amendment to the same Calcic Cambisol [7]. Considering that both SS-derived hydrochars and pyrochars showed Cd and Zn values slightly higher than the limits and that only the hydrochars decreased the germination rate, the presence of heavy metals does not explain our results. A decrease in germination after addition of hydrochars has also been observed by others [14–17] and was attributed to phytotoxic volatile organic compounds adsorbed on the surface of the hydrochar. These compounds are mostly water soluble and can be removed by washing the hydrochars with distillate water [16] but soluble nutrients are expected to be at least partially lost during this process.

The fermentation of hydrochars in an anaerobic biogas reactor has also been proposed to eliminate toxic compounds since it avoids the loss of nutrients [23,24]. However, analysis of our chars prior to incubation did not indicate the presence of HMF, furfural, resorcinol, catechol, phenol, and kresol. Thus, either other non-measured phytotoxic compounds were present in our hydrochars or other factors were responsible. Previous works have also found diverse germination responses among hydrochars produced from different feedstocks [14,16].

Dry pyrolysis performed at temperatures greater than 500 °C in appropriate pyrolysis reactors reduces the presence of volatile compounds, including polycyclic aromatic compounds and other phytotoxics [25]. Thus, the possible negative impact of pyrochars on germination rates should be less prominent than that of hydrochars. Previous works have shown both neutral and positive effects of pyrochars [7,15,17,26]. This is in agreement with our results indicating no major impact of pyrochar addition except for W_Py applied at 25 t ha^{-1}. This effect caused by W_Py could be due to its extremely high aromaticity (H/C$_{at}$ ≤ 0.4; [10]) together with the probable immobilization of nitrogen, especially when adding the highest dose of this amendment (C/N = 922; Table 1).

4.2. Biomass Production and Chars Properties

Despite some hydrochars decreasing the survival rate of *Lolium perenne*, the total biomass production was always higher or equal in the hydrochar amended soils, compared to the control. This is best explained by the size of the remaining plants, since hydrochars tended to increased biomass production per plant (data not shown). However, our pyrochars had no effect on biomass production. The latter is in contrast to Reference [7] who observed a significant increase of *Lolium perenne* yields in pyrochar amended soils compared to the control. However, here it has to be taken into account that in the former study, inorganic N fertilizer was added.

The positive correlation between N_i and N_{org} contents of the chars with the total and shoot biomass production per plant suggests that the availability of nitrogen of the chars represents an important factor determining the additional growth of *Lolium* in the amended soils. Accordingly, the pyrochars showed the lowest N_{org} contents along with absence of N_i, which is in line with the lack of increase of *Lolium* biomass production. The hydrochars produced at 260 °C exhibited higher N_{org} than the pyrochars and contained some N_i which may explain the slight increase of plant biomass production. However, amendment of the hydrochars produced at 200 °C, containing the highest concentrations of N_{org} and N_i, yielded the highest amount of *Lolium* biomass. This is in line with former experiments showing that green-waste-derived hydrochars produced at 200 °C also produced higher plant yields than hydrochars produced at higher temperatures [27].

In contrast to N, the P_T and K_T contents of the chars revealed no correlation with growth parameters. Either the soil already contained sufficient plant-available P and K prior to char application or the P and K of the chars were less plant available. In both cases, our results indicated that factors other than P_T and K_T contents had a stronger impact on plant growth. The correlation of C_{org} contents of the chars with the total and shoot biomass production may be due to the improvement of some soil physical properties, such as a decrease in soil density, an increase of the soil water holding capacity or providing a habitat for soil microorganisms. Note that hydrochars exhibited higher C_{org} contents, and thus, higher biomass production than pyrochars. Since the latter are expected to exhibit a higher biochemical stability than hydrochars, one has to keep in mind that increasing the potential of soils to act as a C sink via addition of charcoal does not necessarily coincide with enhancement of biomass production.

4.3. Root-to-Shoot Ratios and Char Properties

Although pyrochar addition had no impact on *Lolium* biomass production, their presence increased their R:S ratios compared to the control. Considering that increased R:S ratios have been observed when growth is limited by N or P supply [28], our observation may be related to the absence of N_i in our pyrochars. This deficiency may have been increased by adsorption of NO_3^- and NH_4^+ from the soil solution to the char surface as it was suggested to occur for some biochars [29,30].

Hydrochars applied at 25 t ha^{-1} showed lower R:S ratios than the control, whereas the 5 t ha^{-1} application rate produced no significant impact or a lower decrease of this parameter. In general, when nutrient availability increases, plants can develop their aboveground vegetation in detriment to their subsoil part because less effort is required to acquire nutrients [31]. This is in agreement with our results, since char amendment with 25 t ha^{-1} delivered higher nutrients levels than addition of 5 t ha^{-1}.

4.4. Fertilization Potential of Thermally Treated SS

This work indicates that low-temperature HTC of SS results in a product with N fertilization potential, whereas chars yielded from HTC with higher temperatures and pyrolysis do not exhibit the same suitability. This potential depends on the amount of N_i contained in the chars together with the bioavailability of their organic N. Whereas N_i provides fast and immediate N fertilization, the degradation rate of the organic N compounds determines the slow-release N fertilization potential of the amendments. Our previous work [10] showed that organic N in SS occurred mainly in peptide-like structures, part of which may be easily degradable. During both HTC and pyrolysis, these compounds were partially transformed into N-heterocyclic aromatic entities [32], which are microbiologically less accessible. This transformation was more efficient for pyrolysis than for HTC, which suggested a quicker degradation of the N_{org} in hydrochars than in pyrochars. Therefore, not only the total nutrient content values but also their speciation should be considered since their bioavailability from chars is modified with processing conditions.

In addition, it also has to be considered that the same hydrochars/pyrochars may exhibit a different crop response depending on the crop type as well as on the soil properties and the climatic conditions [33,34].

5. Conclusions

The pyrolysis process and conditions together with the composition of the sludge were determinant in the properties of the resulting chars and therefore in their applicability as soil amendment. This study showed that although the problem of reduced germination after HTC application still has to be solved, SS-derived hydrochars have the potential to be used as soil amendment with immediate fertilizing effect. Bearing in mind that plant growth parameters correlated better with N than with P and K contents, the increase of plant yields in the hydrochar amended pots is best explained by the concomitant presence of N_i and easily available organic N forms. However, further and long-term experiments are needed to discern if the presence of less microbially accessible N from the pyrochars may turn into an advantage for crops which need low but constant N fertilization.

From an energetic, and thus, economic point of view, HTC may be advantageous over dry pyrolysis since drying of the SS prior to its thermal treatment can be avoided. In addition, lower temperatures have to be applied during HTC, which considerably reduce the costs for energy. Our studies reveal further that hydrochars produced at 200 °C caused higher plant growth than those yielded at 260 °C, which allows even less energy consumption. In addition, aside from economic considerations, the role of low temperature hydrochar as N-supplier may be of environmental interest since the use of mineral fertilizer could be reduced and the release of greenhouse gases into the atmosphere occurring during disposal of untreated SS can be avoided.

Author Contributions: M.P., H.K. and J.M.D.l.R. conceived the work. J.K. provided the facilities to produce the hydrochars and pyrochars from sewage sludge and carried out the production M.P. carried out the incubation experiment, laboratory analyses and edited the manuscript. All authors contributed to the interpretation of the data and to drafting the manuscript.

Funding: This research was funded by the Spanish Ministry of Economy and Competitiveness, AEI/FEDER funds of the European Union (projects PCGL2012-37041, CGL2015-64811-P and CGL2016-76498-R) and the Spanish Ministry of Education, Culture and Sport (Marina Paneque FPU fellowship, FPU 13/05831).

Acknowledgments: Carlos Aragón and the Experimental Wastewater Treatment plant (CENTA) are acknowledged for providing the sewage sludges. Ulf Lüder is thanked for his assistance during the production of the chars. Marta Velasco-Molina is gratefully acknowledged for her technical assistance.

Conflicts of Interest: The authors declare no conflict of interest.

References

1. Milieu Ltd.; WRc and Risk; Policy Analysts Ltd. (RPA). Environmental, Economic and Social Impacts of the Use of Sewage Sludge on Land. Final Report, Part I: Overview Report. Available online: http://ec.europa.eu/environment/archives/waste/sludge/pdf/part_i_report.pdf (accessed on 1 April 2019).
2. Lehmann, J.; Joseph, S. *Biochar for Environmental Management: Science and Technology*, 2nd ed.; Earthscan: London, UK, 2015.
3. Libra, J.; Ro, K.; Kammann, C.; Funke, A.; Berge, N.; Neubauer, Y.; Titirici, M.; Fuhner, C.; Bens, O.; Kern, J.; et al. Hydrothermal carbonization of biomass residuals: A comparative review of the chemistry, processes and applications of wet and dry pyrolysis. *Biofuels* **2011**, *2*, 89–124. [CrossRef]
4. International Biochar Initiative. Standardized Product Definition and Product Testing Guidelines for Biochar that is Used in Soil: Version Number 2.1. 2015. Available online: https://www.biochar-international.org/wp-content/uploads/2018/04/IBI_Biochar_Standards_V2.1_Final.pdf (accessed on 1 April 2019).
5. European Biochar Foundation. European Biochar Certificate—Guidelines for a Sustainable Production of Biochar: Version 8E 2019. Available online: http://www.european-biochar.org/biochar/media/doc/ebc-guidelines.pdf (accessed on 1 April 2019).
6. Bachmann, H.J.; Bucheli, T.D.; Dieguez-Alonso, A.; Fabbri, D.; Knicker, H.; Schmidt, H.P.; Ulbricht, A.; Becker, R.; Buscaroli, A.; Buerge, D.; et al. Toward the standardization of biochar analysis: The COST Action TD1107 interlaboratory comparison. *J. Agric. Food Chem.* **2016**, *64*, 513–527. [CrossRef] [PubMed]
7. De la Rosa, J.M.; Paneque, M.; Miller, A.Z.; Knicker, H. Relating physical and chemical properties of four different biochars and their application rate to biomass production of *Lolium perenne* on a Calcic Cambisol during a pot experiment of 79 days. *Sci. Total Environ.* **2014**, *499*, 175–184. [CrossRef] [PubMed]
8. Chan, K.Y.; van Zwieten, L.; Meszaros, I.; Downie, A.; Joseph, S. Agronomic values of green waste biochar as a soil amendment. *Aust. J. Soil Res.* **2007**, *45*, 629–634. [CrossRef]
9. Chan, K.Y.; van Zwieten, L.; Meszaros, I.; Downie, A.; Joseph, S. Using poultry litter biochars as soil amendments. *Aust. J. Soil Res.* **2008**, *46*, 437–444. [CrossRef]
10. Paneque, M.; De la Rosa, J.M.; Kern, J.; Reza, M.T.; Knicker, H. Hydrothermal carbonization and pyrolysis of sewage sludges: What happen to carbon and nitrogen? *J. Anal. Appl. Pyrolysis* **2017**, *128*, 314–323. [CrossRef]
11. Frišták, V.; Pipíška, M.; Soja, G. Pyrolysis treatment of sewage sludge: A promising way to produce phosphorus fertilizer. *J. Clean. Prod.* **2018**, *172*, 1772–1778. [CrossRef]
12. Knicker, H. "Black nitrogen"—An important fraction in determining the recalcitrance of charcoal. *Org. Geochem.* **2010**, *41*, 947–950. [CrossRef]
13. Huang, R.; Tang, Y. Evolution of phosphorus complexation and mineralogy during (hydro)thermal treatments of activated and anaerobically digested sludge: Insights from sequential extraction and P K-edge XANES. *Water Res.* **2016**, *100*, 439–447. [CrossRef] [PubMed]
14. Jandl, G.; Eckhardt, K.U.; Bargmann, I.; Kücke, M.; Greef, J.M.; Knicker, H.; Leinweber, P. Hydrothermal carbonization of biomass residues: Mass spectrometric characterization for ecological effects in the soil–plant system. *J. Environ. Qual.* **2013**, *42*, 199–207. [CrossRef]
15. Busch, D.; Kammann, C.; Grünhage, L.; Müller, C. Simple biotoxicity tests for evaluation of carbonaceous soil additives: Establishment and reproducibility of four test procedures. *J. Environ. Qual.* **2012**, *41*, 1023–1032. [CrossRef] [PubMed]
16. Bargmann, I.; Rillig, M.C.; Buss, W.; Kruse, A.; Kuecke, M. Hydrochar and biochar effects on germination of spring barley. *J. Agron. Crop Sci.* **2013**, *199*, 360–373. [CrossRef]
17. Thuille, A.; Laufer, J.; Höhl, C.; Gleixner, G. Carbon quality affects the nitrogen partitioning between plants and soil microorganisms. *Soil Biol. Biochem.* **2015**, *81*, 266–274. [CrossRef]
18. Buss, W.; Mašek, O. Mobile organic compounds in biochar—A potential source of contamination – Phytotoxic effects on cress seed (*Lepidium sativum*) germination. *J. Environ. Manag.* **2014**, *137*, 111–119. [CrossRef] [PubMed]
19. European Commission. *Working Document on Sludge. 3rd Draft*; ENV.E.3/ LM; Directorate-General for the Environment: Brussels, Belgium, 2000.

20. IUSS Working Group WRB. World Reference Base for Soil Resources 2014, Update 2015 International Soil Classification System for Naming Soils and Creating Legends for Soil Maps. World Soil Resources Reports No. 106. FAO: Rome. Available online: http://www.fao.org/3/i3794en/I3794en.pdf (accessed on 1 April 2019).
21. Veihmeyer, F.J.; Hendrickson, A.H. Methods of measuring field capacity and wilting percentages of soils. *Soil Sci.* **1949**, *68*, 75–94. [CrossRef]
22. Oksanen, J.; Blanchet, F.G.; Kindt, R.; Legendre, P.; Minchin, P.R.; O'Hara, R.B.; Simpson, G.L.; Solymos, P.; Stevens, M.H.H.; Wagner, H. Vegan: Community Ecology Package Version 2.4-4. 2015. Available online: http://cran.r-project.org/ (accessed on 27 February 2019).
23. Mumme, J.; Srocke, F.; Heeg, K.; Werner, M. Use of biochars in anaerobic digestion. *Bioresour. Technol.* **2014**, *164*, 189–197. [CrossRef] [PubMed]
24. Lanza, G.; Stang, A.; Kern, J.; Wirth, S.; Gessler, A. Degradability of raw and post-processed chars in a two-year field experiment. *Sci. Total Environ.* **2018**, *628–629*, 1600–1608. [CrossRef] [PubMed]
25. De la Rosa, J.M.; Paneque, M.; Hilber, I.; Blum, F.; Knicker, H.; Bucheli, T.D. Assessment of polycyclic aromatic hydrocarbons in biochar and biochar-amended agricultural soil from Southern Spain. *J. Soils Sediments* **2016**, *16*, 557–565. [CrossRef]
26. Free, H.F.; McGill, C.R.; Rowarth, J.S.; Hedley, M.J. The effect of biochars on maize (*Zea mays*) germination. *N. Z. J. Agric. Res.* **2010**, *53*, 1–4. [CrossRef]
27. Fang, J.; Gao, B.; Chen, J.; Zimmerman, A.R. Hydrochars derived from plant biomass under various conditions: Characterization and potential applications and impacts. *Chem. Eng. J.* **2015**, *267*, 253–259. [CrossRef]
28. Andrews, M.; Sprent, J.I.; Raven, J.A.; Eady, P.E. Relationships between shoot to root ratio, growth and leaf soluble protein concentration of *Pisum sativum*, *Phaseolus vulgaris* and *Triticum aestivum* under different nutrient deficiencies. *Plant Cell Environ.* **1999**, *22*, 949–958. [CrossRef]
29. Spokas, K.A.; Novak, J.M.; Venterea, R.T. Biochar's role as an alternative N fertilizer: Ammonia capture. *Plant Soil* **2012**, *350*, 35–42. [CrossRef]
30. Nelissen, V.; Rütting, T.; Huygens, D.; Staelens, J.; Ruysschaert, G.; Boeckx, P. Maize biochars accelerate short-term soil nitrogen dynamics in a loamy sand soil. *Soil Biol. Biochem.* **2012**, *55*, 20–27. [CrossRef]
31. Ågren, G.I.; Franklin, O. Root:shoot ratios, optimization and nitrogen productivity. *Ann. Bot.* **2003**, *92*, 795–800. [CrossRef] [PubMed]
32. De la Rosa, J.M.; Knicker, H. Bioavailability of N released from N-rich pyrogenic organic matter: An incubation study. *Soil Biol. Biochem.* **2011**, *43*, 2368–2373. [CrossRef]
33. Jeffery, S.; Abalos, D.; Prodana, M.; Bastos, A.C.; Groenigen, J.W.; Hungate, B.A.; Verheijen, F. Biochar boosts tropical but not temperate crop yields. *Environ. Res. Lett.* **2017**, *12*, 053001. [CrossRef]
34. Paneque, M.; De la Rosa, J.M.; Franco-Navarro, J.D.; Colmenero-Flores, J.M.; Knicker, H. Effect of biochar amendment on morphology, productivity and water relations of sunflower plants under non-irrigation conditions. *Catena* **2016**, *147*, 280–287. [CrossRef]

© 2019 by the authors. Licensee MDPI, Basel, Switzerland. This article is an open access article distributed under the terms and conditions of the Creative Commons Attribution (CC BY) license (http://creativecommons.org/licenses/by/4.0/).

Article

Hydrochar-Amended Substrates for Production of Containerized Pine Tree Seedlings under Different Fertilization Regimes

Samieh Eskandari [1,*], Ali Mohammadi [1], Maria Sandberg [1], Rolf Lutz Eckstein [2], Kjell Hedberg [3] and Karin Granström [1]

1. Department of Engineering and Chemical Sciences, Karlstad University, 65188 Karlstad, Sweden
2. Department of Environmental and Life Sciences, Biology, Karlstad University, 65188 Karlstad, Sweden
3. Ulf Ahldén Ingenjörsfirma AB, Älvhagsvägen 11, 19454 Upplands Väsby, Sweden
* Correspondence: samie.eskandari@gmail.com; Tel.: +46-5470-0118-1

Received: 1 June 2019; Accepted: 1 July 2019; Published: 2 July 2019

Abstract: There is a growing body of research that recognizes the potentials of biochar application in agricultural production systems. However, little is known about the effects of biochar, especially hydrochar, on production of containerized seedlings under nursery conditions. This study aimed to test the effects of hydrochar application on growth, quality, nutrient and heavy metal contents, and mycorrhizal association of containerized pine seedlings. The hydrochar used in this study was produced through hydrothermal carbonization of paper mill biosludge at 200 °C. Two forms of hydrochar (powder and pellet) were mixed with peat at ratios of 10% and 20% (v/v) under three levels of applied commercial fertilizer (nil, half and full rates). Application of hydrochar had positive or neutral effects on shoot biomass and stem diameter compared with control seedlings (without hydrochar) under tested fertilizer levels. Analysis of the natural logarithmic response ratios (LnRR) of quality index and nutrient and heavy metal uptake revealed that application of 20% (v/v) hydrochar powder or pellet with 50% fertilizer resulted in same quality pine seedlings with similar heavy metal (Cu, Ni, Pb, Zn and Cr) and nutrient (P, K, Ca and Mg) contents as untreated seedlings supplied with 100% fertilizer. Colonization percentage by ectomycorrhizae significantly increased when either forms of hydrochar were applied at a rate of 20% under unfertilized condition. The results of this study implied that application of proper rates of hydrochar from biosludge with adjusted levels of liquid fertilizer may reduce fertilizer requirements in pine nurseries.

Keywords: containerized production systems; heavy metals; paper mill sludge; biochar-ash pellet; quality index

---

1. Introduction

Sweden is the world's second largest exporter of pulp, paper and wood products; this amounted to ≅ 125 billion Swedish krona (SEK)in 2017 [1]. There is about 28 million hectares of forest land in Sweden and pine trees constitute 40% of the total standing volume [2]. Although approximately 80 million m³ of Swedish forest stands are harvested annually, the total standing volume has considerably increased in the last century [3]. In fact, a total of 400 million containerized tree seedlings are produced by Swedish forest nurseries to restock forests each year [3]. However, intensive annual forest harvests remove essential soil nutrients, which may cause problems for forest productivity.

In Sweden, container-grown seedlings are dominantly produced in peat and peat-based growth media [4]. Peat-based substrates have many advantages such as long-term drainage ability, good aeration for tree seedling roots, good fertilizer absorbance and release capability. However, peat-based media are considered non-sustainable as their extraction have adverse environmental impacts [5,6]. Hence, researchers tend to evaluate various growing substrate alternatives that could fully or partially replace peat [7]. Particularly, nutrient-rich growing media have become a special area of interest for research, as the growing global demand for wood harvest will cause further depletion of soil nutrients. Therefore, sustainable approaches towards forest production and plantation management are urgently needed.

Biochar has been introduced as an environmentally sustainable option to replace peat in nursery conditions [8–11]. Biochar is the carbonaceous residue of rapid biomass combustion in the absence or partial supply of oxygen (pyrolysis) [12]. Biochars as soil amendment not only contribute to storing carbon in the soils but also act as fertilizers [13,14], which subsequently reduces the environmental impact [15,16] and economic cost of plant production [17]. Moreover, biochar can increase substrate pH, improve water-holding capacity and enhance phyto-available nutrients by decreasing nutrient leaching and increasing cation exchange capacity (CEC) and consequently, enhance plant growth and productivity [18–21]. Biochar from paper sludge has also been characterized with high specific surface area [22] and has shown promising effect on simulating growth of plants such as *Lolium perenne* [23]. Biochar from pulp and paper mill sludge has been reported to have higher environmental performance relative to conventional disposal methods for sludge such as landfilling or incineration [24,25]. In many studies, the term 'biochar' is generally understood to mean 'pyrochar', which is a coproduct of fast pyrolysis or gasification. Another type of biochar is hydrochar, which is produced through hydrothermal carbonization (HTC). Therefore, it may have different physical structure and chemical composition, and consequently cause different effects on nutrient availability and plant growth compared to pyrochar [26,27]. Hydrochar is characterized by a low pH, high carbon content, high nutrient levels, good heavy metal absorption capacity and a carbon supply for microorganisms in the soil [18,28,29]. Generally, the effects of biochar on plant growth and physicochemical properties of the growing media vary depending on biochar feedstock, particle size and production process, e.g., temperature and heating duration [27,30]. Potentials of pyrochars and hydrochars as substrate components for production of tree seedlings in nurseries have not been well documented. Pyrochar has been evaluated in forms of either pellet or powder and shown promising results as nursery substrate for pine tree seedlings as well as sequestering carbon as part of normal reforestation [10,31,32]. Few empirical studies investigated the effects of hydrochar on tree seedling and soil nutrient cycling. Some studies reported adverse impacts of hydrochar derived from beet root chips on plant productivity and seed germination even when applied as low as 20–25% of volume of the growth mixture [33,34]. However, other studies offered contradictory findings about the effects of hydrochar on a fast-growing tree species: Baronti et al. [35] and George et al. [27] showed that biomass productivity and nitrogen use efficiency increased in poplar tree seedlings treated with hydrohar derived from maize (*Zea mays* L.) silage feedstock. These contradictory results concerning the effects of hydrochar on plant productivity call for careful choice of feedstock, application rate and the target species to ensure optimum growth benefits.

As hydrochar is usually friable and dusty, pelletization decreases dust formation, unifies its shape and size and facilitate transportation and distribution. Other benefits of reducing the loss of nutrients (e.g., nitrate and phosphate) and water, reducing bulk density and providing a beneficial environment for microbes as well as improved total porosity and aeration porosity in containers have been also reported by Di Lonardo et al. [20] and Dumroese et al. [10]. Thus far, however, there has been no discussion about the use of hydrochar pellets in nursery substrate constituent. Moreover, despite the potential significance of hydrochar, the effects of this byproduct on plant and microorganism interactions, e.g., mycorrhizal associations, is poorly understood. Ectomycorrhizal (ECM) associations play a key role in nutrient uptake in many woody plants, e.g., pine trees, such that growth and survival of these plants in forest ecosystems considerably rely on ECM fungi [36]. The existing body of research on biochar suggests positive effects of pyrochar on mycorrhizal symbiosis of the host plants, however, some substantial differences between hydrochar and pyrochar require assessment of hydrochar for any potential negative effects [33].

Therefore, the aim of this paper is to study the effects of hydrochar, derived from paper mill biosludge, on growth, quality, mycorrhizal associations and nutrient/heavy metal uptake of pine tree seedlings. We analyzed whether effects varied significantly between hydrochar forms (powder or pellets) or hydrochar proportions mixed with peat (10% or 20% hydrochar v/v). The effects of hydrochar addition on pine tree seedling was evaluated under three fertilization regimes (no fertilizer, 50% fertilizer and 100% fertilizer). We hypothesized that the growth, quality and mycorrhizal colonization of pine tree seedlings grown in substrate mixed with hydrochar would improve. We also expected pine tree seedlings grown with hydrochar to require less fertilizer to achieve similar or higher growth, mycorrhizal colonization and associated nutrient uptake relative to seedlings grown without hydrochar but with optimum rates of fertilizer (100% fertilizer). To the best of our knowledge, this current study is the first paper to explore the potentials of hydrochar powder and pellets for being used as a growing media component in production of containerized pine tree seedlings.

2. Materials and Methods

2.1. Hydrochar Preparation and Properties: Powder and Pellets

The hydrochar used in this study was produced through hydrothermal carbonization (HTC) of biosludge from a market kraft pulp mill. The HTC process started with combusting biosludge in Innventia's Parr-reactor (Relzow, Germany) for 2 h at 200 °C [37]. The attained solid and liquid were filtered and the resultant solid was then reslurried in deionized water at 70 °C and was filtered for a second time. We used the second filtered char as the experimental hydrochar. We enriched the experimental hydrochar by dry blending it with 5% wood ash (w/w) in a rotating mixer for 1 h. The main reason of adding wood ash to hydrochar was to improve the nutrient supply and to counteract the low pH of the raw hydrochar (pH 4.9). Moreover, having ash is in compliance with the current common practice of spreading ash into Swedish forests. Samples of hydrochar and ash were dried at 105 °C overnight and then ground to < 2 mm; the dry materials were then acid digested with ultra-wave digestion technique (Milestone© with unique Single Reaction Chamber (SRC)). Element contents were determined by inductively coupled plasma optical emission spectroscopy (ICP-OES). The pH of hydrochar and ash was determined according to the methods outlined by Chen et al. [38]. Carbon and nitrogen content was quantified using Elementar Vario EL with TCD detector. Properties of ash and hydrochar are summarized in Table 1. Hydrochar pellets were produced at the discipline of Environmental and Energy Systems, Karlstad University (Karlstad, Sweden) by feeding the finely ground ash-enriched hydrochar powder into a single pellet press unit. The measured technical parameters for densification process of hydrochar included temperature (100 °C), compressive force (4 kN), holding time (30 s) and the speed of girder moving (5 mm/min). The length (18.32 mm) and diameter (8.85 mm) of hydrochar pellets were also measured.

Table 1. The pH, nutrient and heavy metal concentrations of the peat, hydrochar and wood ash used in this study.

Parameters	Peat	Hydrochar	Wood Ash
pH	5.17 ± 0.17	5.59 ± 0.12	11.95 ± 0.00
Total Carbon (%)	48.82 ± 0.04	49.20 ± 1.51	3.16 ± 0.88
Total N (%)	1.04 ± 0.01	2.26 ± 0.02	0.05 ± 0.01
P (g/kg)	0.28 ± 0.00	2.77 ± 0.66	8.68 ± 0.08
Ca (g/kg)	7.92 ± 0.08	2.72 ± 0.04	216.66 ± 4.26
K (g/kg)	0.39 ± 0.00	1.34 ± 0.02	19.33 ± 0.12
Mg (g/kg)	2.39 ± 0.01	1.34 ± 0.01	9.55 ± 0.13
Na (g/kg)	0.23 ± 0.00	0.81 ± 0.01	6.93 ± 0.09
Zn (g/kg)	0.01 ± 0.00	0.43 ± 0.00	1.19 ± 0.00
Cu (g/kg)		0.07 ± 0.00	0.06 ± 0.00
Ni (g/kg)		0.02 ± 0.00	0.01 ± 0.00
Pb (g/kg)		0.02 ± 0.00	0.02 ± 0.00
Cd (g/kg)		0.01 ± 0.00	
Cr (g/kg)		0.04 ± 0.00	0.06 ± 0.00

2.2. Experimental Design and Growth Condition

The experiment consisted of 12 unique treatment combinations: two factor levels of the factor "hydrochar type" (powder—PW; pellets—PL), two factor levels of the factor "hydrochar proportion" (10% and 20% v/v) and three factor levels of the factor "liquid fertilizer" (none, half and full rate). Square plastic containers (60 mL) were hand filled with a mixture of peat and either hydrochar powder or pellets at rates of 10% or 20% on a volume basis. Ten and 20 percent volume rates corresponded to mass-based proportions of 23% and 35% for the hydrochar powder and 35% and 49% for the hydrochar pellet treatment, respectively. Element contents of peat summarized in Table 1 was determined through the same approaches as for hydrochar and ash. The pH of peat was measured in a 1:5 solid water suspension (Table 1). Fertilizer regimes were determined and adjusted to our pot volume according to recommended Nitrogen (N) rates by Jackson et al. [39] for pine seedlings. Three pine seeds (*Pinus sylvestris*) were sown in each pot. Three weeks after sowing, seedlings were thinned to one per pot and were then fertilized once a week with a commercial liquid fertilizer (N:P:K, 100:13:65 + 4 Mg w/v; Wallco, Sweden) for 20 weeks. This resulted in a full rate of 2.4 mg N per seedling per week (100% fertilizer treatment) and a half rate of 1.2 mg N per seedling per week (50% fertilizer treatment). Treatments assigned to be unfertilized received only deionized water at fertilization times. There were 10 replicate pots in each treatment combination, including a control with neither forms of hydrochar applied. Irrigation frequency was determined using weight loss method [40] (Table A1). Therefore, the mass of each empty pot and its oven-dried growing medium (60 °C for 72 h) was recorded. Pots filled with assigned growing medium were watered with deionised water to their capacity (until saturated) and then left to freely drain for 60 min. The water content of each pot at container capacity was calculated by subtracting the weight of empty pot and oven-dry medium from the weight of the pot at container capacity (after 60 min of drainage). Pots were weighed daily and manually irrigated to container capacity when the average mass of the water in three pot replicates reached 75% of the water content at container capacity. The required amount of fertilizer was dissolved in the calculated irrigating water in designated fertilized treatments once a week. In order to adjust for substrate shrinkage and seedling biomass, recalculation of container capacity mass was done every two months.

The experiment was conducted at a fully controlled plant growth room at Karlstad University, Karlstad (59°24′12.59″ N and 13°34′32.39″ E). Pots were placed randomly in the growth room with constant temperature maintained at 23 °C throughout the first three weeks after sowing (germination stage). Temperature were then controlled between 23 °C and 16 °C for day and night, respectively. Relative humidity and photoperiod of the room was set at 60% and 18 h (~40,000 lux light intensity), respectively. Pine tree seedlings were grown in the aforementioned conditions for a period of six months and were rearranged monthly to minimize the effects of bench location.

2.3. Seedling Measurements

At 6 months after sowing, seedlings were harvested and their height and stem diameter at the root collar were measured. Seedlings were then dissected into root and shoot sections and the fresh mass of each section determined. Afterwards, they were dried in a fan-forced oven at 60 °C for 72 h and their dry mass were recorded. Determination of the nutrient/heavy metal composition of triplicate shoot samples (dried and ground to <2 mm) was done with ICP-OES following digestion using an ultra-wave digestion technique (Milestone© with unique Single Reaction Chamber (SRC)). Nitrogen content of plant samples was measured using Elementar Vario EL with TCD detector.

To compare the robustness and biomass distribution equilibrium in the pine tree seedlings, the quality index of Dickson (QID) was calculated using a formula first described by Dickson et al. [41] (Equation (1)):

$$QID = \frac{\text{Total dry biomass (g)}}{\left(\frac{\text{Height (cm)}}{\text{Stem diameter (mm)}} + \frac{\text{Shoot dry weight (g)}}{\text{Root dry weight (g)}}\right)} \quad (1)$$

2.4. Mycorrhizal Colonization Test

A second experiment was designed to examine potential effects of hydrochar on mycorrhizal symbiosis of pine tree seedlings. We inoculated hydrochar-amended substrates (mixed with 20% powder or pellet by volume) with a specific commercial ectomycorrhizal inoculant for coniferous trees (ectovit®-Mycorrhiza for Trees). These treatments were either fully fertilized to the recommended rate by Jackson et al. [39] or remained unfertilized (receiving only deionized water at fertilization times). A non-inoculated control treatment was included for each hydrochar type (PW and PL) in both fertilized and unfertilized regimes. We used 60 mL pots and 10 replicates in the same growth condition as in the first experiment. Seedlings were irrigated according to the weight loss method [40].

At six months after sowing, the root systems of the three replicates in each treatment were carefully washed free of soil with deionized water. Roots were cut into 2 cm segments and stored in 50% ethanol at 4 °C for further analysis. In order to quantify mycorrhizal root colonization, the gridline intersect method for ectomycorrhizal associations was used with a stereo microscope [42].

2.5. Data Analysis

The statistical analysis of the experiments was designed based on the hypotheses elaborated earlier in this paper. Firstly, in order to investigate the potential positive/negative effects of hydrochar type and application rates on growth of pine tree seedlings under the tested fertilizer levels, we used the natural logarithmic response ratios (LnRR), which is frequently used in meta-analysis (e.g., Brose et al. [43]) but also in laboratory experiments (e.g., Loydi et al. [44]). Each dependent variable, e.g., shoot dry mass and stem diameter, was standardized by the mean of the associated control treatment (without hydrochar addition) at each fertilization scenario. LnRR was calculated according to Goldberg and Scheiner [45] (Equation (2)):

$$LnRR = Ln \frac{(PT)}{(PC)} \qquad (2)$$

where PT is the value of the treated sample and PC is the mean value of the control treatment. The effects of hydrochar type and hydrochar proportion were considered significantly different from the control at each fertilization regime, when the 95% confidence interval (CI) did not overlap with zero. This analysis allowed to directly compare pine seedling growth responses to 10 and 20% hydrochar powder and pellet addition in growing media under the tested nutritional regimes. To examine the significance of differences among treatments, the LnRR was then subjected to a General Linear Model (GLM) using IBM SPSS Statistics for Windows, Version 25.0.0.2 (Released 2017. Armonk, NY: IBM Corp).

Secondly, in order to test our second hypothesis about lower fertilizer requirements of hydrochar-treated seedlings compared to full-fertilized untreated ones (common nursery practice), LnRR of QID and nutrient/heavy metal uptake were standardized with the mean of the untreated seedlings grown with 100% fertilizer and were then subjected to a GLM using IBM SPSS Statistics for Windows, Version 25.0.0.2 (Released 2017. Armonk, NY: IBM Corp). This analysis allowed for examining whether addition of hydrochar may produce the same quality of seedlings but using less fertilizer relative to the control seedlings. We also examined the LnRR of mycorrhizal colonization percentage standardized with the mean of the full-fertilized control treatment. The discrepancy between untreated and hydrochar treated seedling were considered significantly different when the 95% confidence interval did not overlap with zero.

3. Results

3.1. Effects of Hydrochar on Growth Parameters

3.1.1. Shoot Dry Mass

Hydrochar application had either positive ($P < 0.05$) or neutral effects on shoot dry mass of pine seedlings in tested under different nutritional scenarios (Figure 1a). In the unfertilized treatment, shoot dry mass in pine seedlings receiving hydrochar powder was ~3 times higher than in their counterparts treated with hydrochar pellets (fertilizer × hydrochar type $P < 0.05$; mean ± standard error (SE): 0.225 ± 0.025 versus 0.080 ± 0.006) (Figure 1a). Higher rates of hydrochar application (20%) improved seedling shoot biomass in unfertilized and half fertilized conditions (fertilizer × hydrochar rate $P < 0.05$). The interaction among hydrochar type, application rate and fertilizer level was not significant ($P > 0.05$) (Figure 1a).

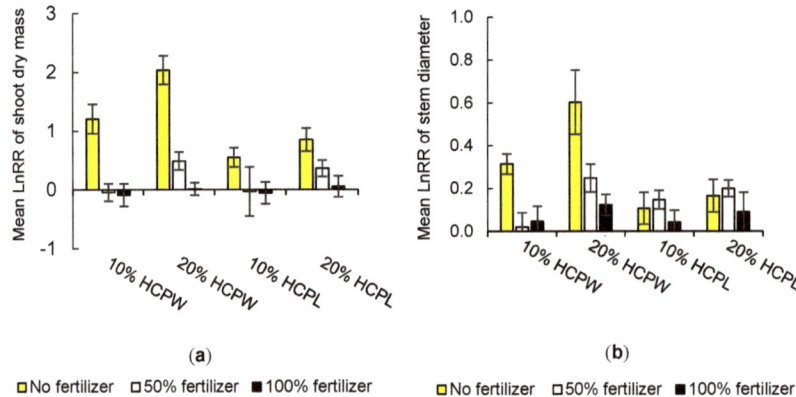

Figure 1. Mean (±95 CI, n = 10) natural logarithmic response ratios (standardized with the untreated controls at the same fertilizer level) of shoot dry mass (**a**) and stem diameter (**b**) of pine tree seedlings grown in peat mixed with 10% and 20% (v/v) hydrochar powder (HCPW) and hydrochar pellet (HCPL) in unfertilized (yellow bars), half (open bars) and full fertilized (black bars) conditions. Effects of hydrochar type and application rate were considered significant when 95% CI did not overlap with zero.

3.1.2. Stem Diameter

The results showed that replacing 20% of the growing media volume with hydrochar powder significantly increased the stem diameter compared to their untreated counterparts in all experimental fertilization scenarios ($P < 0.05$) (Figure 1b). Stem diameter increased when the 20% of hydrochar was mixed with the substrate under unfertilized and 50% fertilizer rate condition (fertilizer rate × hydrochar rate $P = 0.052$) (Figure 1b). Hydrochar powder increased stem diameter by 40% relative to hydrochar pellet under unfertilized condition (Fertilizer rate × hydrochar type $P < 0.05$; mean ± SE: 1.21 ± 0.02 versus 0.86 ± 0.06). However, there was no significant interaction between hydrochar type, rate and fertilizer addition ($P = 0.08$) (Figure 1b).

3.1.3. Quality Index of Dickson (QID)

To test whether hydrochar-treated seedlings needed less fertilizer than untreated ones, natural logarithmic response ratio of QID was standardized with the mean of untreated full-fertilized control. The results suggested that seedlings treated with 20% hydrochar powder or pellets, which received a 50% fertilizer dose, had similar QID than pine seedlings without any hydrochar addition that received 100% fertilizer (Figure 2, 95% CI overlapping with zero). When seedling were fertilized with 100% fertilizer, those grown with 20% hydrochar powder had 75% higher QID values than their untreated counterparts (mean ± SE: 0.07 ± 0.012 versus 0.04 ± 0.004, respectively) (Figure 2). There was no significant interaction among hydrochar type, hydrochar proportion and fertilizer addition ($P > 0.05$) (Figure 2).

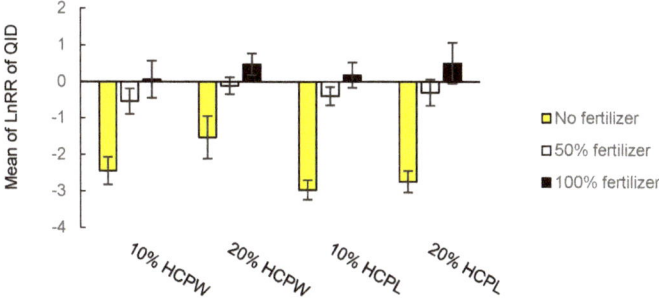

Figure 2. Mean (±95 CI, n = 4) natural logarithmic response ratios (standardized with the untreated full-fertilized control) of Quality Index of Dickson (QID) of pine tree seedlings grown in peat mixed with 10% and 20% (v/v) hydrochar powder (HCPW) and hydrochar pellet (HCPL) in unfertilized (yellow bars), half (open bars) and full fertilized (black bars) conditions. Effects of hydrochar type and application rate were considered significant, relative to controls receiving 100% fertilizer, when 95% CI did not overlap with zero in each nutritional regime.

3.2. Effects of Hydrochar on Above-Ground Nutrient and Heavy Metal Uptake

Shoot N, P, K, Ca and Mg uptake of pine tree seedlings grown with hydrochar under unfertilized condition was significantly lower than uptake of seedlings receiving 100% fertilizer ($P < 0.05$) (Figure 3). Nitrogen uptake reflected QID responses (Figure 3a). Using 20% hydrochar powder or pellet with 50% fertilizer resulted in pine seedling with similar N content to untreated seedlings grown under 100% fertilizer. Adding hydrochar under 100% fertilizer had mostly neutral impact on N uptake (Figure 3a). Shoot P uptake of seedlings treated with hydrochar powder and half or full rates of liquid fertilizer was similar to full-fertilized control ones (Figure 3b). Application of hydrochar pellets decreased shoot P acquisition in 50% fertilizer regime compared to untreated ones with 100% fertilizer rate ($P < 0.05$) (Figure 3b). Comparison of K uptake of hydrochar treated seedlings under 50% and 100% fertilizer with untreated seedlings receiving 100% fertilizer showed mostly positive or neutral impacts of hydrochar addition (Figure 3c). Addition of hydrochar powder and pellet with half and full rates of chemical fertilizer did mostly not significantly affect shoot Ca and Mg uptake. Only seedlings grown under fully fertilized conditions in peat amended with 20% (v/v) hydrochar had 63% higher Ca uptake than untreated seedlings (mean ± SE: 6.14 ± 0.14 versus 3.77 ± 0.60, respectively). There was no significant interaction between hydrochar type, rate and fertilizer level for shoot N, P, K, Ca and Mg uptake ($P > 0.05$).

Concerning heavy metals, our results suggested that the acquisition of Cu, Ni, Pb, Zn and Cr in seedling grown with 20% hydrochar in 50% fertilizer was not significantly different from what was observed in untreated seedlings (Figure 4). However, under the 100% fertilizer regime, application of both hydrochar powder and pellets resulted in significantly higher shoot Cu and Zn content in pine seedlings in comparison with untreated seedlings (Figure 4). Cadmium uptake also increased by 2–5 times when 20% of pot volume was mixed with hydrochar pellet or powder under 50% fertilizer application, respectively ($P < 0.05$) (Figure 4e). However, application of hydrochar pellet resulted in less Cd uptake by seedling compared to hydrochar powder in all tested nutritional regimes (fertilizer rate × hydrochar type $P = 0.004$) (Figure 4e). The shoot heavy metal uptake was not significantly affected by the interaction between hydrochar type, rate and fertilizer ($P > 0.05$).

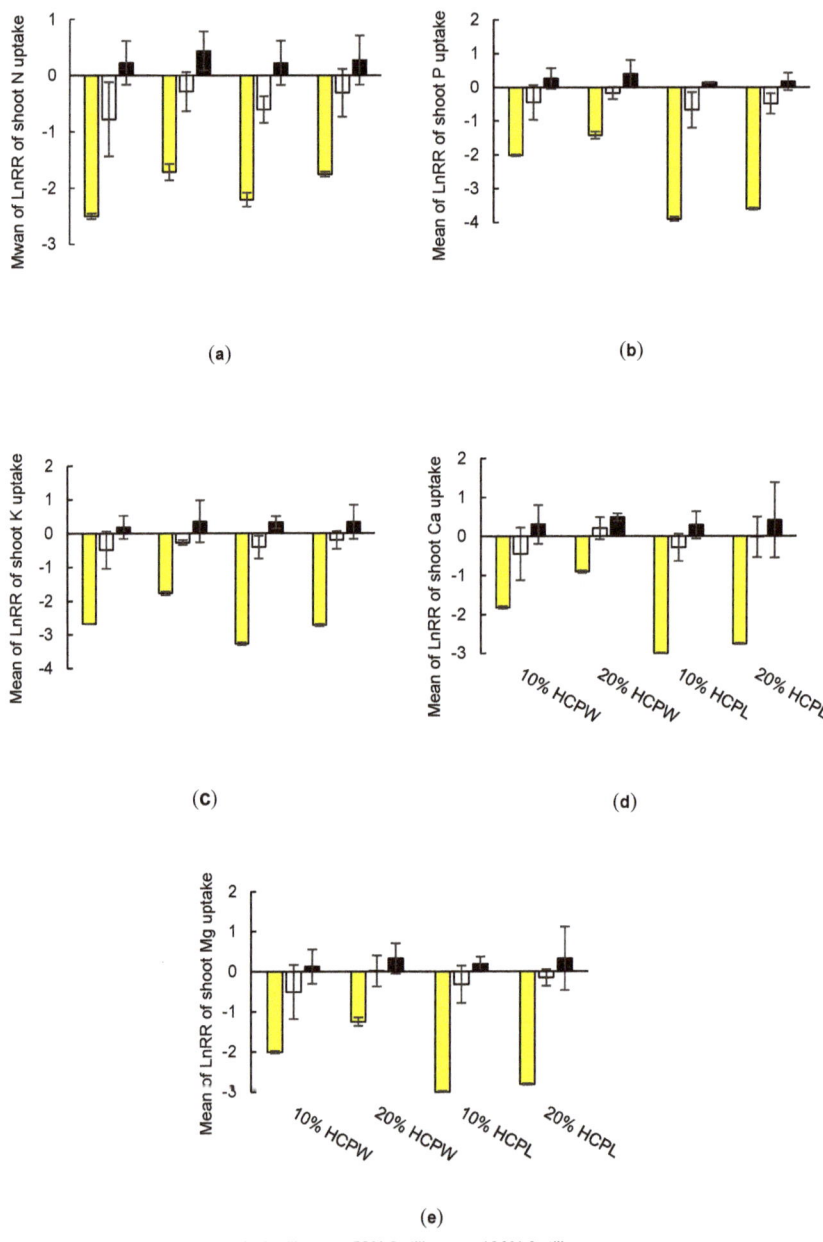

Figure 3. Mean (±95 CI, n = 3) natural logarithmic response ratios (standardized with the untreated fully-fertilized control) of shoot N (**a**), P (**b**), K (**c**), Ca (**d**) and Mg (**e**) uptake of pine tree seedlings grown in peat mixed with 10% and 20% (v/v) hydrochar powder (HCPW) and hydrochar pellet (HCPL) in unfertilized (yellow bars), half (open bars) and full fertilized (black bars) conditions. Effects of hydrochar type and application rate were considered significant, relative to controls receiving 100% fertilizer, when 95% CI did not overlap with zero in each nutritional regime.

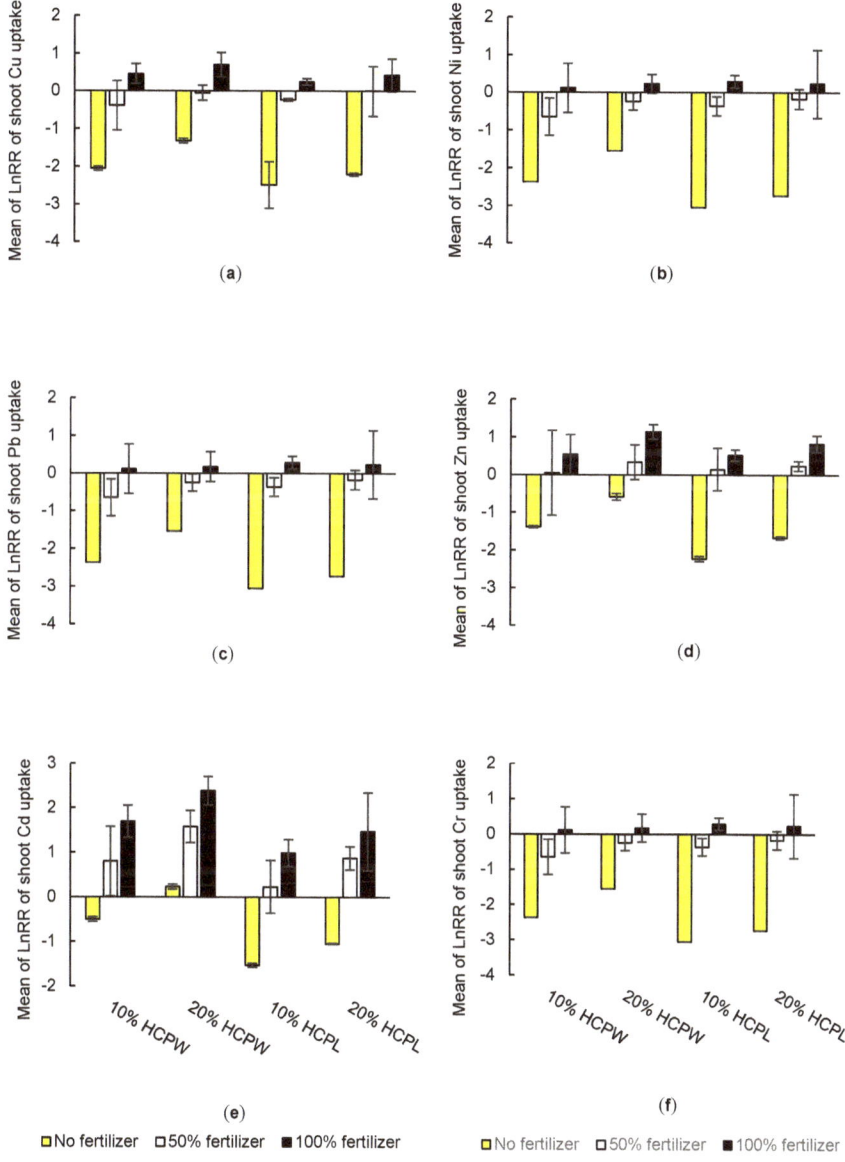

Figure 4. Mean (±95 CI, n = 3) natural logarithmic response ratios (standardized with the untreated fully-fertilized control) of shoot Cu (**a**), Ni (**b**), Pb (**c**), Zn (**d**), Cd (**e**) and Cr (**f**) uptake of pine tree seedlings grown in peat mixed with 10% and 20% (v/v) hydrochar powder (HCPW) and hydrochar pellet (HCPL) in unfertilized (yellow bars), half (open bars) and full fertilized (black bars) conditions. Effects of hydrochar type and application rate were considered significant, relative to controls receiving 100% fertilizer, when 95% CI did not overlap with zero in each nutritional regime.

3.3. Mycorrhizal Colonization Response to Hydrochar

The results of the second experiment showed that percentage of root length colonized by ectomycorrhizal fungi in seedlings grown with hydrochar amendment under unfertilized condition was 18% higher than in those grown without hydrochar but with 100% fertilizer ($P < 0.05$; mean ± SE: 39.6 ± 2.62 versus 21.5 ± 2.1) (Figure 5). However, neither powder nor pellet of hydrochar affected the colonization percentage of pine seedling roots under full-fertilized condition ($P > 0.05$) (Figure 5). The two-way interaction between fertilizer rate and hydrochar type was not significant ($P > 0.05$).

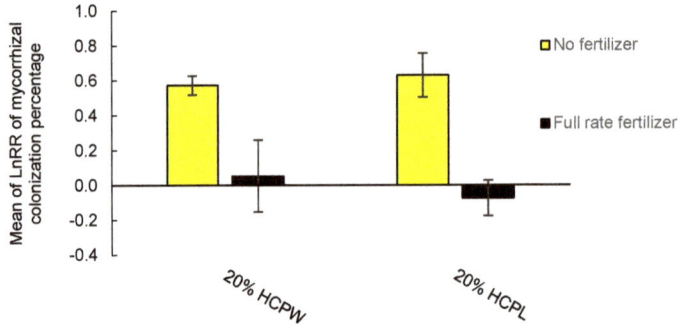

Figure 5. Mean (±95 CI, n = 3) natural logarithmic response ratios (standardized with the untreated fully-fertilized control) of mycorrhizal colonization percentage of pine tree seedlings grown in peat mixed with 20% (v/v) hydrochar powder (HCPW) and hydrochar pellet (HCPL) in unfertilized (yellow bars) and full fertilized (black bars) regimes. Effects of hydrochar type and application rate were considered significant, relative to controls receiving 100% fertilizer, when 95% CI did not overlap with zero in each nutritional regime.

4. Discussion

4.1. Shoot Dry Mass and Stem Diameter

This study is the first report of the effects of ash-enriched hydrochar powder/pellet applied as substrate constituents on growth and quality of pine tree seedlings. Amending growing media with hydrochar had positive or neutral effects on shoot biomass of seedlings compared to those grown in non-amended substrates under all experimental fertilization regimes (0%, 50% and 100% fertilizer), which supports the first hypothesis of this study (Figure 1). Although there is no published study on the potentials of hydrochar for production of containerized pine tree seedlings, our results are in accordance with those of George et al. [27] who reported 82% increase in above-ground biomass of hydrochar treated poplar seedlings in the first year and neutral effect in the second year of their study. We suggest that seedlings treated with hydrochar enriched with 5% ash in this study benefitted from improved nutrient availability in the rhizosphere that consequently enhanced shoot biomass. George et al. [27] showed that >24% of the total N in hydrochar treated poplar tree seedlings came from the char proving that hydrochar may behave as a direct source of nutrients for poplar seedlings. Our results contradict those of Belda et al. [30] who found that hydrochar from forest waste decreased stem dry weight of myrtle (*Myrtus communis* L.) and mastic (*Pistacia lentiscus*) seedlings by up to 75% and 49%, respectively. While limited studies are available on the use of hydrochar for container grown seedlings, the effects of pyrochar on pine seedlings have recently been investigated [10], showing similar growth of seedlings grown with ≤50% pyrochar (v/v) and in those grown in untreated peat. Under higher rates of nitrogen fertilizer, morphological traits were similar between pine seedlings grown with either 25% pyrochar powder or pyrolyzed wood pellets and in those grown in raw peat without biochar addition [10]. This is consistent with the results of the present study (Figure 1). Therefore, the growth response of plants may be idiosyncratic, depending on the biochar feedstock, biochar production processes and plant species. Toufiq et al. [28] showed that phytotoxic effects of hydrochar depended on feedstock and production conditions and do not necessarily occur in every hydrochar application.

Lower effects on seedling shoot biomass grown in peat amended with hydrochar pellets compared to those grown with hydrochar power under unfertilized condition, might be related to the fact that pellets release growth-affecting nutrients much more slowly. This may have important implications for potential long-term benefits of hydrochar pellets in forest soils after transplantation.

We found that amending peat with 20% hydrochar powder (v/v) increased the stem diameter compared to untreated control under all fertilizer regimes. Stem diameter is important for the survival of containerized pine and spruce tree seedlings after transplantation [46,47]. Approximately 20% of pine tree seedlings die during the first couple of growing seasons [48]. There is a strong positive relationship between initial stem-base diameter of spruce seedlings and their survival rate [47] so that containerized spruce with ~8 mm stem diameter showed low mortality rate through Pine Weevils (*Hylobius* spp.) and may be an alternative for insecticide-treated seedlings. Our finding, while preliminary, suggests that hydrochar application, especially hydrochar powder, in containerized pine tree seedlings may also enhance the survival rates after transplantation.

4.2. Quality of The Pine Tree Seedlings

Another aim of the present study was to estimate the potential of ash-enriched hydrochar for compensating fertilizer requirements of containerized pine tree seedlings. This was done using QID of different hydrochar application under the various fertilization scenarios. QID is one of the most comprehensive indices evaluating seedling quality [49]. We found that application of 20% (v/v) hydrochar as either powder or pellet and a 50% fertilizer dose resulted in the same quality of pine seedlings as seedlings growing in non-amended substrate but receiving full rate of fertilizer (optimum growth condition) (Figure 2). This result can be explained by boosted N, P, Ca and Mg availability in hydrochar-amended substrates (Figure 3).

These results suggest the possibility of decreasing chemical fertilizer application in pine tree nurseries by partially mixing the growing media with hydrochar powder or pellet. This probability had been previously brought to attention by Steinbeiss et al. [50], who believed that hydrochar, from yeast and glucose, could be added into soil as a fertilizer while also acting as a decadal carbon pool. The fertilizer effects of hydrochar might be attributed to stimulation of soil microorganisms that participate in nutrient recycling, improved water retention and CEC of the growing media due to large surface area of hydrochar [50]. However, our results should be interpreted with caution since we found only a relatively small reduction in irrigation frequency requirement (3–10%) (Table A1) and we did not analyze microbial activity or composition of the experimental growing media used in this study.

Under the full fertilizer regime, we found 75% increment in the quality index of the pine seedlings when grown with 20% hydrochar powder (v/v) in comparison with untreated seedlings. This is corroborated by Rezende et al. [51] and Aung et al. [49], who also found that biochar treatment significantly increased the quality index in *Tectona grandis*, *Quercus serrata* and *Prunus sargentii*. These researchers explained the observed increase in QID with improved water retention in growing substrates amended with biochar in a containerized production system. However, this might not be the case here as no significant difference was recorded in irrigation frequency demand of seedling grown with 100% fertilizer and 20% hydrochar powder and those with only full rate of fertilizer (Table A1). Indeed, it may be that these seedlings benefitted from better nutrient availability through either improved nutrient recycling by simulated beneficial microbes or direct supplement from hydrochar powder.

4.3. Nutrient and Heavy Metal Contents

The results of the current study showed that nutrient uptake of seedlings grown with either forms of ash-enriched hydrochar without addition of chemical fertilizer was significantly lower than in those grown in untreated peat with 100% fertilizer. Similarly, Sarauer and Coleman [21] found that N concentration in Douglas-fir seedlings was significantly lower in biochar treated seedlings fertilized with $\frac{1}{4}$ rate fertilizer relative to untreated ones with full rate of fertilizer. Therefore, it can be concluded that application of 10–20% hydrochar as powder or pellet without added fertilizer will not provide the seedlings with sufficient nutrients to achieve optimal growth and nutrient acquisition. Substrates amended with 20% hydrochar and 50% fertilizer produced seedling with similar N, P, K, Ca and Mg contents as non-amended but fully fertilized ones, which may result in similar seedling quality in both groups. Therefore, mixing 20% volume of the growing media with ash-enriched hydrochar from biosludge may provide sufficient plant-available nutrients to compensate for halved levels of fertilizer. Prior studies have noted that amending peat and perlite media with biochar can improve nutrient availability in these substrates [52]. Similarly, enhanced phytonutrient availability and phyto-stimulant ability of hydrochar-peat mixtures have been shown by Álvarez, et al. [53]. Furthermore, wood ash application to peat media can also boost foliar P and K concentrations [54].

Sludge of pulp and paper mill effluent contains significant amounts of heavy metals [55]. Due to environmental issues, considering data about heavy metal content is crucial before introducing hydrochar to either forestry or agriculture sectors. Hydrochar treated seedlings with 50% fertilizer that showed the same quality index as full-fertilized control ones also had similar heavy metal contents except for Cd. This may provide some support for the conceptual premise that the bioavailability of heavy metals existing in biosludge reduce when the sludge are hydrothermally carbonized. Devi and Saroha [55] believed that the bioavailability and eco-toxicity of the heavy metals in biochar derived from pulp and paper mill sludge declined because bioavailable and mobile heavy metals became relatively stable when pyrolyzed. Similarly, Liu et al. [56] found that exchangeable and reducible fraction of heavy metals in sewage sludge decreased by hydrothermal carbonization. The immobilized heavy metals in hydrochar considerably reduce the terrestrial ecotoxicity [24] and aquatic ecotoxicity impacts [25] of soil application of biochar.

Accumulation of heavy metals were previously suggested to cause negative effects of hydrochar on plants [28]. However, our results, while preliminary, proposes that hydrochar from biosludge, in combination with proper levels of fertilizer may safely be used in forestry and agriculture. However, the potential of using hydrochar in agricultural systems warrants further research.

However, Cd uptake in hydrochar-treated seedling receiving 50% fertilizer was significantly higher than in untreated seedlings receiving 100% fertilizer (Figure 4e). Likewise, unfertilized seedlings in peat amended with 20% hydrochar powder had greater Cd values than those grown fully fertilized without hydrochar. Considering the heavy metal contents of the peat, hydrochar and ash summarized in Table 1, accessible Cd for pine tree seedling was from the hydrochar source. The reason for this is not clear but it may be related to relatively high leaching potential of Cd from the hydrochar at low pH conditions of the experimental growing media (pH ~ 5). According to Devi and Saroha [55], Cd in biochar from paper mill sludge is mainly associated with the oxidizable fraction, which represents potentially bioavailable metals. They also found that potential ecological risk index for Cd in biochar derived from paper mill sludge pyrolyzed at 200 °C was significantly higher than for other heavy metals, i.e., Cr, Cu, Ni, Zn, Pb.

Our results showed increased accumulation of heavy metals by using hydrochar powder/pellet combined with 100% fertilizer. This result must be interpreted with caution because the interaction between root system and hydrochar-peat mixture is very complex, and therefore, it may partly be explained by either increased leaching of heavy metals from hydrochar in response to acidic exudates released by the developing root system and microbial community or direct uptake from hydrochar sources. Further work is required to test these relationships.

4.4. Colonization with Ectomycorrhizal Fungi

This is the first study describing ectomycorrizal responses of pine tree seedlings to hydrochar addition. Our results confirmed that application of either hydrochar powder or pellets, at a rate of 20% (v/v), boosted the percentage of root length colonized by ectomycorrhizal fungi of pine tree seedlings grown with no added fertilizer. Rillig et al. [33] reported higher root colonization percentage, spore germination and arbuscule formation by arbuscular mycorrhizae when *Taraxacum* was grown on soil mixed with 20% (v/v) hydrochar. This might be due to changes in physicochemical properties of the soil, e.g., nutrient status and pH, and signaling interaction in the root zone [33,57]. On the other hand, George et al. [58] found reduced mycorrhizal colonization and arbuscule formation when hydrochar was applied at rates of 5–10%. Indeed, different hydrothermal carbonization conditions and feedstock used to produce hydrochar might be responsible for contradictory results on mycorrhizal associations [28]. Unchanged mycorrhizal colonization percentage in hydrochar amended seedlings grown with 100% fertilizer relative to their untreated counterparts was not surprising, since pine tree seedlings might have reduced carbon transfer to the fungus due to high nutrient availability in the rhizosphere.

4.5. Environmental and Economic Implications of Using Hydrochar

Based on the results of the current study, peat substrate can successfully be replaced with hydrochar up to ratio of 20% (v/v). This substitution of peat, and even fertilizer, with hydrochar is important from environmental and economic point of views, particularly for nitrogen (N) fertilizer, the production of which is a greenhouse gas and energy intensive process. Moreover, peat use reduction due to replacement with hydrochar will result in less peat extraction and, therefore, will decrease the associated environmental impacts. Results of our previous studies showed that pyrochar and hydrochar production from paper mill sludge can significantly increase the environmental performance of sludge management and contribute to sustainable forest ecosystems [24,25]. Hydrochar-amended substrates can contribute to climate change mitigation with carbon sequestration as part of the normal reforestation. Potential contribution to climate change mitigation have been supported with studies of the life cycle carbon footprint of biochar systems [16,59,60] Moreover, potentials of using hydrochar pellets in containerized production systems add to the economic viability of HTC and densification processes. To date, no studies assessing the environmental and economic impacts of using hydrochar powder/pellets in nursery production or forest landscape have been published. Therefore, further study is recommended for scientific evidence of the sustainability and productivity of hydrochar use to assess the economic market and to identify and manage the associated risks of broad application in forestry sector.

5. Conclusions

The present study is the first study assessing the potentials of ash-enriched hydrochar application in containerized pine seedling production. We examined the effects of hydrochar addition in two forms (powder and pellets) and two mixing rates (10% and 20% v/v) under different nutritional regimes (no fertilizer, 50% fertilizer and 100% fertilizer). The most important finding was that hydrochar derived from biosludge did not adversely affect the growth of containerized pine tree seedlings. Moreover, we could show that pine tree seedlings grown in peat mixed with 20% hydrochar and 50% fertilizer rate had similar quality index values and nutrient and heavy metal uptake (except for Cd) to those that did not receive hydrochar but supplied with 100% fertilizer. Percentage of root length colonized with ectomycorrhizae also showed positive responses to hydrochar application. The results of our study imply that application of proper rates of hydrochar from biosludge with adjusted levels of liquid fertilizer may reduce fertilizer requirements in pine nurseries through recycling nutrients from forest waste materials. Furthermore, these findings have significant implications for the understanding of how forest industry can approach a circular bioeconomy. However, more information about the effects of hydrochar on chemistry of growing media and microbial activity would help to establish a greater degree of accuracy on this matter.

Author Contributions: Conceptualization, S.E. and M.S.; Data curation, S.E., A.M. and M.S.; Formal analysis, S.E.; Funding acquisition, K.G.; Investigation, S.E. and K.H.; Methodology, S.E.; Project administration, K.G. and M.S.; Resources, A.M. and K.H.; Supervision, K.G. and R.L.E.; Visualization, S.E.; Writing—original draft preparation, S.E.; Writing—review and editing, R.L.E.

Funding: This research was funded by the Swedish agency for economic and regional growth, grant number 20201239, project name FOSBE, and by a European union grant through the interreg sweden-norway program, grant number 20200023, project name IMTRIS.

Acknowledgments: The authors would like to acknowledge Kajsa Fougner (Åforsk), Bergvik Skog AB and The Norwegian Institute of Bioeconomy Research (NIBIO) for their technical help in this study.

Conflicts of Interest: The authors declare no conflict of interest. The funder had no role in the design of the study; in the collection, analyses, or interpretation of data; in the writing of the manuscript, or in the decision to publish the results".

Appendix A

Table A1. Irrigation frequency for 6 months-old pine tree seedlings grown with 10% and 20% hydrochar powder and pellet standardized with values of control pines (without hydrochar) with 100% fertilizer.

	Percentage of Peat Replaced (v/v)						
	0	10			20		
Fertilizer rate	100%	0%	50%	100%	0%	50%	100%
Peat	100						
Hydrochar Powder		66	94	96	72	97	100
Hydrochar Pellet		61	92	92	61	90	93

References

1. The Swedish Forest Industries' Statistics. Economic Importance. Available online: http://www.forestindustries.se/forest-industry/statistics/economic-importance/ (accessed on 4 April 2019).
2. Nilsson, P. Forest Land. Available online: https://www.slu.se/en/Collaborative-Centres-and-Projects/the-swedish-national-forest-inventory/forest-statistics/forest-statistics/skogsmark (accessed on 21 August 2018).
3. Mattsson, A. Reforestation challenges in Scandinavia. *Reforesta* **2016**, 67–85. [CrossRef]
4. Bohlin, C.; Holmberg, P. Peat: Dominating growing medium in Swedish horticulture. In Proceedings of the International Symposium on Growing Media and Hydroponics, Alnarp, Sweden, 8–14 September 2001; pp. 177–181.
5. Cleary, J.; Roulet, N.T.; Moore, T.R. Greenhouse Gas Emissions from Canadian Peat Extraction, 1990–2000: A Life-cycle Analysis. *Ambio J. Hum. Environ.* **2005**, *34*, 456–461. [CrossRef]
6. Alexander, P.; Bragg, N.; Meade, R.; Padelopoulos, G.; Watts, O. Peat in horticulture and conservation: The UK response to a changing world. *Mires Peat* **2008**, *3*, 8.
7. Dombrowsky, M.; Dixon, M. Sustainable growing substrates for potted greenhouse gerbera production. In Proceedings of the International Symposium on Responsible Peatland Management and Growing Media Production, Quebec City, Canada, 13–17 June 2011; pp. 61–68.
8. Tian, Y.; Sun, X.; Li, S.; Wang, H.; Wang, L.; Cao, J.; Zhang, L. Biochar made from green waste as peat substitute in growth media for *Calathea rotundifola* cv. *Fasciata*. *Sci. Hortic.* **2012**, *143*, 15–18. [CrossRef]
9. Vaughn, S.F.; Eller, F.J.; Evangelista, R.L.; Moser, B.R.; Lee, E.; Wagner, R.E.; Peterson, S.C. Evaluation of biochar-anaerobic potato digestate mixtures as renewable components of horticultural potting media. *Ind. Crop. Prod.* **2015**, *65*, 467–471. [CrossRef]
10. Dumroese, K.R.; Pinto, J.R.; Heiskanen, J.; Tervahauta, A.; McBurney, K.G.; Page-Dumroese, D.S.; Englund, K. Biochar Can Be a Suitable Replacement for Sphagnum Peat in Nursery Production of Pinus ponderosa Seedlings. *Forests* **2018**, *9*, 232. [CrossRef]
11. Matt, C.P.; Keyes, C.R.; Dumroese, R.K. Biochar effects on the nursery propagation of 4 northern Rocky Mountain native plant species. *Nativ. Plants J.* **2018**, *19*, 14–26. [CrossRef]
12. Huber, G.W.; Iborra, S.; Corma, A. Synthesis of transportation fuels from biomass: Chemistry, catalysts, and engineering. *Chem. Rev.* **2006**, *106*, 4044–4098. [CrossRef]
13. Glaser, B.; Haumaier, L.; Guggenberger, G.; Zech, W. The 'Terra Preta' phenomenon: A model for sustainable agriculture in the humid tropics. *Naturwissenschaften* **2001**, *88*, 37–41. [CrossRef]
14. Marris, E. Black is the new green. *Nature* **2006**, *442*, 624–626. [CrossRef]

15. Mohammadi, A.; Cowie, A.; Anh Mai, T.L.; de la Rosa, R.A.; Kristiansen, P.; Brandão, M.; Joseph, S. Biochar use for climate-change mitigation in rice cropping systems. *J. Clean. Prod.* **2016**, *116*, 61–70. [CrossRef]
16. Mohammadi, A.; Cowie, A.L.; Anh Mai, T.L.; Brandão, M.; Anaya de la Rosa, R.; Kristiansen, P.; Joseph, S. Climate-change and health effects of using rice husk for biochar-compost: Comparing three pyrolysis systems. *J. Clean. Prod.* **2017**, *162*, 260–272. [CrossRef]
17. Mohammadi, A.; Cowie, A.L.; Cacho, O.; Kristiansen, P.; Anh Mai, T.L.; Joseph, S. Biochar addition in rice farming systems: Economic and energy benefits. *Energy* **2017**, *140*, 415–425. [CrossRef]
18. Hu, B.; Wang, K.; Wu, L.; Yu, S.-H.; Antonietti, M.; Titirici, M.-M. Engineering Carbon Materials from the Hydrothermal Carbonization Process of Biomass. *Adv. Mater.* **2010**, *22*, 813–828. [CrossRef] [PubMed]
19. Laird, D.; Fleming, P.; Wang, B.; Horton, R.; Karlen, D. Biochar impact on nutrient leaching from a Midwestern agricultural soil. *Geoderma* **2010**, *158*, 436–442. [CrossRef]
20. Di Lonardo, S.; Baronti, S.; Vaccari, F.P.; Albanese, L.; Battista, P.; Miglietta, F.; Bacci, L. Biochar-based nursery substrates: The effect of peat substitution on reduced salinity. *Urban For. Urban Green.* **2017**, *23*, 27–34. [CrossRef]
21. Sarauer, J.L.; Coleman, M.D. Biochar as a growing media component for containerized production of Douglas-fir. *Can. J. For. Res.* **2018**, *48*, 581–588. [CrossRef]
22. Paneque, M.; De la Rosa, J.M.; Franco-Navarro, J.D.; Colmenero-Flores, J.M.; Knicker, H. Effect of biochar amendment on morphology, productivity and water relations of sunflower plants under non-irrigation conditions. *Catena* **2016**, *147*, 280–287. [CrossRef]
23. De la Rosa, J.M.; Paneque, M.; Miller, A.Z.; Knicker, H. Relating physical and chemical properties of four different biochars and their application rate to biomass production of *Lolium perenne* on a Calcic Cambisol during a pot experiment of 79days. *Sci. Total Environ.* **2014**, *499*, 175–184. [CrossRef]
24. Mohammadi, A.; Sandberg, M.; Venkatesh, G.; Eskandari, S.; Dalgaard, T.; Joseph, S.; Granström, K. Environmental analysis of producing biochar and energy recovery from pulp and paper mill biosludge. *J. Ind. Ecol.* **2019**. [CrossRef]
25. Mohammadi, A.; Sandberg, M.; Venkatesh, G.; Eskandari, S.; Dalgaard, T.; Joseph, S.; Granström, K. Environmental performance of end-of-life handling alternatives for paper-and-pulp-mill sludge: Using digestate as a source of energy or for biochar production. *Energy* **2019**, *182*, 594–605. [CrossRef]
26. Kambo, H.S.; Dutta, A. A comparative review of biochar and hydrochar in terms of production, physico-chemical properties and applications. *Renew. Sustain. Energy Rev.* **2015**, *45*, 359–378. [CrossRef]
27. George, E.; Ventura, M.; Panzacchi, P.; Scandellari, F.; Tonon, G. Can hydrochar and pyrochar affect nitrogen uptake and biomass allocation in poplars? *J. Plant Nutr. Soil Sci.* **2017**, *180*, 178–186. [CrossRef]
28. Toufiq, M.R.; Andert, J.; Wirth, B.; Busch, D.; Pielert, J.; Lynam, J.G.; Mumme, J. Hydrothermal carbonization of biomass for energy and crop production. *Appl. Bioenergy* **2014**, *1*, 11–29. [CrossRef]
29. Fornes, F.; Belda, R.M. Biochar versus hydrochar as growth media constituents for ornamental plant cultivation. *Sci. Agric.* **2018**, *75*, 304–312. [CrossRef]
30. Belda, R.M.; Lidón, A.; Fornes, F. Biochars and hydrochars as substrate constituents for soilless growth of myrtle and mastic. *Ind. Crop. Prod.* **2016**, *94*, 132–142. [CrossRef]
31. Dumroese, R.K.; Heiskanen, J.; Englund, K.; Tervahauta, A. Pelleted biochar: Chemical and physical properties show potential use as a substrate in container nurseries. *Biomass Bioenergy* **2011**, *35*, 2018–2027. [CrossRef]
32. Robertson, S.J.; Rutherford, P.M.; Lopez-Gutierrez, J.C.; Massicotte, H.B. Biochar enhances seedling growth and alters root symbioses and properties of sub-boreal forest soils. *Can. J. Soil Sci.* **2012**, *92*, 329–340. [CrossRef]
33. Rillig, M.C.; Wagner, M.; Salem, M.; Antunes, P.M.; George, C.; Ramke, H.-G.; Titirici, M.-M.; Antonietti, M. Material derived from hydrothermal carbonization: Effects on plant growth and arbuscular mycorrhiza. *Appl. Soil. Ecol.* **2010**, *45*, 238–242. [CrossRef]
34. Busch, D.; Kammann, C.; Grunhage, L.; Muller, C. Simple biotoxicity tests for evaluation of carbonaceous soil additives: Establishment and reproducibility of four test procedures. *J. Environ. Qual.* **2012**, *41*, 1023–1032. [CrossRef]
35. Baronti, S.; Alberti, G.; Camin, F.; Criscuoli, I.; Genesio, L.; Mass, R.; Vaccari, F.P.; Ziller, L.; Miglietta, F. Hydrochar enhances growth of poplar for bioenergy while marginally contributing to direct soil carbon sequestration. *GCB Bioenergy* **2017**, *9*, 1618–1626. [CrossRef]

36. Nara, K. Ectomycorrhizal networks and seedling establishment during early primary succession. *New Phytol.* **2006**, *169*, 169–178. [CrossRef] [PubMed]
37. Ahlroth, M.; Bialik, M.; Jensen, A. Hydrothermal Carbonization of Pulp-and Paper Mill Effluent Sludge. *Bioresource Technol.* **2016**, *200*, 15–489.
38. Chen, H.; Zhou, Y.; Zhao, H.; Li, Q. A comparative study on behavior of heavy metals in pyrochar and hydrochar from sewage sludge. *Energy Sources Part A Recovery Util. Environ. Eff.* **2018**, *40*, 565–571. [CrossRef]
39. Jackson, D.P.; Dumroese, K.R.; Barnett, J.P. Nursery response of container *Pinus palustris* seedlings to nitrogen supply and subsequent effects on outplanting performance. *For. Ecol. Manag.* **2012**, *265*, 1–12. [CrossRef]
40. Dumroese, R.K.; Pinto, J.R.; Montville, M.E. Using container weights to determine irrigation needs: A simple method. *Nativ. Plants J.* **2015**, *16*, 67–71. [CrossRef]
41. Dickson, A.; Leaf, A.L.; Hosner, J.F. Quality appraisal of white spruce and white pine seedling stock in nurseries. *For. Chron.* **1960**, *36*, 10–13. [CrossRef]
42. Brundrett, M.; Bougher, N.; Dell, B.; Grove, T.; Malajczuk, N. *Working with Mycorrhizas in Forestry and Agriculture*; Australian Centre for International Agricultural Research Canberra: Canberra, Australia, 3 January 1996.
43. Brose, P.H.; Dey, D.C.; Phillips, R.J.; Waldrop, T.A. A Meta-Analysis of the Fire-Oak Hypothesis: Does Prescribed Burning Promote Oak Reproduction in Eastern North America? *For. Sci.* **2012**, *59*, 322–334. [CrossRef]
44. Loydi, A.; Donath, T.W.; Eckstein, R.L.; Otte, A. Non-native species litter reduces germination and growth of resident forbs and grasses: Allelopathic, osmotic or mechanical effects? *Biol. Invasions* **2015**, *17*, 581–595. [CrossRef]
45. Goldberg, D.E.; Scheiner, S.M. ANOVA and ANCOVA: Field competition experiments. In *Design and Analysis of Ecological Experiments*, 2nd ed.; Scheiner, S.M., Gurevitch, J., Eds.; Oxford University Press Inc: Oxford, UK, 2001; pp. 77–98.
46. Selander, J.; Immonen, A.; Raukko, P. Resistance of naturally regenerated and nursery-raised Scots pine seedlings to the large pine weevil. *Folia For.* **1990**, *766*, 19.
47. Thorsen, Å.A.; Mattsson, S.; Weslien, J. Influence of Stem Diameter on the Survival and Growth of Containerized Norway Spruce Seedlings Attacked by Pine Weevils (*Hylobius* spp.). *Scand. J. For. Res.* **2001**, *16*, 54–66. [CrossRef]
48. Köster, E.; Pumpanen, J.; Köster, K. Biochar as a possible new tool for afforestation practices. In Proceedings of the EGU General Assembly Conference Abstracts, Vienna, Austria, 8–13 April 2018; p. 6082.
49. Aung, A.; Han, S.H.; Youn, W.B.; Meng, L.; Cho, M.S.; Park, B.B. Biochar effects on the seedling quality of Quercus serrata and Prunus sargentii in a containerized production system. *For. Sci. Technol.* **2018**, *14*, 112–118. [CrossRef]
50. Steinbeiss, S.; Gleixner, G.; Antonietti, M. Effect of biochar amendment on soil carbon balance and soil microbial activity. *Soil Biol. Biochem.* **2009**, *41*, 1301–1310. [CrossRef]
51. Rezende, F.A.; Santos, V.A.H.F.D.; Maia, C.M.B.D.F.; Morales, M.M. Biochar in substrate composition for production of teak seedlings. *Pesqui. Agropec. Bras.* **2016**, *51*, 1449–1456. [CrossRef]
52. Locke, J.C.; Altland, J.E.; Ford, C.W. Gasified Rice Hull Biochar Affects Nutrition and Growth of Horticultural Crops in Container Substrates. *J. Environ. Hortic.* **2013**, *31*, 195–202. [CrossRef]
53. Álvarez, M.L.; Gascó, G.; Plaza, C.; Paz-Ferreiro, J.; Méndez, A. Hydrochars from Biosolids and Urban Wastes as Substitute Materials for Peat. *Land Degrad. Dev.* **2017**, *28*, 2268–2276. [CrossRef]
54. Hytönen, J. Wood ash fertilisation increases biomass production and improves nutrient concentrations in birches and willows on two cutaway peats. *Balt* **2016**, *22*, 98–106.
55. Devi, P.; Saroha, A.K. Risk analysis of pyrolyzed biochar made from paper mill effluent treatment plant sludge for bioavailability and eco-toxicity of heavy metals. *Bioresour. Technol.* **2014**, *162*, 308–315. [CrossRef]
56. Liu, T.; Liu, Z.; Zheng, Q.; Lang, Q.; Xia, Y.; Peng, N.; Gai, C. Effect of hydrothermal carbonization on migration and environmental risk of heavy metals in sewage sludge during pyrolysis. *Bioresour. Technol.* **2018**, *247*, 282–290. [CrossRef]
57. Warnock, D.D.; Lehmann, J.; Kuyper, T.W.; Rillig, M.C. Mycorrhizal responses to biochar in soil—Concepts and mechanisms. *Plant Soil* **2007**, *300*, 9–20. [CrossRef]
58. George, C.; Wagner, M.; Kücke, M.; Rillig, M.C. Divergent consequences of hydrochar in the plant—Soil system: Arbuscular mycorrhiza, nodulation, plant growth and soil aggregation effects. *Appl. Soil Ecol.* **2012**, *59*, 68–72. [CrossRef]

59. Mohammadi, A.; Cowie, A.; Mai, T.L.A.; de la Rosa, R.A.; Brandão, M.; Kristiansen, P.; Joseph, S. Quantifying the Greenhouse Gas Reduction Benefits of Utilising Straw Biochar and Enriched Biochar. *Energy Procedia* **2016**, *97*, 254–261. [CrossRef]
60. Cowie, A.; Woolf, D.; Gaunt, J.; Brandão, M.; Anaya de la Rosa, R.; Cowie, A. Biochar, carbon accounting and climate change. In *Biochar for Environmental Management: Science, Technology, and Implementation*; Lehmann, J., Joseph, S., Eds.; Routledge: Milton Park, UK, 2015; pp. 763–794.

© 2019 by the authors. Licensee MDPI, Basel, Switzerland. This article is an open access article distributed under the terms and conditions of the Creative Commons Attribution (CC BY) license (http://creativecommons.org/licenses/by/4.0/).

Article

ACC Deaminase Producing PGPR *Bacillus amyloliquefaciens* and *Agrobacterium fabrum* along with Biochar Improve Wheat Productivity under Drought Stress

Muhammad Zafar-ul-Hye [1], Subhan Danish [1], Mazhar Abbas [2], Maqshoof Ahmad [3] and Tariq Muhammad Munir [4,*]

1. Department of Soil Science, Faculty of Agricultural Sciences and Technology, Bahauddin Zakariya University, Multan 60800, Pakistan
2. Institute of Horticultural Sciences, Faculty of Agriculture, University of Agriculture, Faisalabad 38000, Pakistan
3. Department of Soil Science, University College of Agriculture & Environment Sciences, The Islamia University of Bahawalpur, Bahawalpur 63100, Pakistan
4. Department of Geography, University of Calgary, 2500 University Drive NW, Calgary, AB T2N 1N4, Canada
* Correspondence: tmmunir@ucalgary.ca; Tel.: +1-403-971-5693

Received: 16 June 2019; Accepted: 27 June 2019; Published: 29 June 2019

Abstract: Drought stress retards wheat plant's vegetative growth and physiological processes and results in low productivity. A stressed plant synthesizes ethylene which inhibits root elongation; however, the enzyme 1-Aminocyclopropane-1-Carboxylate (ACC) deaminase catabolizes ethylene produced under water stress. Therefore, the ACC deaminase producing plant growth promoting rhizobacteria (PGPR) can be used to enhance crop productivity under drought stress. Biochar (BC) is an organically active and potentially nutrient-rich amendment that, when applied to the soil, can increase pore volume, cation exchange capacity and nutrient retention and bioavailability. We conducted a field experiment to study the effect of drought tolerant, ACC deaminase producing PGPR (with and without timber waste BC) on plant growth and yield parameters under drought stress. Two PGPR strains, *Agrobacterium fabrum* or *Bacillus amyloliquefaciens* were applied individually and in combination with 30 Mg ha^{-1} BC under three levels of irrigation, i.e., recommended four irrigations (4I), three irrigations (3I) and two irrigations (2I). Combined application of *B. amyloliquefaciens* and 30 Mg ha^{-1} BC under 3I, significantly increased growth and yield traits of wheat: grain yield (36%), straw yield (50%), biological yield (40%). The same soil application under 2I resulted in greater increases in several of the growth and yield traits: grain yield (77%), straw yield (75%), above- and below-ground biomasses (77%), as compared to control; however, no significant increases in chlorophyll a, b or total, and photosynthetic rate and stomatal conductance in response to individual inoculation of a PGPR strain (without BC) were observed. Therefore, we suggest that the combined soil application of *B. amyloliquefaciens* and BC more effectively mitigates drought stress and improves wheat productivity as compared to any of the individual soil applications tested in this study.

Keywords: activated carbon; biofertilizers; gas exchange attributes; wheat; water stress; yield attributes

1. Introduction

Wheat is a staple and cash crop globally recognized for its nutritional and economic importance [1,2]. Wheat grain (flour) constitutes 20% of daily human diet and contains protein (8–12%) and a high amount of carbohydrates (55%). Drought is a worldwide, most critical abiotic factor due to which sustainable wheat crop productivity is at risk [3–5]. Drought severity is predicted to successively increase under climate change scenarios of atmospheric and soil warmings and altered precipitation patterns [6–11]. Consistent and prolonged warming and drought conditions combined with associated abiotic and biotic changes [12] may drastically retard crop productivity and risk food security [13,14]. Drought stress reduces nutrient uptake, which can cause poor development of roots, low transpiration and photosynthetic rates, closure of leaf stomata and desiccation resulting in wilting of plants [15–17]. Like other abiotic stresses, the drought also stimulates stress ethylene synthesis through an elevated level of 1-Aminocyclopropane-1-carboxylic acid (ACC; an ethylene precursor) via the methionine pathway, in higher plants [18,19]. Accumulation of stress ethylene in-turn inhibits roots elongation and consequently, shoot growth in plants [20].

Water management strategies and genetic engineering are useful tools to adapt to or mitigate drought stress. While irrigation water is being managed in irrigation-dependent cropping systems, genetic engineering to cope with water stress remains limited. However, a vital biological approach to combat drought impacts is the soil inoculation of plant growth promoting rhizobacteria (PGPR). The PGPR are frequently reported to efficiently elongate plant roots in the pot [21,22] and mitigate drought impacts in field or greenhouse conditions [23,24], and mobilize the immobile nutrients that lead to significant increases in plant vegetative growth [25] and crop yield [26,27]. PGPR produces ACC deaminase enzyme, which catabolizes stress ethylene through cleavage of ACC into α-ketobutyrate and ammonium ion (NH_4^+) under drought stress, e.g., [28], thus reducing the level of stress ethylene [29,30].

Biochar (BC) is an organically active soil amendment with very high soil pore volume and cation exchange capacity and has been reported to reduce drought stress in plants [31–34]. Biochar is a nutrient-rich, black carbon soil amendment [35] that is produced through pyrolysis of waste feedstock at high temperature [36] under anaerobic or partially anaerobic condition [37–39].

While individual soil application of ACC-deaminase containing PGPR or BC has been frequently investigated for combating drought effects in pot experiments, controlled field experimentation for evaluation of cumulative mitigating effects remains limited. Therefore, the objective of this research was to observe the efficiency of combined application of ACC-deaminase producing PGPR and timber waste BC in granting resistance to field-scale wheat crop against drought impacts. We hypothesized that soil inoculation of drought-tolerant ACC-deaminase containing PGPR along with timber waste BC amendment would be a more efficient technique to mitigate adverse drought effects on wheat growth and yield traits.

2. Materials and Methods

We conducted this experiment in the research area of the Department of Soil Science, Bahauddin Zakariya University, Multan, Pakistan, in November 2016. A total of 54 same size plots (9 m^2) were prepared and randomly divided into six triplicate treatments (T) (6 × 3 = 18) with each applied at three levels of irrigation (I), (i.e., 4I, 3I and 2I) following a randomized complete block design (RCBD; 18 × 3 = 54 plots). The experimental area was cropped with wheat and maize (rotation) during the last five years.

Recommended nitrogen (N), phosphorus (P) and potassium (K) fertilizers (RNPKF) were applied at the rates of 120, 60 and 60 kg ha^{-1} [40,41]. Full doses of P (as diammonium phosphate) and K (as sulphate of potash), and a 1/3rd dose of N were incorporated to topsoil at the seedbed preparation stage, and the remaining two splits of N were top-dressed after 30 and 60 days of seeding. We used standard crop management practices such as irrigation, fertilization, weeding, hoeing and plant protection to grow wheat crop during the study season. Timber-waster biochar (BC) was applied at a rate of 1.5%, i.e., 30 Mg ha^{-1}. Treatments included: Control (No PGPR + No BC + RNPKF), *A. fabrum*, *B. amyloliquefaciens*, 30 Mg ha^{-1} BC, *A. fabrum* + 30 Mg ha^{-1} BC and *B. amyloliquefaciens* + 30 Mg ha^{-1} BC.

The two most competent drought-tolerant ACC-deaminase producing PGPR strains, *Agrobacterium fabrum* (NR_074266.1) and *Bacillus amyloliquefaciens* (FN597644.1), as documented by Danish and Zafar-ul-Hye [22], were provided from the collection of Soil and Environmental Microbiology

Laboratory, Bahauddin Zakariya University Multan, Pakistan. Both strains were initially tested and found eligible to grow in Dworkin and Foster (DF) minimal salt medium at −0.78 Mpa osmotic potential, generated by 20% polyethylene glycol 6000 (PEG) [22]. For experimental purpose, DF minimal salt medium without agar was used to prepare inoculum of desired PGPR strains [42]. For measuring ACC-deaminase produced by PGPR strains (*A. fabrum* = 349.6 ± 21.4 and *B. amyloliquefaciens* = 313.2 ± 34.3 μmol α-ketobutyrate mg^{-1} protein h^{-1}), we followed El-Tarabily [43]. Glickmann and Dessaux methods [44] was used for assessment of indole acetic acid with (*A. fabrum* = 58.8 ± 3.27 and *B. amyloliquefaciens* = 17.3 ± 2.34 μg/mL) and without 0.5 gL^{-1} L-tryptophan (*A. fabrum* = 2.43 ± 0.34 and *B. amyloliquefaciens* = 1.12 ± 0.60 μg/mL) using Salkowski reagent. Vazquez et al. [45] and Sheng and He [46] methodologies were followed for determination of P (*A. fabrum* = 16.2 ± 1.48 and *B. amyloliquefaciens* = 20.9 ± 2.48 μg/mL) and K solubilizing activities (*A. fabrum* = 26.7 ± 1.49 and *B. amyloliquefaciens* = 23.4 ± 1.92 μg/mL) [22].

Initially, timber waste was sun-dried and then pyrolyzed at 389 °C in a pyrolyzer for 80 min [22]. pH and EC*e* of BC were also assessed following Danish and Zafar-ul-Hye [22]. Biochar was digested with di-acid (HNO$_3$: HClO$_4$) mixture [47] for determination of total P using a UV-VIS spectrophotometer (Model 6305, Jenway, UK) at 430 nm wavelength following Tandon et al. [48]. For the development of colour, ammonium molybdate and ammonium metavanadate were used [49]. The K and sodium were determined on flame photometer (Model EEL 410, Watford, UK) [49]. For assessing nitrogen, H$_2$SO$_4$ digestion [49] was carried out, followed by Kjeldahl's distillation [49]. Volatile matter and ash content of BC were measured by heating BC in a muffle furnace at 450 and 550 °C respectively. Fixed carbon was calculated following Ronsse et al. [50] (Table 1).

Table 1. Characteristics of soil and timber waste biochar (BC).

Soil	Unit	Value	Biochar	Unit	Value
Sand	%	55	pH	-	7.26
Silt	%	25	EC$_e$	dS m^{-1}	1.22
Clay	%	30	Volatile Matter	%	8.96
Texture	Sandy Clay Loam		Ash Content	%	28.9
pH$_s$	-	8.52	Fixed Carbon	%	62.1
EC$_e$	dS m^{-1}	3.69	Total N	%	0.21
Organic Matter	%	0.45	Total P	%	0.62
Total N	%	0.02	Total K	%	1.61
Extractable P	mg kg^{-1}	5.26	Total Na	%	0.19
Extractable K	mg kg^{-1}	170			

The hydrometer method was used for the textural class analysis of soil [51]. Using the United States Department of Agriculture triangle (USDA triangle), the textural class was assessed as "sandy clay loam". The Walkley method [52] was used for the determination of soil organic matter. The following equation was used to calculate organic soil N:

$$\text{Organic N (\%) = Soil Organic Matter}/20 \quad (1)$$

Extractable soil P was determined by the Olsen and Sommers methodology [53]. The Chapman and Pratt [47] protocol was followed for the determination of extractable K. The physiochemical characteristics of soil are provided in Table 1.

Wheat seeds (Glaxay-2013) were purchased from the Government of Punjab certified seed dealer. Weak seeds were initially screened out manually. Seeds sterilization was done with sodium hypochlorite (5%). Finally, seeds were ethanol (95%) washed thrice, followed by three times sterilized distilled water washings [54]. For inoculation, 100 g of sterilized seeds were inoculated with 1 mL of PGPR inoculum having optical density 0.5 at 535 nm wavelength along with 10% sugar (glucose) solution. After sticking of inoculum and sugar solution uniformly [55], seeds were top dressed in BC. Before inoculation of seeds, the BC was sterilized for 20 min at 121 °C in an autoclave [56]. For the control treatment, seed top dressing was also done with BC along with 10% sugar solution [57].

In each of the 18 plots (9 m^2), six rows of seeds were sown using the drill method. Four irrigations were applied according to the production technology of wheat recommended and published by the Directorate of Agricultural Information Punjab [58]. There was no precipitation event during the study period, therefore, no precipitation-induced soil moisture variations were monitored. To create a mild drought, three irrigations were applied (one irrigation was skipped at the tillering stage). However, severer drought stress was induced by using two irrigations (two irrigations were skipped; one at the tillering stage and other at the milky stage). The irrigation schedule was:

1st = 25 days after sowing (Crown root Initiation)
2nd = 55 days after sowing (Tillering stage)
3rd = 80 days after sowing (Heading stage)
4th = 110 days after sowing (Milky stage/soft dough)

After 65 days of sowing (vegetative phase), we collected vegetative samples from four random spots in each plot for the determination of chlorophyll contents, gas exchange attributes, electrolyte leakage and nutrient concentrations in the shoot. At the vegetative phase, samples were collected only from 4I (control) and 3I (mild drought) treatments (no 2I (severe drought) treatment was available at this point of time). Skipping one irrigation created mild drought treatment as compared to skipping two irrigations (2nd and 4th) which created severe drought treatment. The drought and control treatments were sampled at maturity point of time for estimating yield attributes.

We followed Kumar et al. [59] for root sampling and Newman [60] for root length measurement at 120 days after seeding. Briefly, an augar of 10 cm internal diameter was used and the core samples were taken at 10 cm depth intervals to a total depth of 90 cm. Random sampling locations within each plot included sampling at row and midway between rows for collecting four, 90 cm depth samples. Soil/root cores were placed on a 32 cm mesh screen and gently washed in water [59]. Root length was measured by the line intercept technique of Newman [59,60]. For yield attributes and grain analyses, harvesting was done at the time of maturity when soil and plants were fully dried. The plant height, spike length, grains spike^{-1}, spikelets spike^{-1}, 1000-grains weight, grains yield, straw yield and biological yield (aboveground + root biomass) data were collected at the time of maturity (approx. 140 days).

Leaf samples were digested in sulfuric acid for analysis of nitrogen on Kjeldahl's distillation apparatus [49]. Leaf P concentration was determined using the yellow colour development method and spectrophotometric absorbance at 420 nm [49]. Total K concentration in shoot and grain were found out by digesting the samples in di-acid (HNO_3-$HClO_4$) mixture [49] and using a flamephotometer (Model EEL 410, Watford, UK).

CI-340 Photosynthesis system (Bio Science Inc., WA, USA), Infra-Red Gas Analyzer (IRGA–EGM-4, PP Systems, USA) was used for assessment of net transpiration rate, stomatal conductance and photosynthetic rate [61]. On a sunny day, all readings were taken at a saturating intensity of light between 10:17 and 11:56 AM [62].

For the determination of photosynthetic pigments, the methodology of Arnon [63] was followed. Leaf samples were initially ground in a mortar by adding acetone (80%) solution. After that, absorbance was taken on a spectrophotometer at 645 and 663 nm wavelengths. Final chlorophyll contents were calculated by using equations;

$$\text{Chlorophyll a (mg/g)} = 12.7 \text{ (OD 663)} - 2.69 \text{ (OD 645) V}/(1000 \times W) \quad (2)$$

$$\text{Chlorophyll b (mg/g)} = 22.9 \text{ (OD 645)} - 4.68 \text{ (OD 663) V}/(1000 \times W) \quad (3)$$

$$\text{Total Chlorophyll (mg/g)} = \text{Chlorophyll a} + \text{Chlorophyll b} \quad (4)$$

where OD = Optical density (nm), V = Final volume made (mL), W = Fresh weight of sample (g).

The Lutts et al. [64] method was adopted for the determination of electrolyte leakage (EL). All the leaves samples were washed with deionized (DI) water for the removal of dust particles. After that, discs of uniform size were cut with a steel cylinder of 1 cm diameter. Finally, one gram of equal discs was dipped in a test tube containing 20 mL DI water and incubated at 25 °C for 24 h. First electrical conductivity (EC1) was determined using a pre-calibrated EC meter. Second EC (EC2) was noted after heating the test tubes at 120 °C for 20 min in a water bath. We calculated the final value of electrolyte leakage (EL) by using the equation:

$$\text{Electrolyte Leakage (\%)} = (EC1/EC2) \times 100 \quad (5)$$

Maximum increase (%) was calculated by using the formula:

$$\text{Maximum Increase (\%)} = (\text{Highest Value} - \text{Control treatment value}/\text{Control treatment value}) \times 100 \quad (6)$$

Statistical analysis was performed using standard statistical procedures as described by Steel and Torrie [65]. Two factorial ANOVA was applied on Statistix 8.1 software for determination of treatments significance under various levels of irrigations. Tukey's test at $p \leq 0.05$ was applied for comparison of the treatments.

3. Results

3.1. Plant Height, Root Length and Spike Length

Both the individual and interactive effects of T and I were significant on plant height and root length. For spike length, the main effects were significantly different while the interactive effects (T × I) remained nonsignificant. Application of BC, *A. fabrum* + BC and *B. amyloliquefaciens* + BC significantly improved plant height compared to control, with 4I and 2I (Table 2). The treatments *A. fabrum*, *B. amyloliquefaciens*, BC, *A. fabrum* + BC and *B. amyloliquefaciens* + BC differed significantly from control at 3I for plant height. A maximum increase of 0.31-fold in plant height was observed in *A. fabrum* + BC at 4I while 0.81-fold in *B. amyloliquefaciens* + BC with 2I from control. However, plant height was the maximum (0.42-fold) from control, in responses to *A. fabrum* + BC and *B. amyloliquefaciens* + BC treatments. For root length, the BC, *A. fabrum* + BC and *B. amyloliquefaciens* + BC differed significantly from control at 4I and 3I. The *B. amyloliquefaciens*, BC, *A. fabrum* + BC and *B. amyloliquefaciens* + BC were significantly better from control for root length with 2I (Table 2). Maximum increases, i.e., 0.49, 1.11 and 0.90-fold in root length were noted over control in *B. amyloliquefaciens* + BC with 4I, 3I and 2I, respectively. In the case of spike length, all the treatments were statistically alike but different from control (Table 2).

3.2. Grain, Straw and Biological Yield

Both the individual and interactive effects of T and I were significantly different for grain, straw and biological yield of wheat. The *A. fabrum* + BC and *B. amyloliquefaciens* + BC differed significantly from control for grain yield with 4I (Figure 1). Applications of *A. fabrum*, *B. amyloliquefaciens*, BC, *A. fabrum* + BC and *B. amyloliquefaciens* + BC differed significantly from control for grain yield at 3I. However, the BC, *A. fabrum* + BC and *B. amyloliquefaciens* + BC showed significantly better results over control for grain yield at 2I. The maximum increases, i.e., 0.29, 0.36 and 0.77-fold in grain yield were noted from control in *B. amyloliquefaciens* + BC with 4I, 3I and 2I, respectively. For straw yield, the *B. amyloliquefaciens*, BC, *A. fabrum* + BC and *B. amyloliquefaciens* + BC differed significantly from control with 3I and 2I (Figure 2). Maximum increases of 0.25, 0.50 and 0.75-fold in straw yield were noted from control in *B. amyloliquefaciens* + BC. In case of biological yield, the *B. amyloliquefaciens*, BC, *A. fabrum* + BC and *B. amyloliquefaciens* + BC differed significantly from control with 4I and 2I. From control, the *A. fabrum*, *B. amyloliquefaciens*, BC, *A. fabrum* + BC and *B. amyloliquefaciens* + BC differed significantly at

3I for biological yield (Figure 3). The maximum increases of 0.28, 0.40 and 0.77-fold in biological yield were noted from control in *B. amyloliquefaciens* + BC with 4I, 3I and 2I, respectively.

Table 2. Effect of *Agrobacterium fabrum*, *Bacillus amyloliquefaciens* with/without biochar (30 Mg ha^{-1}) on plant height, root length and spike length of wheat cultivated in drought-stressed field conditions.

Treatments	Plant Height (cm)				Root Length (cm)				Spike Length (cm)			
	IE (T × I)			ME (T)	IE (T × I)			ME (T)	IE (T × I)			ME (T)
	4I	3I	2I		4I	3I	2I		4I	3I	2I	
Control	59.0 d-f	48.1 g,h	33.0 i	46.7 D	8.82 d-f	5.65 h,i	4.46 i	6.31 D	5.75	4.86	4.34	4.98 B
A. fabrum	65.5 c-f	58.3 e,f	40.1 h,i	54.6 C	10.1 b-d	7.08 f-h	6.22 g-i	7.79 C	6.68	6.22	4.87	5.92 A
B. amyloliquefaciens	66.2 b-e	60.0 d-f	39.6 h,i	55.3 B,C	10.3 b-d	7.46 f-h	6.36 g,h	8.03 C	6.69	6.27	4.78	5.91 A
BC	71.2 a-c	60.9 d-f	45.6 h	59.2 B	11.5 a-c	9.67 c-e	7.81 e-g	9.67 B	6.67	6.46	4.96	6.03 A
A. fabrum + BC	76.8 a	68.3 a-d	56.5 f,g	67.2 A	12.8 a	11.2 a-c	8.46 d-f	10.8 A	7.13	6.33	5.10	6.19 A
B. amyloliquefaciens + BC	75.5 a,b	68.3 a-d	59.7 d-f	67.8 A	13.1 a	11.9 a,b	8.49 d-f	11.2 A	7.12	6.55	5.05	6.24 A
ME (I)	69.0 A	60.7 B	45.8 C		11.1 A	8.83 B	6.97 C		6.67 A	6.1 B	4.85 C	

Means sharing different letters are significantly different ($p \leq 0.05$). ME = indicates main effect; IE = interactive effect; 4I = 4 irrigations; 3I = 3 irrigations; 2I = 2 irrigations.

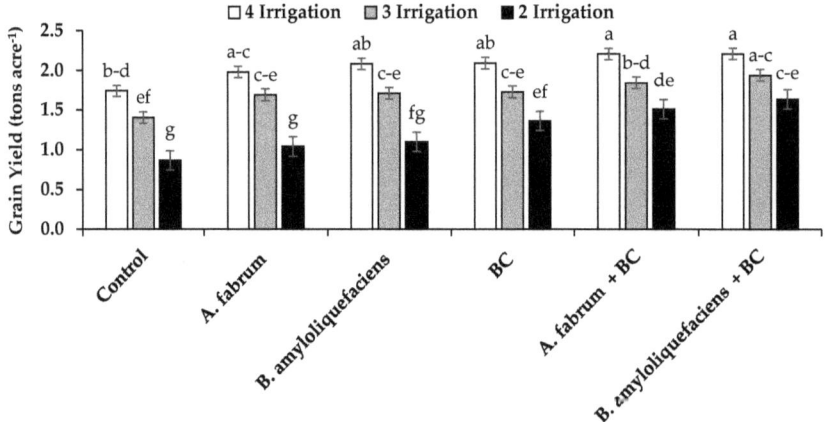

Figure 1. Effect of *Agrobacterium fabrum*, *Bacillus amyloliquefaciens* with/without biochar (30 Mg ha^{-1}) on grains yield (tons acre^{-1}) in wheat cultivated in drought-stressed field conditions.

3.3. Spikelets Spike^{-1}, Grains Spik^{-1} and 1000 Grain Weight

Main effects of T and I differed significantly for spikelets spike^{-1}, grains spike^{-1} and 1000 grain weight but the interaction (T × I) was significantly different only for 1000 grain weight. From control, the applications of *B. amyloliquefaciens*, BC, *A. fabrum* + BC and *B. amyloliquefaciens* + BC differed significantly from control for spikelets spike^{-1} (Table 3). The treatment *B. amyloliquefaciens* + BC differed significantly over BC and *B. amyloliquefaciens* for spikelets spike^{-1}. Similarly, *A. fabrum* + BC differed significantly as compared to *A. fabrum* but did not differ significantly as compared to BC for spikelets spike^{-1}. A maximum increase of 0.24-fold in spikelets spike^{-1} was noted from control in *B. amyloliquefaciens* + BC. In the case of grains spike^{-1}, the BC, *A. fabrum* + BC and *B. amyloliquefaciens* + BC were statistically alike but differed significantly from control (Table 3). Inoculation of *B. amyloliquefaciens* also differed significantly from control for grains spike^{-1}. A maximum increase of 0.51-fold in grains spike^{-1} was noted from control in *B. amyloliquefaciens* + BC. For 1000 grain weight, the *A. fabrum* + BC differed significantly from control with 4I. It was noted that *A. fabrum* + BC and *B. amyloliquefaciens* + BC differed significantly from control at 3I for 1000 grain weight (Table 3). However, the applications of *B. amyloliquefaciens*, BC, *A. fabrum* + BC and *B. amyloliquefaciens* + BC differed significantly from control with 2I for 1000 grain weight. A maximum increase of 0.20-fold in 1000 grain weight was noted as compared to control in *A. fabrum* + BC with 4I. With 3I, the application of *B. amyloliquefaciens* + BC gave a maximum increase of 0.29-fold as compared to control in 1000 grains weight. However, the BC and *A. fabrum* + BC gave a maximum rise of 0.46-fold as compared to control in 1000 grain weight with 2I.

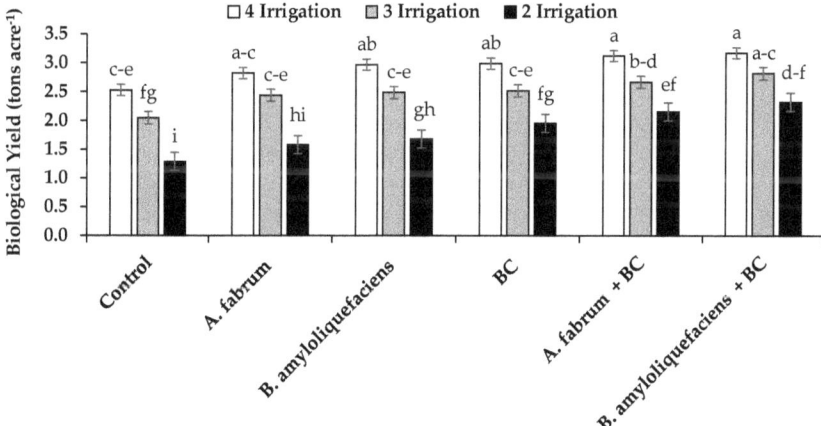

Figure 2. Effect of *Agrobacterium fabrum*, *Bacillus amyloliquefaciens* with/without biochar (30 Mg ha^{-1}) on straw yield (tons acre^{-1}) in wheat cultivated in drought-stressed field conditions.

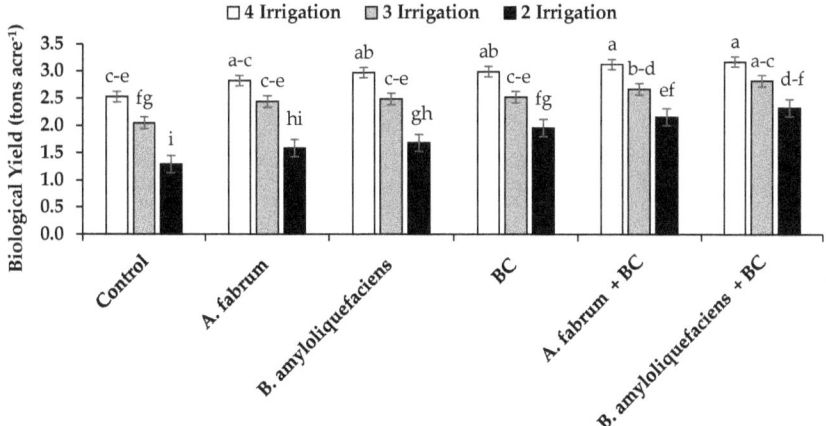

Figure 3. Effect of *Agrobacterium fabrum*, *Bacillus amyloliquefaciens* with/without biochar (30 Mg ha^{-1}) on biological yield (tons acre^{-1}) in wheat cultivated in drought-stressed field conditions.

Table 3. Effect of *Agrobacterium fabrum*, *Bacillus amyloliquefaciens* with/without biochar (30 Mg ha^{-1}) on spikelet's spike^{-1}, grains spike^{-1} and 1000 grains weight of wheat cultivated in drought-stressed field conditions.

Treatments	Spikelets spike^{-1}				Grains Spike^{-1}				1000 Grains Weight (g)			
	IE (T × I)			ME	IE (T × I)			ME	IE (T × I)			ME
	4I	3I	2I	(T)	4I	3I	2I	(T)	4I	3I	2I	(T)
Control	15.3	13.0	10.7	13.0 D	37.7	27.7	22.0	29.1 C	35.3 b–d	28.2 f,g	20.2 h	27.9 C
A. fabrum	16.0	13.7	13.0	14.2 C,D	38.3	33.7	26.7	32.9 B,C	34.2 c–f	30.7 c–g	25.5 g,h	30.2 B,C
B. amyloliquefaciens	16.0	13.3	12.7	14.0 C	39.3	35.0	29.7	34.7 B	35.2 b–e	31.0 c–g	26.8 g	31.0 B
BC	16.3	14.3	13.0	14.6 B,C	47.3	42.0	32.7	40.7 A	36.3 a–c	31.1 c–g	29.4 d–f	32.3 B
A. fabrum + BC	17.0	15.3	14.3	15.6 A,B	45.3	40.7	39.0	41.7 A	42.5 a	34.7 b–e	29.4 d–f	35.5 A
B. amyloliquefaciens + BC	17.3	15.7	15.3	16.1 A	49.0	43.0	39.3	43.8 A	40.8 a,b	36.5 a–c	28.9 e–f	35.4 A
ME (I)	16.3 A	14.2 B	13.2 C		42.8 A	37.0 B	31.6 C		37.4 A	32.0 B	26.7 C	

Means sharing different letters are significantly different ($p \leq 0.05$). ME = indicates main effect; IE = interactive effect; 4I = 4 irrigations; 3I = 3 irrigations; 2I = 2 irrigations.

3.4. N, P and K Concentration in Grains

Both the main and interactive effects of T and I were significant for N, P and K concentrations in wheat grains. All the treatments were statistically alike with 4I for grains N concentration. Applications of BC, *A. fabrum* + BC and *B. amyloliquefaciens* + BC performed significantly better from control with 3I and 2I for grains' N concentration (Table 4). The maximum increases of 0.13, 0.37 and 0.57-fold in grains' N concentration were noted in *B. amyloliquefaciens* + BC with 4I, 3I and 2I, respectively. In the case of grains' P concentration, *B. amyloliquefaciens*, BC, *A. fabrum* + BC and *B. amyloliquefaciens* + BC remained statistically alike but only *A. fabrum* + BC and *B. amyloliquefaciens* + BC differed significantly from control with 4I. The *A. fabrum*, *B. amyloliquefaciens*, BC, *A. fabrum* + BC and *B. amyloliquefaciens* + BC were significantly better from control with 3I and 2I for grains P concentration (Table 4). Both the *A. fabrum* + BC and *B. amyloliquefaciens* + BC showed a maximum increase of 0.32-fold in the grains' P concentration from control with 4I. However, with 3I and 2I, the *B. amyloliquefaciens* + BC gave the maximum increases of 0.91 and 1.64-fold in grains P concentration from control, respectively. For the grains' K concentration, the *A. fabrum*, *B. amyloliquefaciens*, BC, *A. fabrum* + BC and *B. amyloliquefaciens* + BC were statistically similar to each other while, the BC, *A. fabrum* + BC and *B. amyloliquefaciens* + BC differed significantly from control with 4I (Table 4). The *A. fabrum* + BC and *B. amyloliquefaciens* + BC differed significantly with 3I from control for the grains' K concentration. However, the *A. fabrum*, *B. amyloliquefaciens*, BC, *A. fabrum* + BC and *B. amyloliquefaciens* + BC differed significantly over control for the grains' K concentration with 2I. The maximum increases of 0.22, 0.27 and 0.61-fold in the grains' K concentration were noted in *B. amyloliquefaciens* + BC with 4I, 3I and 2I, respectively.

3.5. N, P and K Concentration in Shoot

Both the individual and interactive effects of T and I differed significantly for shoot nitrogen compared to only individual effects of T and I were significant for P and K concentrations in wheat. All treatments were statistically alike with 4I for shoot nitrogen concentration. The *A. fabrum* + BC, *B. amyloliquefaciens* + BC and BC differed significantly from control at 3I for shoot nitrogen concentration (Table 5). *A. fabrum* and *B. amyloliquefaciens* were non-significant over control for shoot nitrogen concentration. A maximum increase of 0.32-fold in shoot nitrogen concentration was noted from control with 3I in both the *B. amyloliquefaciens* + BC and *A. fabrum* + BC treatments. For P concentration in the shoot, B. amyloliquefaciens + BC differed significantly from control. A. fabrum and B. amyloliquefaciens and BC also differed substantially from control (Table 5). Maximum increases of 0.44-fold in shoot P concentration were noted from control in B. amyloliquefaciens + BC treatment. However, wheat cultivation with 4I gave 0.44-fold higher P shoot concentration from 3I. For shoot K concentration, A. fabrum + BC and B. amyloliquefaciens + BC remained statistically alike but significantly better from control. Both the A. fabrum and B. amyloliquefaciens inoculations also differed significantly from control for shoot K concentration. We observed that BC was significantly different from A. fabrum, B. amyloliquefaciens and control treatments for shoot K concentration (Table 5). Maximum increases of 0.51-fold in shoot K concentration were noted from control in A. fabrum + BC. However, wheat cultivation with 4I gave 0.13-fold higher K shoot concentration from 3I treatment.

Table 4. Effect of *Agrobacterium fabrum*, *Bacillus amyloliquefaciens* with/without biochar (30 Mg ha^{-1}) on nitrogen, phosphorus and potassium concentration in wheat grain cultivated in drought-stressed field conditions *.

Treatments	Grains Nitrogen (%)				Grains Phosphorus (%)				Grains Potassium (%)			
	IE (T × I)			ME (T)	IE (T × I)			ME (T)	IE (T × I)			ME (T)
	4I	3I	2I		4I	3I	2I		4I	3I	2I	
Control	2.58 a-e	1.99 h-j	1.54 k	2.04 D	0.66 d-g	0.43 i	0.25 j	0.44 D	0.50 b-e	0.45 e	0.31 f	0.42 D
A. fabrum	2.78 a,b	2.21 f-i	1.79 j,k	2.26 C	0.71 c-f	0.60 f-h	0.45 i	0.59 C	0.55 a-c	0.47 d,e	0.43 e	0.48 C
B. amyloliquefaciens	2.73 a-c	2.33 e-h	1.86 i-k	2.31 C	0.75 a-e	0.61 e-h	0.48 h,i	0.61 B,C	0.56 a-c	0.48 c-e	0.46 e	0.50 C
BC	2.74 a-c	2.55 b-f	2.13 g-j	2.47 B	0.78 a-d	0.68 c-g	0.54 g-i	0.67 B	0.58 a	0.50 b-e	0.46 e	0.51 B,C
A. fabrum + BC	2.87 a,b	2.68 a-d	2.33 d-h	2.63 A,B	0.87 a,b	0.73 b-f	0.62 e-h	0.74 A	0.60 a	0.54 a-d	0.49 c-e	0.55 A,B
B. amyloliquefaciens + BC	2.92 a	2.76 a-c	2.42 c-g	2.70 A	0.87 a	0.82 a-c	0.66 d-g	0.78 A	0.61 a	0.57 a,b	0.50 b-e	0.56 A
ME (I)	2.77 A	2.42 B	2.01 C		0.77 A	0.65 B	0.50 C		0.57 A	0.50 B	0.44 C	

Followed Danish and Zafar-ul-Hye, 2019 [22] for comparisons. Means sharing different letters are significantly different at $p \leq 0.05$. ME = indicates main effect; IE = interactive effect; 4I = 4 irrigations; 3I = 3 irrigations; 2I = 2 irrigations.

3.6. Gas Exchange Attributes

Main effects of T and I differed significantly but the interactive effect (T × I) was non-significant for photosynthetic rate and stomatal conductance. For the photosynthetic rate, the BC, A. fabrum + BC and B. amyloliquefaciens + BC were statistically alike but differed significantly from control (Table 6). Applications of A. fabrum and B. amyloliquefaciens were non-significant from control for the photosynthetic rate. A maximum increase of 0.48-fold in the photosynthetic rate was observed from control in B. amyloliquefaciens + BC treatment. However, wheat cultivation with 4I showed 0.35-fold higher photosynthetic rate from 3I. In case of transpiration rate, A. fabrum + BC and B. amyloliquefaciens + BC were statistically similar to each other but differed significantly from control. A. fabrum, B. amyloliquefaciens and BC proved significantly better treatments from control for transpiration rate (Table 6). A maximum increase of 0.81-fold in the rate of transpiration was noted from control in B. amyloliquefaciens + BC. However, wheat cultivation with 4I showed 0.32-fold higher transpiration rate from 3I. For stomatal conductance, the BC, A. fabrum + BC and B. amyloliquefaciens + BC treatments remained statistically alike but were significantly different from A. fabrum and control (Table 6). The A. fabrum and B. amyloliquefaciens were statistically similar to control for stomatal conductance. The maximum increases of 0.42-fold in stomatal conductance were noted from control in B. amyloliquefaciens + BC treatment. However, wheat cultivation with 4I showed 0.24-fold higher stomatal conductance from 3I.

Table 5. Effect of *Agrobacterium fabrum*, *Bacillus amyloliquefaciens* with/without biochar (30 Mg ha^{-1}) on nitrogen, phosphorus and potassium concentration in wheat shoot cultivated in drought-stressed field conditions *.

Treatments	Shoot Nitrogen (%)			Shoot Phosphorus (%)			Shoot Potassium (%)		
	IE (T × I)		ME (T)	IE (T × I)		ME (T)	IE (T × I)		ME (T)
	4I	3I		4I	3I		4I	3I	
Control	1.88 a	1.33 c	1.60 B	0.42	0.26	0.34 C	1.85	1.56	1.71 D
A. fabrum	1.85 a	1.58 b,c	1.71 A,B	0.46	0.36	0.41 B	2.20	1.97	2.09 C
B. amyloliquefaciens	1.85 a	1.57 b,c	1.71 A,B	0.45	0.38	0.42 B	2.20	2.01	2.11 C
BC	1.88 a	1.59 b	1.73 A,B	0.48	0.42	0.45 A,B	2.56	2.17	2.37 B
BC + A. fabrum	1.93 a	1.75 a,b	1.84 A	0.51	0.43	0.47 A	2.69	2.47	2.58 A
BC + B. amyloliquefaciens	1.94 a	1.75 a,b	1.85 A	0.53	0.45	0.49 A	2.68	2.44	2.56 A
ME (I)	1.89 A	1.59 B		0.47 A	0.39 B		2.37 A	2.10 B	

* Followed Danish and Zafar-ul-Hye, 2019 [22] for comparisons. Means sharing different letters are significantly different at $p \leq 0.05$. Means sharing no letters are non-significant at $p \leq 0.05$. ME = indicates main effect; IE = interactive effect; 4I = 4 irrigations; 3I = 3 irrigations.

Table 6. Effect of *Agrobacterium fabrum*, *Bacillus amyloliquefaciens* with/without biochar (30 Mg ha^{-1}) on gas exchange attributes of wheat cultivated in drought-stressed field conditions *.

Treatments	Photosynthetic Rate (μmol (CO$_2$) m^{-2} s^{-1})			Transpiration Rate (mmol (H$_2$O) m^{-2} s^{-1})			Stomatal Conductance (μmol (CO$_2$) m^{-2} s^{-1})		
	IE (T × I)		ME (T)	IE (T × I)		ME (T)	IE (T × I)		ME (T)
	4I	3I		4I	3I		4I	3I	
Control	14.5	9.07	11.8 C	4.35	2.97	3.66 D	150.7	105.3	128.0 C
A. fabrum	16.1	10.7	13.4 C	4.90	4.23	4.56 C	148.3	125.7	137.0 C
B. amyloliquefaciens	15.9	10.1	13.0 B,C	5.41	4.17	4.79 B,C	166.7	127.7	147.2 B,C
BC	17.4	13.5	15.5 A,B	6.27	4.64	5.46 B	181.3	145.0	163.2 A,B
A. fabrum + BC	18.6	15.6	17.1 A	7.35	5.85	6.60 A	193.0	156.3	174.7 A
B. amyloliquefaciens + BC	19.1	15.8	17.5 A	7.86	5.43	6.64 A	193.3	172.3	182.8 A
ME (I)	16.9 A	12.5 B		6.02 A	4.55 B		172.2 A	138.7 B	

* Followed Danish and Zafar-ul-Hye, 2019 [22] for comparisons. Means sharing different letters are significantly different at $p \leq 0.05$. Means sharing no letters are non-significant at $p \leq 0.05$. ME = indicates main effect; IE = interactive effect; 4I = 4 irrigations; 3I = 3 irrigations.

3.7. Chlorophyll Content

The main effects of T and I were significantly different but interaction (T × I) was non-significant for chlorophyll a, chlorophyll b and total chlorophyll contents in wheat leaves. In the case of chlorophyll a, the *A. fabrum* + BC and *B. amyloliquefaciens* + BC were statistically similar, while both differed significantly from control (Table 7). The BC also differed significantly from control for chlorophyll a content. The *A. fabrum* and *B. amyloliquefaciens* did not vary significantly from control for chlorophyll a content. A maximum increase of 0.40-fold in chlorophyll a was noted in *B. amyloliquefaciens* + BC treatment over control. However, wheat cultivation with 4I showed 0.15-fold higher chlorophyll a content from 3I. For chlorophyll b, the *B. amyloliquefaciens* + BC and *A. fabrum* + BC treatments differed significantly from control. The BC also differed significantly from control for chlorophyll b. While, *A. fabrum* and *B. amyloliquefaciens*, did not differ significantly from control for chlorophyll b, maximum increase of 0.42-fold in chlorophyll b was noted from control in *B. amyloliquefaciens* + BC treatment. However, wheat cultivation with 4I showed 0.20-fold higher chlorophyll b content from 3I. In case of total chlorophyll, *A. fabrum* + BC and *B. amyloliquefaciens* + BC differed significantly from control. Inoculation of *A. fabrum* and *B. amyloliquefaciens* did not vary significantly but BC was significant from control for total chlorophyll (Table 7). A maximum increase of 0.41-fold in total chlorophyll was noted over control due to *B. amyloliquefaciens* + BC application. However, wheat cultivation with 4I showed 0.17-fold higher total chlorophyll content from 3I.

Table 7. Effect of *Agrobacterium fabrum*, *Bacillus amyloliquefaciens* with/without biochar (30 Mg ha^{-1}) on photosynthetic pigments synthesis in wheat leaves cultivated in drought-stressed field conditions *.

Treatments	Chlorophyll a (mg g^{-1})			Chlorophyll b (mg g^{-1})			Total Chlorophyll (mg g^{-1})			Electrolyte Leakage (%)		
	IE (T × I)		ME (T)	IE (T × I)		ME (T)	IE (T × I)		ME (T)	IE (T × I)		ME (T)
	4I	3I		4I	3I		4I	3I		4I	3I	
	No. of Irrigations (I)											
Control	0.87	0.68	0.77 C	0.42	0.34	0.38 C	1.29	1.02	1.15 C	41.0	59.3	50.2 A
A. fabrum	0.91	0.78	0.85 B,C	0.47	0.39	0.43 B,C	1.38	1.17	1.28 B,C	40.3	55.3	47.8 A,B
B. amyloliquefaciens	0.90	0.78	0.84 B,C	0.48	0.38	0.43 B,C	1.37	1.16	1.27 B,C	41.3	54.0	47.7 A,B
BC	0.99	0.85	0.92 B	0.48	0.42	0.45 B	1.47	1.27	1.37 B	41.0	47.0	44.0 A,B
A. fabrum + BC	1.16	0.98	1.07 A	0.53	0.45	0.49 A,B	1.68	1.44	1.56 A	39.0	41.0	40.0 B
B. amyloliquefaciens + BC	1.10	1.06	1.08 A	0.59	0.49	0.54 A	1.69	1.55	1.62 A	37.0	42.3	39.7 B
ME (I)	0.99 A	0.86 B		0.49 A	0.41 B		1.48 A	1.27 B		39.9 B	49.8 A	

* Followed Danish and Zafar-ul-Hye, 2019 [22] for comparisons. Means sharing different letters are significantly different at $p \leq 0.05$. Means sharing no letters are non-significant at $p \leq 0.05$. ME = indicates main effect; IE = interactive effect; 4I = 4 irrigations; 3I = 3 irrigations.

3.8. Electrolyte Leakage

Main effects of T and I were significantly different from control for electrolyte leakage. The *A. fabrum* + BC and *B. amyloliquefaciens* + BC treatments differed significantly from control for electrolyte leakage (Table 7). The *A. fabrum*, *B. amyloliquefaciens* and BC were statistically similar to control for electrolyte leakage. The *B. amyloliquefaciens* + BC exhibited significant reduction, i.e., 0.21-fold in electrolyte leakage compared to control. However, with 4I application wheat plants showed a significant reduction (0.20-fold) in electrolyte leakage from 3I.

4. Discussion

The sole application of BC under 2I significantly improved the root length and grain yield of wheat as compared to control. Biochar is frequently reported to have very high pore volume, water holding and cation exchange capacities, e.g., [66], and such properties stimulate root growth and facilitate better water and nutrient uptakes resulting in improved vegetative and reproductive growth [67,68]. Significantly greater K concentrations in the shoot and grain and improved plant yield in this field and in an earlier pot study [22] have validated the reportedly productive characteristics of BC. However, the specific objective of this study was to investigate and present the cumulative role of drought tolerant ACC-deaminase producing PGPR and BC in mitigating drought stress in wheat crop under field conditions.

Combined application of ACC-deaminase producing PGPR *Agrobacterium fabrum* or *Bacillus amyloliquefaciens* and timber-waste BC significantly improved the growth and yield of field grown wheat under mild (3I) and severe drought (2I) conditions. Our field study results validate earlier pot study results of improved growth and yield in response to comparable drought conditions [22–24]. Both the PGPR strains, *A. fabrum* and *B. amyloliquefaciens* along with BC significantly enhanced root length and plant height compared to those under control condition. Similar results were also observed in a previous pot study where *A. fabrum* and *B. amyloliquefaciens* significantly improved morphological growth attributes in wheat under drought stress [22]. As the *A. fabrum, B. amyloliquefaciens* were capable of producing ACC-deaminase, improvement in root length and plant height might be due to a reduction in ethylene level.

According to Mayak et al. [29], raised level of 1-aminocyclopropane-1-carboxylic acid (ACC) in plants exposed to drought, raises ethylene concentration in root and shoot of plants. Roots secrete accumulated ACC into rhizosphere which is cleaved by PGPR secreted ACC-deaminase into NH_3 and α-ketobutyrate, and ultimately ethylene level decreases. The decrease in ethylene concentration results in better root coverage, which results in improvements in the uptake of water and nutrients due to the enhanced rhizospheric area [69].

Significant improvements in grain yield, photosynthetic rate, transpiration rate, stomatal conductance chlorophyll a, chlorophyll b and total chlorophyll validated the enhanced functioning of the *A. fabrum* and *B. amyloliquefaciens* when applied in combination with BC, as compared to using the same rhizobacteria without BC [25]. Secretion of growth hormone, i.e., IAA by the *A. fabrum* and *B. amyloliquefaciens* and greater water holding capacity of BC in addition to ACC-deaminase production are the allied factors responsible for the improvement in wheat growth. The findings of previous pot studies also support this argument [22–24]. Xie et al. [70] described IAA as a co-factor, playing a crucial role in crop growth enhancement. Moreover, increases in surface area and length of lateral and adventitious roots due to high IAA secretion by PGPR play a vital role in better nutrient uptake [71].

This study finds that both the *A. fabrum* and *B. amyloliquefaciens* were solubilizing P and K, which may explain why grain and shoot P and K concentrations were significantly improved [72,73] with and without BC. Also, the increases in N, P and K contents in shoot and grain in responses to BC (without rhizobacteria) might be due to the retention of N and presences of P and K in BC. Improvement in cation exchange sites through BC addition also increases the retentions of mobile nutrients like N [74], thus, enhancing its bioavailability by decreasing leaching and volatilization losses [75]. Significant improvements in total chlorophyll, chlorophyll a, and chlorophyll b in the current study were probably due to better uptake of N.

Singh et al. [76] stated that greater K concentration in BC ash also contributed to better K uptake. Improvement in K concentration might have maintained the cell turgor pressure and regulated the stomatal conductance by osmoregulation [77,78]. Novak et al. [79] and Lehmann et al. [80] also observed a significant improvement in water holding capacity of soil where BC was applied. The greater surface area and pore spaces of BC facilitate the retention of water when used in soil [33,79–81]. According to Singh et al. [76], the organic carbon in BC significantly facilitates PGPR for improvement in their growth. Danish and Zafar-ul-Hye [22] also documented the synergistic effects of PGPR and BC against drought. They argued that root elongation and retention of water and nutrients by PGPR and BC respectively create a favourable environment in rhizosphere for plants to perform better under drought. Specifically, significant increases in growth and yield of wheat through the co-application of both ACC-deaminase PGPR (*A. fabrum* and *B. amyloliquefaciens*) along with BC might be due to better survivability, activity and proliferation of PGPR in combination with the water and nutrients holding potentials of BC under 3I and 2I.

5. Conclusions

Combined application of PGPR and biochar more effectively mitigates drought impacts as compared to individual PGPR inoculation or BC application, in field-grown wheat crop. Specifically, soil application of drought-tolerant ACC-deaminase producing PGPR *Agrobacterium fabrum* or *Bacillus amyloliquefaciens*, in addition to timber waste BC (30 tons ha^{-1}), significantly promotes growth and yield traits of wheat under field drought conditions.

Author Contributions: S.D. conducted research, prepared biochar, collected data and wrote manuscript; M.Z.H. designed and supervised experiment and wrote manuscript. M.A. and M.A. assisted in investigation and methodology validation. T.M.M. supported research, reviewed/edited article and collaborated funding.

Acknowledgments: First author is grateful to Higher Education Commission (HEC) of Pakistan to provide Indigenous Scholarship (PIN: 315-8403-2AV3-049) for his Ph.D. study in Pakistan.

Conflicts of Interest: The authors declare no conflict of interest.

References

1. Bos, C.; Juillet, B.; Fouillet, H.; Turlan, L.; Daré, S.; Luengo, C.; N'tounda, R.; Benamouzig, R.; Gausserès, N.; Tomé, D.; et al. Postprandial metabolic utilization of wheat protein in humans. *Am. J. Clin. Nutr.* **2005**, *81*, 87–94. [CrossRef] [PubMed]
2. Shiferaw, B.; Smale, M.; Braun, H.-J.; Duveiller, E.; Reynolds, M.; Muricho, G. Crops that feed the world 10. Past successes and future challenges to the role played by wheat in global food security. *Food Secur.* **2013**, *5*, 291–317. [CrossRef]
3. Ahmad, Z.; Waraich, E.A.; Akhtar, S.; Anjum, S.; Ahmad, T.; Mahboob, W.; Hafeez, O.B.A.; Tapera, T.; Labuschagne, M.; Rizwan, M. Physiological responses of wheat to drought stress and its mitigation approaches. *Acta Physiol. Plant.* **2018**, *40*, 80. [CrossRef]
4. Kilic, H.; Yağbasanlar, T. The effect of drought stress on grain yield, yield components and some quality traits of durum wheat (*Triticum turgidum* ssp. Durum) cultivars. *Not. Bot. Horti Agrobot. Clujnapoca* **2010**, *38*, 164–170.
5. Waraich, E.A.; Ahmad, R.; Ashraf, M. Role of mineral nutrition in alleviation of drought stress in plants. *Aust. J. Crop Sci.* **2011**, *5*, 764–777.
6. Griffin, M.T.; Montz, B.E.; Arrigo, J.S. Evaluating climate change induced water stress: A case study of the lower cape fear basin, nc. *Appl. Geogr.* **2013**, *40*, 115–128. [CrossRef]
7. Munir, T.; Perkins, M.; Kaing, E.; Strack, M. Carbon dioxide flux and net primary production of a boreal treed bog: Responses to warming and water-table-lowering simulations of climate change. *Biogeosciences* **2015**, *12*, 1091–1111. [CrossRef]
8. Bechtold, M.; De Lannoy, G.; Koster, R.; Reichle, R.; Mahanama, S.; Bleuten, W.; Bourgault, M.; Brümmer, C.; Burdun, I.; Desai, A.; et al. Peat-clsm: A specific treatment of peatland hydrology in the nasa catchment land surface model. *J. Adv. Model. Earth Syst.* **2019**. [CrossRef]
9. Saikia, J.; Sarma, R.K.; Dhandia, R.; Yadav, A.; Bharali, R.; Gupta, V.K.; Saikia, R. Alleviation of drought stress in pulse crops with acc deaminase producing rhizobacteria isolated from acidic soil of northeast india. *Sci. Rep.* **2018**, *8*, 3560. [CrossRef]
10. Zhang, S.; Kang, H.; Yang, W. Climate change-induced water stress suppresses the regeneration of the critically endangered forest tree nyssa yunnanensis. *PLoS ONE* **2017**, *12*, e0182012. [CrossRef]
11. Strack, M.; Munir, T.M.; Khadka, B. Shrub abundance contributes to shifts in dissolved organic carbon concentration and chemistry in a continental bog exposed to drainage and warming. *Ecohydrology* **2019**. [CrossRef]
12. Preston, K.L.; Rotenberry, J.T.; Redak, R.A.; Allen, M.F. Habitat shifts of endangered species under altered climate conditions: Importance of biotic interactions. *Glob. Chang. Biol.* **2008**, *14*, 2501–2515. [CrossRef]
13. Allen, M.R.; Barros, V.R.; Broome, J.; Cramer, W.; Christ, R.; Church, J.A.; Clarke, L.; Dahe, Q.; Dasgupta, P.; Dubash, N.K. IPCC Fifth Assessment Synthesis Report-Climate Change 2014 Synthesis Report. 2014. Available online: https://ar5-syr.ipcc.ch/ (accessed on 28 June 2019).
14. Mishra, V.; Cherkauer, K.A. Retrospective droughts in the crop growing season: Implications to corn and soybean yield in the midwestern united states. *Agric. Meteorol.* **2010**, *150*, 1030–1045. [CrossRef]
15. Fahad, S.; Bajwa, A.A.; Nazir, U.; Anjum, S.A.; Farooq, A.; Zohaib, A.; Sadia, S.; Nasim, W.; Adkins, S.; Saud, S. Crop production under drought and heat stress: Plant responses and management options. *Front. Plant Sci.* **2017**, *8*, 1147. [CrossRef] [PubMed]
16. Reddy, S.K.; Liu, S.; Rudd, J.C.; Xue, Q.; Payton, P.; Finlayson, S.A.; Mahan, J.; Akhunova, A.; Holalu, S.V.; Lu, N. Physiology and transcriptomics of water-deficit stress responses in wheat cultivars tam 111 and tam 112. *J. Plant Physiol.* **2014**, *171*, 1289–1298. [CrossRef] [PubMed]
17. Munir, T.M.; Khadka, B.; Xu, B.; Strack, M. Mineral nitrogen and phosphorus pools affected by water table lowering and warming in a boreal forested peatland. *Ecohydrology* **2017**, *10*, e1893. [CrossRef]
18. Wang, W.; Vinocur, B.; Altman, A. Plant responses to drought, salinity and extreme temperatures: Towards genetic engineering for stress tolerance. *Planta* **2003**, *218*, 1–14. [CrossRef] [PubMed]

19. Zafar-Ul-Hye, M.; Shahjahan, A.; Danish, S.; Abid, M.; Qayyum, M.F. Mitigation of cadmium toxicity induced stress in wheat by acc-deaminase containing pgpr isolated from cadmium polluted wheat rhizosphere. *Pak. J. Bot.* **2018**, *50*, 1727–1734.
20. Sharp, R.E.; LeNoble, M.E. Aba, ethylene and the control of shoot and root growth under water stress. *J. Exp. Bot.* **2002**, *53*, 33–37. [CrossRef]
21. Danish, S.; Zafar-ul-Hye, M.; Hussain, M.; Shaaban, M.; Núñez-Delgado, A.; Hussain, S.; Qayyum, M.F. Rhizobacteria with ACC-deaminase activity improve nutrient uptake, chlorophyll contents and early seedling growth of wheat under PEG-induced osmotic stress. *Intl. J. Agric. Biol.* **2019**, *21*, 1212–1220.
22. Danish, S.; Zafar-ul-Hye, M. Co-application of acc-deaminase producing pgpr and timber-waste biochar improves pigments formation, growth and yield of wheat under drought stress. *Sci. Rep.* **2019**, *9*, 5999. [CrossRef] [PubMed]
23. Stromberger, M.E.; Abduelafez, I.; Byrne, P.; Elamari, A.A.; Manter, D.K.; Weir, T. Genotype-specific enrichment of 1-aminocyclopropane-1-carboxylic acid deaminase-positive bacteria in winter wheat rhizospheres. *Soil Sci. Soc. Am. J.* **2017**, *81*, 796–805. [CrossRef]
24. Salem, G.; Stromberger, M.E.; Byrne, P.F.; Manter, D.K.; El-Feki, W.; Weir, T.L. Genotype-specific response of winter wheat (*Triticum aestivum* L.) to irrigation and inoculation with acc deaminase bacteria. *Rhizosphere* **2018**, *8*, 1–7. [CrossRef]
25. Kumputa, S.; Vityakon, P.; Saenjan, P.; Lawongsa, P. Carbonaceous greenhouse gases and microbial abundance in paddy soil under combined biochar and rice straw amendment. *Agronomy* **2019**, *9*, 228. [CrossRef]
26. Basak, B.B.; Biswas, D.R. Co-inoculation of potassium solubilizing and nitrogen fixing bacteria on solubilization of waste mica and their effect on growth promotion and nutrient acquisition by a forage crop. *Biol. Fertil. Soils* **2010**, *46*, 641–648. [CrossRef]
27. Mwajita, M.R.; Murage, H.; Tani, A.; Kahangi, E.M. Evaluation of rhizosphere, rhizoplane and phyllosphere bacteria and fungi isolated from rice in kenya for plant growth promoters. *SpringerPlus* **2013**, *2*, 606. [CrossRef]
28. Belimov, A.A.; Safronova, V.I.; Sergeyeva, T.A.; Egorova, T.N.; Matveyeva, V.A.; Tsyganov, V.E.; Borisov, A.Y.; Tikhonovich, I.A.; Kluge, C.; Preisfeld, A. Characterization of plant growth promoting rhizobacteria isolated from polluted soils and containing 1-aminocyclopropane-1-carboxylate deaminase. *Can. J. Microbiol.* **2001**, *47*, 642–652. [CrossRef]
29. Mayak, S.; Tirosh, T.; Glick, B.R. Plant growth-promoting bacteria confer resistance in tomato plants to salt stress. *Plant Physiol. Biochem.* **2004**, *42*, 565–572. [CrossRef]
30. Zahir, Z.A.; Munir, A.; Asghar, H.N.; Shaharoona, B.; Arshad, M. Effectiveness of rhizobacteria containing acc deaminase for growth promotion of peas (*Pisum sativum*) under drought conditions. *J. Microbiol. Biotechnol.* **2008**, *18*, 958–963.
31. Gundale, M.J.; DeLuca, T.H. Charcoal effects on soil solution chemistry and growth of koeleria macrantha in the ponderosa pine/douglas-fir ecosystem. *Biol. Fertil. Soils* **2006**, *43*, 303–311. [CrossRef]
32. Hartmann, M.; Fliessbach, A.; Oberholzer, H.R.; Widmer, F. Ranking the magnitude of crop and farming system effects on soil microbial biomass and genetic structure of bacterial communities. *FEMS Microbiol. Ecol.* **2006**, *57*, 378–388. [CrossRef] [PubMed]
33. De Jesus Duarte, S.; Glaser, B.; Pellegrino Cerri, C.E. Effect of biochar particle size on physical, hydrological and chemical properties of loamy and sandy tropical soils. *Agronomy* **2019**, *9*, 165. [CrossRef]
34. Wacal, C.; Ogata, N.; Basalirwa, D.; Handa, T.; Sasagawa, D.; Acidri, R.; Ishigaki, T.; Kato, M.; Masunaga, T.; Yamamoto, S.; et al. Growth, seed yield, mineral nutrients and soil properties of sesame (*Sesamum indicum* L.) as influenced by biochar addition on upland field converted from paddy. *Agronomy* **2019**, *9*, 55. [CrossRef]
35. De la Rosa, J.M.; Rosado, M.; Paneque, M.; Miller, A.Z.; Knicker, H. Effects of aging under field conditions on biochar structure and composition: Implications for biochar stability in soils. *Sci. Total Environ.* **2018**, *613*, 969–976. [CrossRef] [PubMed]
36. Paneque, M.; De la Rosa, J.M.; Kern, J.; Reza, M.; Knicker, H. Hydrothermal carbonization and pyrolysis of sewage sludges: What happen to carbon and nitrogen? *J. Anal. Appl. Pyrolysis* **2017**, *128*, 314–323. [CrossRef]
37. Lehmann, J. Bio-energy in the black. *Front. Ecol. Environ.* **2007**, *5*, 381–387. [CrossRef]
38. Singh, B.; Singh, B.P.; Cowie, A.L. Characterisation and evaluation of biochars for their application as a soil amendment. *Aust. J. Soil Res.* **2010**, *48*, 516–525. [CrossRef]

39. Hagemann, N.; Subdiaga, E.; Orsetti, S.; De la Rosa, J.M.; Knicker, H.; Schmidt, H.-P.; Kappler, A.; Behrens, S. Effect of biochar amendment on compost organic matter composition following aerobic composting of manure. *Sci. Total Environ.* **2018**, *613*, 20–29. [CrossRef]
40. Sarfraz, M.; Mehdi, S.; Abid, M.; Akram, M. External and internal phosphorus requirement of wheat in bhalike soil series of pakistan. *Pak. J. Bot.* **2008**, *40*, 2031–2040.
41. Ahmad, I.; Bibi, F.; Ullah, H.; Munir, T.M. Mango fruit yield and critical quality parameters respond to foliar and soil applications of zinc and boron. *Plants* **2018**, *7*, 97. [CrossRef]
42. Dworkin, M.; Foster, J.W. Experiments with some microorganisms which utilize ethane and hydrogen. *J. Bacteriol.* **1958**, *75*, 592–603. [PubMed]
43. El-Tarabily, K.A. Promotion of tomato (*Lycopersicon esculentum* mill.) plant growth by rhizosphere competent 1-aminocyclopropane-1-carboxylic acid deaminase-producing streptomycete actinomycetes. *Plant Soil* **2008**, *308*, 161–174. [CrossRef]
44. Glickmann, E.; Dessaux, Y. A critical examination of the specificity of the salkowski reagent for indolic compounds produced by phytopathogenic bacteria. *Appl. Environ. Microbiol.* **1995**, *61*, 793–796. [PubMed]
45. Vazquez, P.; Holguin, G.; Puente, M.E.; Lopez-Cortes, A.; Bashan, Y. Phosphate-solubilizing microorganisms associated with the rhizosphere of mangroves in a semiarid coastal lagoon. *Biol. Fertil. Soils* **2000**, *30*, 460–468. [CrossRef]
46. Sheng, X.F.; He, L.Y. Solubilization of potassium-bearing minerals by a wild-type strain of bacillus edaphicus and its mutants and increased potassium uptake by wheat. *Can. J. Microbiol.* **2006**, *52*, 66–72. [CrossRef] [PubMed]
47. Chapman, H.; Pratt, P. *Methods of Analysis for Soils, Plants and Waters*; University of California, Division of Agricultural Sciences: Oakland, CA, USA, 1961; pp. 169–176.
48. Tandon, H.; Cescas, M.; Tyner, E. An acid-free vanadate-molybdate reagent for the determination of total phosphorus in soils 1. *Soil Sci. Soc. Am. J.* **1968**, *32*, 48–51. [CrossRef]
49. Jones, J.B., Jr.; Wolf, B.; Mills, H.A. *Plant Analysis Handbook. A Practical Sampling, Preparation, Analysis, and Interpretation Guid*; Micro-Macro Publishing, Inc.: Athens, GA, USA, 1991.
50. Ronsse, F.; Van Hecke, S.; Dickinson, D.; Prins, W. Production and characterization of slow pyrolysis biochar: Influence of feedstock type and pyrolysis conditions. *Gcb Bioenergy* **2013**, *5*, 104–115. [CrossRef]
51. Gee, G.; Bauder, J. Particle-size analysis. *Agronomy* **1986**, *4*, 255–293.
52. Walkley, A. An examination of methods for determining organic carbon and nitrogen in soils. *J. Agric. Sci.* **1935**, *25*, 598–609. [CrossRef]
53. Olsen, S.R.; Sommers, L.E. Phosphorus. In *Method of Soil Analysis*, 2nd ed.; American Society of Agronomy: Madison, WI, USA, 1982; pp. 403–430.
54. Ahmad, I.; Akhtar, M.J.; Zahir, Z.A.; Naveed, M.; Mitter, B.; Sessitsch, A. Cadmium-tolerant bacteria induce metal stress tolerance in cereals. *Environ. Sci. Pollut. R.* **2014**, *21*, 11054–11065. [CrossRef]
55. Ahmad, M.T.; Asghar, N.; Saleem, M.; Khan, M.Y.; Zahir, Z.A. Synergistic effect of rhizobia and biochar on growth and physiology of maize. *Agron. J.* **2015**, *107*, 2327–2334. [CrossRef]
56. Nadeem, S.M.; Zahir, Z.A.; Naveed, M.; Arshad, M. Rhizobacteria containing acc-deaminase confer salt tolerance in maize grown on salt-affected fields. *Can. J. Microbiol.* **2009**, *55*, 1302–1309. [CrossRef] [PubMed]
57. Shaharoona, B.; Arshad, M.; Zahir, Z.A. Effect of plant growth promoting rhizobacteria containing acc-deaminase on maize (*Zea mays* L.) growth under axenic conditions and on nodulation in mung bean (vigna radiata L.). *Lett. Appl. Microbiol.* **2006**, *42*, 155–159. [CrossRef] [PubMed]
58. Sheikh, A.; Rehman, T.; Yates, C. Logit models for identifying the factors that influence the uptake of new 'no-tillage'technologies by farmers in the rice–wheat and the cotton–wheat farming systems of pakistan's punjab. *Agric. Syst.* **2003**, *75*, 79–95. [CrossRef]
59. Kumar, K.; Prihar, S.; Gajri, P. Determination of root distribution of wheat by auger sampling. *Plant Soil* **1993**, *149*, 245–253. [CrossRef]
60. Newman, E. A method of estimating the total length of root in a sample. *J. Appl. Ecol.* **1966**, *3*, 139–145. [CrossRef]
61. Munir, T.; Khadka, B.; Xu, B.; Strack, M. Partitioning forest-floor respiration into source based emissions in a boreal forested bog: Responses to experimental drought. *Forests* **2017**, *8*, 75. [CrossRef]

62. Tekalign, T.; Hammes, P. Growth and productivity of potato as influenced by cultivar and reproductive growth: I. Stomatal conductance, rate of transpiration, net photosynthesis, and dry matter production and allocation. *Sci. Hortic.* **2005**, *105*, 13–27. [CrossRef]
63. Arnon, D.I. Copper enzymes in isolated chloroplasts:Polyphenoloxidase in beta vulgaris. *Plant Physiol.* **1949**, *24*, 1–15. [CrossRef]
64. Lutts, S.; Kinet, J.M.; Bouharmont, J. Nacl-induced senescence in leaves of rice (*oryza sativa* L.) cultivars differing in salinity resistance. *Ann. Bot.* **1996**, *78*, 389–398. [CrossRef]
65. Steel, R.G.D.; Torrie, J.H. *Principles and Procedures of Statistics, a Biometrical Approach*; McGraw-Hill Kogakusha, Ltd.: New York, NY, USA, 1980.
66. Horel, Á.; Tóth, E.; Gelybó, G.; Dencső, M.; Farkas, C. Biochar amendment affects soil water and CO_2 regime during capsicum annuum plant growth. *Agronomy* **2019**, *9*, 58. [CrossRef]
67. Sánchez-Monedero, M.A.; Cayuela, M.L.; Sánchez-García, M.; Vandecasteele, B.; D'Hose, T.; López, G.; Martínez-Gaitán, C.; Kuikman, P.J.; Sinicco, T.; Mondini, C. Agronomic evaluation of biochar, compost and biochar-blended compost across different cropping systems: Perspective from the european project fertiplus. *Agronomy* **2019**, *9*, 225. [CrossRef]
68. Amonette, J.E.; Joseph, S. Characteristics of biochar: Microchemical properties. In *Biochar for Environmental Management*; Routledge: London, UK, 2012; pp. 65–84.
69. Glick, B.R.; Liu, C.; Ghosh, S.; Dumbroff, E.B. Early development of canola seedlings in the presence of the plant growth-promoting rhizobacterium pseudomonas putida gr12-2. *Soil Biol. Biochem.* **1997**, *29*, 1233–1239. [CrossRef]
70. Xie, H.; Pasternak, J.J.; Glick, B.R. Isolation and characterization of mutants of the plant growth-promoting rhizobacterium pseudomonas putida gr12-2 that overproduce indoleacetic acid. *Curr. Microbiol.* **1996**, *32*, 67–71. [CrossRef]
71. Mohite, B. Isolation and characterization of indole acetic acid (iaa) producing bacteria from rhizospheric soil and its effect on plant growth. *J. Soil Sci. Plant Nutr.* **2013**, *13*, 638–649. [CrossRef]
72. Pérez-Fernández, M.; Alexander, V. Enhanced plant performance in *cicer arietinum* l. Due to the addition of a combination of plant growth-promoting bacteria. *Agriculture* **2017**, *7*, 40. [CrossRef]
73. Ma, Y.; Látr, A.; Rocha, I.; Freitas, H.; Vosátka, M.; Oliveira, R.S. Delivery of inoculum of rhizophagus irregularis via seed coating in combination with pseudomonas libanensis for cowpea production. *Agronomy* **2019**, *9*, 33. [CrossRef]
74. De la Rosa, J.M.; Paneque, M.; Hilber, I.; Blum, F.; Knicker, H.E.; Bucheli, T.D. Assessment of polycyclic aromatic hydrocarbons in biochar and biochar-amended agricultural soil from southern spain. *J. Soils Sediments* **2016**, *16*, 557–565. [CrossRef]
75. Chan, K.Y.; Zwieten, L.V.; Meszaros, I.; Downie, A.; Joseph, S. Using poultry litter biochars as soil amendments. *Aust. J. Soil Res.* **2008**, *46*, 437–444. [CrossRef]
76. Singh, A.; Singh, A.P.; Singh, S.K.; Rai, S.; Kumar, D. Impact of addition of biochar along with pgpr on rice yield, availability of nutrients and their uptake in alluvial soil. *J. Pure Appl. Microbiol.* **2016**, *10*, 2181–2188.
77. Shabala, S. Regulation of potassium transport in leaves: From molecular to tissue level. *Ann. Bot.* **2003**, *95*, 627–634. [CrossRef] [PubMed]
78. Wilkinson, S.; Davies, W.J. Aba-based chemical signalling: The co-ordination of responses to stress in plants. *Plant Cell Environ.* **2002**, *25*, 195–210. [CrossRef] [PubMed]
79. Novak, J.M.; Lima, I.; Xing, B.; Gaskin, J.W.; Steiner, C.; Das, K.C.; Ahmedna, M.; Rehrah, D.; Watts, D.W.; Busscher, W.J.; et al. Characterization of designer biochar produced at different temperatures and their effects on a loamy sand. *Ann. Environ. Sci.* **2009**, *3*, 195–2006.
80. Lehmann, J.; Rillig, M.C.; Thies, J.; Masiello, C.A.; Hockaday, W.C.; Crowley, D. Biochar effects on soil biota—A review. *Soil Biol. Biochem.* **2011**, *43*, 1812–1836. [CrossRef]
81. Abbas, T.; Rizwan, M.; Ali, S.; Adrees, M.; Zia-ur-Rehman, M.; Qayyum, M.F.; Ok, Y.S.; Murtaza, G. Effect of biochar on alleviation of cadmium toxicity in wheat (*Triticum aestivum* L.) grown on cd-contaminated saline soil. *Environ. Sci. Pollut. R.* **2018**, *25*, 25668–25680. [CrossRef] [PubMed]

© 2019 by the authors. Licensee MDPI, Basel, Switzerland. This article is an open access article distributed under the terms and conditions of the Creative Commons Attribution (CC BY) license (http://creativecommons.org/licenses/by/4.0/).

Article

Addition of Biochar to a Sandy Desert Soil: Effect on Crop Growth, Water Retention and Selected Properties

Khaled D. Alotaibi [1,*] and Jeff J. Schoenau [2]

1 Department of Soil Science, King Saud University, P.O. Box 2460, Riyadh 11451, Saudi Arabia
2 Department of Soil Science, University of Saskatchewan, 51 Campus Drive, Saskatoon, SK S7N 5A8, Canada; jeff.schoenau@usask.ca
* Correspondence: khalotaibi@ksu.edu.sa

Received: 8 April 2019; Accepted: 15 June 2019; Published: 20 June 2019

Abstract: Agricultural and environmental applications of biochar (BC) to soils have received increasing attention as a possible means of improving productivity and sustainability. Most previous studies have focused on tropical soils and more recently temperate soils. However, benefits of BC addition to desert soils where many productivity constraints exist, especially water limitations, have not been widely explored. Thus, three experiments were designed using a desert soil from Saudi Arabia to address three objectives: (1) to evaluate the effect of BCs produced from date palm residues added at 8 t ha^{-1} on wheat growth, (2) to determine the effect of BC addition and BC aging in soil on water retention, and (3) to reveal the effect of BC on selected soil physical (bulk density, BD; total porosity; TP) and chemical (pH; electrical conductivity, EC; organic matter, OM; cation exchange capacity, CEC) properties. The feedstock (FS) of date palm residues were pyrolyzed at 300, 400, 500, and 600 °C, referred to here as BC300, BC400, BC500, and BC600, respectively. The BC products produced at low temperatures were the most effective in promoting wheat growth when applied with the NPK fertilizer and in enhancing soil water retention, particularly with aging in soil, whereas high-temperature BCs better improved the selected soil physical properties. The low-temperature BCs increased the yield approximately by 19% and improved water retention by 46% when averaged across the incubation period. Higher water retention observed with low-temperature BCs can be related to an increased amount of oxygen-containing functional groups in the low-temperature BCs, rendering BC surfaces less hydrophobic. Only the BC300 treatment showed a consistent positive impact on pH, OM, and CEC. Pyrolysis temperature of date palm residue along with aging are key factors in determining the potential benefit of BC derived from date palm residues added to sandy desert soil.

Keywords: biochar; desert soil; crop growth; water retention; nutrient; soil chemical properties; soil physical properties

1. Introduction

Biochar (BC) is a carbonaceous residue produced through the thermal breakdown of organic materials under limited conditions of oxygen. The use of BC as a soil amendment is not a new concept, but it has received growing attention in the last few years, generally due to its role in mitigating greenhouse gas emissions by sequestering carbon in the soils, reducing nitrous oxide emissions, improving soil properties and quality, and increasing nutrient use efficiency and crop production [1–4]. BC can influence crop growth and yield directly via its effects on pH and nutrient retention and supply or indirectly through improvement of soil physical and chemical properties of importance for crop production [3,5–7]. Impact of BC addition on crop yield is more pronounced in infertile and degraded soils compared to that in the soils of high fertility [7]. However, the extent to which BC can influence crop yield is largely dependent on BC production temperature and the feedstock (FS) [5–7]. In a recent review, it was reported that the fertilizing value of BC and its nutrient availability, especially N and P, decrease with increasing pyrolysis temperature [5]. Therefore, the effect of BC on crop yield and plant nutrition was found to be variable, and its use as a sole amendment was not always effective, having its supplementation with mineral fertilizers sometimes necessary to promote crop growth [8–16]. Some studies found BC to be the most effective when it is applied with mineral fertilizers [8,14,17–19]. Therefore, positive impact of BC on soil fertility and crop yield is not always certain, and it highly depends on FS, BC production temperature, and most importantly soil type [3,5–7,20].

BC benefits to soil also include improvement of soil physical properties. For instance, soil bulk density (BD) and porosity was found to be improved after BC application [20–26]. BC was also reported to increase the soil ability to retain water, and this is mostly attributed to the high total porosity (TP) of BC, retaining water in small pores and thereby increasing water holding capacity and infiltration [5,20,21]. However, enhancement of soil physical properties following BC application is more evident in infertile and course-textured soils [7,23].

Key soil chemical properties, such as pH, OM content, and CEC are also influenced by BC addition, affecting soil fertility and productivity [3,7,27–29]. In most cases, BC is alkaline, and therefore its effect on increasing acidic soil pH was demonstrated in several studies whereas its impact on alkaline soil pH is expected to be minimal [5–7]. Reviewed studies reported increases in CEC and organic matter in soil treated with BC; however, the magnitude of this effect varies according to the BC production condition, application rate, and soil type [3–7].

BC use is most frequently reported to be effective in tropical zones where soils are acidic and nutrient leaching potential is high [30]. Under these soils conditions, BC can have liming and nutrient availability effects, as BC was found to increase pH values of acidic soils and thereby improve nutrient bioavailability and use efficiency [7,31]. A meta-analysis using data from 103 studies found a better performance of BC addition in acidic soils than in neutral soils, and in sandy than in loam and silt soils, suggesting liming and physical property improvement effects of BC [32].

Less attention has been given to effects of BC in semi-arid [33–36] and especially arid environments [26,37–40] with few studies in true desert soils.

Arid soils of Saudi Arabia are characterized in general by very low content of organic matter (<1%) and nutrients and sandy texture with poor water retention capacity and nutrient use efficiency [41]. The soils are alkaline, with a pH value in many cases found to be greater than 8.0 and have great abundance of calcium carbonate higher than 30% in some soils [41]. The use of BC from locally abundant feedstock, such as date palm residues as amendment can be a possible means to alleviate the productivity constraints of these desert soils. Therefore, the aim of the current study was to evaluate the effect of BC produced from date palm residues at different temperatures on wheat growth, soil water retention, and selected physical and chemical properties of a typical desert soil from Saudi Arabia.

2. Materials and Methods

2.1. Soil and Biochar

Soils were collected from a private farm near the City of Thadiq, approximately 120 km northwest of the City of Riyadh in Saudi Arabia. The particular site from where the soils were collected was abandoned and left uncultivated for ten years. Before abandonment, the site was under continuous cultivation of wheat and alfalfa for more than twenty years. Several soil samples were collected from 0–20 cm depth and thoroughly mixed to provide a composite sample. Soil was brought back to the laboratory, air-dried, and sieved (<2 mm). Prior to the experiments, the processed soil was analyzed for its basic characteristics. The soil was found to have an organic matter content of 0.48% OM. determined according to Nelson and Sommers (1996) [42], and a pH value of 8.4, EC of 0.47 dS m^{-1}, and 19.0% by weight of $CaCO_3$, determined according to the method of Loeppert and Suarez (1996) [43]. Soil pH and EC were measured in a 1:2.5 soil/distilled water suspension. The soil texture was determined as described by Gee and Bauder (1986) [44] and the textural class was loamy sand, having 79%, 14%, and 7% of sand, silt, and clay, respectively.

Date palm tree residues used as the FS for BC production included frond midrib and frond base residues. These residues were collected randomly from a local farm, chopped at the site, brought to the laboratory, and air-dried at 40 °C. A subsample was ground and prepared for direct application to soil. Another subsample was pyrolyzed for BC production. The pyrolysis process was carried out as described by Usman et al. (2015) [45]. Briefly, the fine-ground date palm residue was placed in a stainless steel container and closed tightly to exclude oxygen. The closed container was transferred to an electrical muffle furnace where it remained for 4 h at the desired temperature. The date palm residues were pyrolyzed at 300, 400, 500, and 600 °C, generating four BCs referred to as BC300, BC400, BC500, and BC600, respectively. The BCs were left to cool down to room temperature, followed by basic characterization analyses. This included analysis of pH, EC, total C, and total N. The pH was measured in 1:10 BC:water suspension, and the EC was measured in an extract of the same suspension. The total C and N were analyzed using a CHNS analyzer (Series II, PerkinElmer, Waltham, MA, USA). The basic characteristics of date palm residues and its BC are presented in Table 1.

Table 1. Basic characteristics of date palm residue BC used in the current study.

FS/BC	Parameters			
	C (%)	N (%)	pH	EC (dS m^{-1})
FS	42.8	1.5	7.20	3.00
BC300	61.4	0.75	7.67	5.61
BC400	65	0.93	8.22	6.67
BC500	71.7	1.0	8.40	7.53
BC600	73.0	1.1	8.75	8.07

2.2. Experimental Design

2.2.1. Experiment1: Wheat Growth

This experiment was designed to include treatments of FS and four BCs (BC300, BC400, BC500, and BC600) without and with a fertilizer, a fertilized control (FC), and an unfertilized control (UFC). The FS and BCs were applied at one rate of 8 t ha^{-1}. This rate of BC application is considered practical if BC is to be applied at a large scale under field conditions. The FS and BCs were weighed and mixed with 100 g soil followed by addition of NPK fertilizer as urea, triple superphosphate, and potassium sulphate applied at 200 kg N ha^{-1}, 100 kg P ha^{-1}, and 50 kg K ha^{-1}, respectively. In case of FS and BCs applied alone, no fertilizer was applied. The mixture was spread onto the surface of 900 g of air-dried sieved soil in a 1L plastic pot with a 12 cm height and a 12 cm diameter, bringing the total soil weight to 1 kg per pot. The FC treatment received 200 kg N ha^{-1}, 100 kg P ha^{-1}, and 50 kg K ha^{-1}, as urea, triple superphosphate, and potassium sulphate, respectively, whereas the UFC received no fertilizers or BCs. Each treatment was replicated four times. Pots were left to equilibrate for 24 h, and then seven seeds of wheat (*Triticum aestivum* L. cultivar YecoraRojo) were sown in each pot. Soil moisture was brought to 80% of field capacity and maintained at this moisture level by daily watering for the entire period of the study. The pots were placed in a controlled environment growth chamber set at 22–25 °C. After plant emergence, seedlings were thinned to three plants per pot. The plants were left to grow for six weeks after which the total aboveground biomass was cut at the soil surface, harvested, and dried at 60 °C, and dry matter weight was recorded. A finely ground subsample of the plant material was first digested as described by Thomas et al. (1976) [46] followed by colorimetric determination to determine N and P contents of the digested plant material.

2.2.2. Experiment2: Water Retention

Five hundred grams of air-dried and sieved soil was placed into a plastic container. Then, BC was added and mixed manually with the soil. The soil mixed with BC was repacked into PVC columns with a 5.5 cm internal diameter and a 20 cm height, giving a soil depth of 15 cm. The columns were sealed at the bottom with glass wool to hold the soil during leaching. A headspace of approximately 3 cm was left above the soil surface to allow for water addition. The treatments included the four BCs used in this study applied at one rate of 50 t ha^{-1} without a fertilizer, in addition to untreated control. Each treatment was replicated three times. The rate of 50 t BC ha^{-1} used in this experiment was selected according to initial trials where low rates of BC did not show a clear effect on water retention. This is within the range of BC rates used in the literature where a rate of >20 t ha^{-1} is required in several studies to detect significant responses to BC addition [7,23,30]. The cylinders with soil were placed in a rack and allowed to stand. Then, the soil was brought to 50% of the field capacity and left to equilibrate for 24 h, after which the leaching events were conducted. Leaching was carried out by adding distilled water using a perforated container and allowing water to percolate through the soil overnight, and leachate was collected and its volume was recorded. The percentage of the retained water was calculated by subtraction of the volume of recovered leachate from the volume of the added water [47]. The leaching occurred five times at the beginning of the experiment and after 14, 30, 45, and 60 days, in order to investigate the effect of BC aging in soil on water retention.

2.2.3. Experiment3: Changes in Selected Physical and Chemical Properties

In this experimental setup, treatments included BC300, BC400, BC500, and BC600 applied without a fertilizer at the same rate (8 t ha^{-1}) as in experiment 1 in addition to the untreated control, but left for a longer time to allow for effects on physical and chemical properties to be fully revealed. The treatments were thoroughly mixed with the entire soil placed in a 1L plastic pot with a 12 cm height and a 12 cm diameter and incubated for eight months at 80% of field capacity and under controlled conditions. At the end of the eight-month incubation, soil BD and TP were determined using the core method [48]. For chemical analysis, soils were removed from the pots, air-dried, and analyzed for pH, EC, OM, and CEC. The OM concentration was analyzed according to the method of Nelson and Sommers (1996) [42]. The CEC was determined according to Rhoades (1982) [49].

2.3. Statistical Analysis

Data generated from experiments 1 and 3 were analyzed using a one-way analysis of variance (ANOVA) where treatment effects on plant and soil variables were tested. Data obtained from experiment 2 were analyzed using a two-way ANOVA where the effects of treatment, time of incubation, and their interaction on water retention were analyzed. Treatments effects were considered significant at $p < 0.05$ at which the mean comparisons was also performed using the Student–Newman–Keuls (SNK) test. Statistical analysis was performed using CoStat software package (CoHort Software, 2008). Each variable data were reported as the mean ± standard error of the treatment replicates. Statistical analysis outputs are either included in the tables or the figures.

3. Results

3.1. Wheat Yield and N and P Uptake Response to Biochar Addition

The impact of date palm residue BC amendment on wheat yield clearly varied according to the BC production temperature (Figure 1). This variation was also observed in either the presence or the absence of fertilizer addition (Figure 1). In general, the yield was higher in the presence of NPK compared to that using the same treatments without NPK. BC produced at 300 or 400 °C and added with NPK provided the highest yield compared to that with the NPK alone treatment and also compared to the BC produced at 500 or 600 °C treatments (Figure 1). The BC300 and BC400 treatments also had higher yield than the FS when NPK was added. The yield in the FS with a fertilizer did not exceed that in NPK alone treatment. Without a fertilizer, compared to the UFC, the FS and BC 300 treatments did not significantly increase yield, while yield was decreased in treatments with BC produced at higher temperature (BC400, BC500, and BC600). Compared to the FC, the fertilized BC300 and BC400 treatments led to higher yield while other treatments were not significantly different. A similar pattern was evident among treatments in fertilized versus unfertilized. In general, the most effective treatments were the BC300 and BC400 when applied with a fertilizer; both treatments resulted in the greatest yield that was significantly higher than any other treatment.

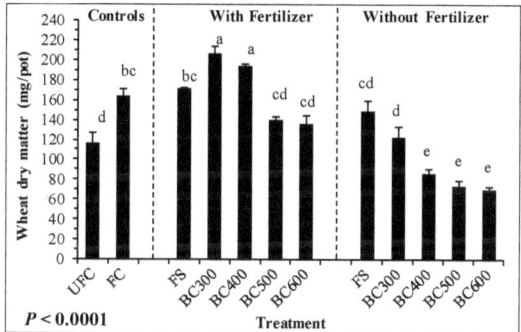

Figure 1. Response of total biomass yield of wheat grown for 5 weeks to application with and without a fertilizer of 8 t ha^{-1} of four date palm residue biochars (BCs) that were produced with different pyrolysis temperatures (BC300, BC400, BC500, and BC600) and the BC feedstock (FS). Treatments also included an unfertilized control (UFC) and a fertilized control (FC). Bars sharing the same letter among treatments are not significantly different according to the Student–Newman–Keuls (SNK) test ($p \leq 0.05$). Errors bars represent the standard error of the mean ($n = 4$).

Treatments had a significant impact on plant N and P uptake (Figures 2 and 3) that followed a similar pattern to that of the impact on yield. The BC300 treatment with a fertilizer showed the greatest N and P uptake (Figures 2 and 3). Higher-temperature BCs (BC500 and BC600) treatments reduced plant N and P uptake to the background level either when applied alone or in combination with the NPK fertilizer. Application of BCs produced at high temperatures may limit N and P through retention processes and/or result in reduced uptake due to a phytotoxic effect.

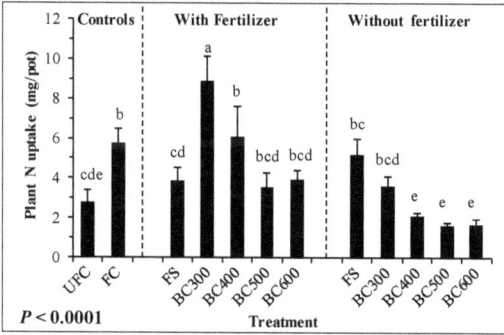

Figure 2. Plant N uptake in soil receiving 8 t ha^{-1} of four date palm residue BCs produced with different pyrolysis temperatures (BC300, BC400, BC500, and BC600) and the BC FS with and without a fertilizer added. Treatments also included a UFC and an FC. Bars sharing the same letter among treatments are not significantly different according to the SNK test ($p \leq 0.05$). Errors bars represent the standard error of the mean ($n = 4$).

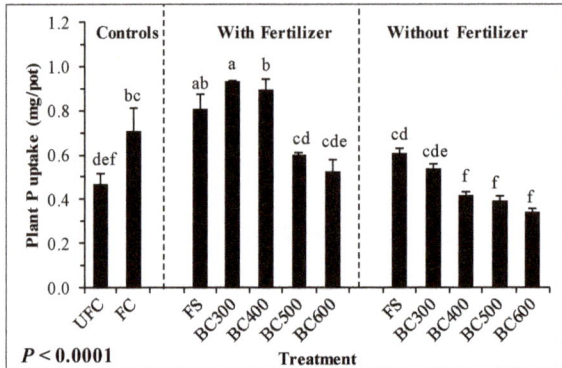

Figure 3. Plant P uptake in soil receiving 8 t ha^{-1} of four date palm residue BCs produced with different pyrolysis temperatures (BC300, BC400, BC500, and BC600) and the BC FS with and without a fertilizer added. Treatments also included a UFC and an FC. Bars sharing the same letter among treatments are not significantly different according to the SNK test ($p \leq 0.05$). Errors bars represent the standard error of the mean ($n = 4$).

3.2. Impact of Date Palm Residues Biochars on Water Retention

The effects of treatment, time of incubation, and their interaction on water retention were statistically significant (Figure 4). The positive effect of BC treatments on water retention in the soil increased with increasing time of incubation and was greatest for the low pyrolysis temperature BC (BC300). This indicated that, as BC ages in soil, it can be more effective in promoting water retention, especially the low pyrolysis temperature BC. When averaged across the time of incubation periods, the BC300 treatment showed the greatest amount of retained water, significantly higher than that observed in BC400 treatment (Figure 5A) and higher than the BC500 and BC600 treatments. The treatments in order of retained water amounts were BC300>BC400>BC500≥BC600>unamended control. Averaged across all the treatments, water retention was greatest at the end of the incubation (Figure 5B), with Day45≥Day60>Day30>Day14>Day0.

3.3. Impact of Date Palm Residues Biochars on Soil Bulk Density and Total Porosity

The statistical analysis showed a significant treatment effect on soil BD (Figure 6A) and TP (Figure 6B). All the treatments significantly reduced the BD in comparison to the control; however, the effect of the treatments varied in their magnitude (Figure 6A). The lowest decreases in BD were observed with FS, BC300, and BC400 treatments, all of which did not significantly differ from each other. The greatest decreases in BD were shown in soil treated with BC500 and BC500, both of which did not significantly differ from each other but were significantly different from any other treatment (Figure 6A). The lowest value of TP was observed with the control treatment; however, it was the only BC600 treatment that provided a significantly higher TP value than the control (Figure 6B).

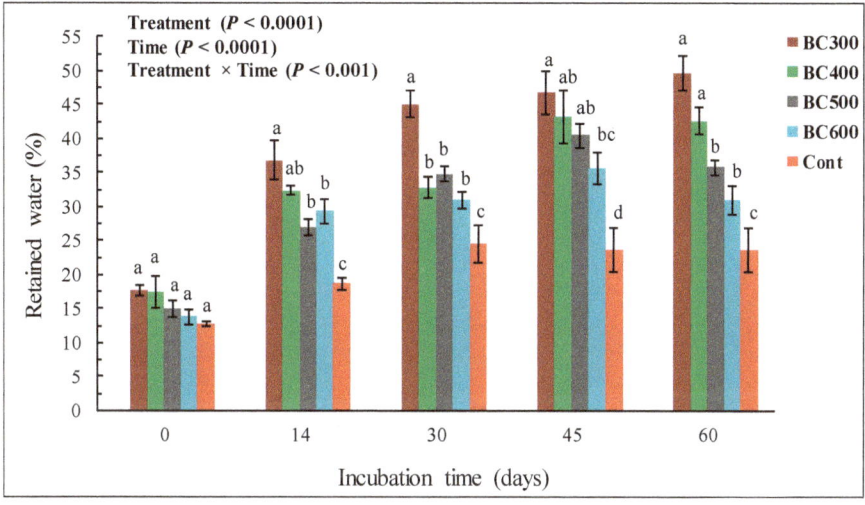

Figure 4. Percentage of retained water in sandy desert soil treated with 50 t ha^{-1} of four date palm residue BCs produced with different pyrolysis temperatures (BC300, BC400, BC500, and BC600) over time. Bars sharing the same letter among treatments are not significantly different according to the SNK test ($p \leq 0.05$). Errors bars represent the standard error of the mean ($n = 3$).

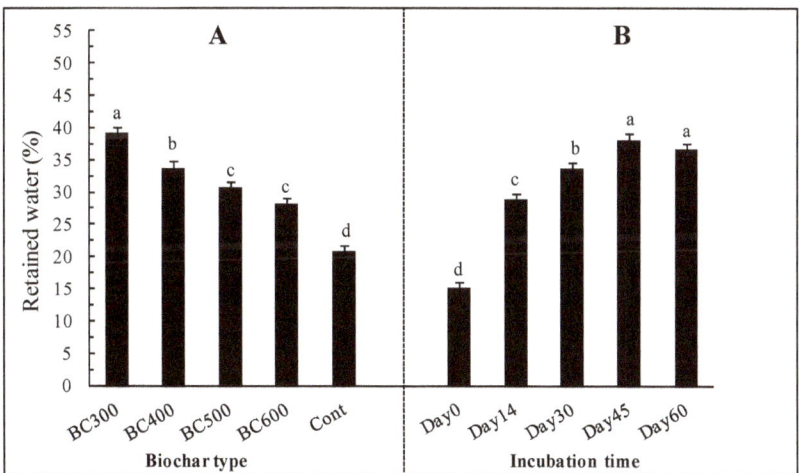

Figure 5. Percentage of retained water in sandy desert soil treated with 50 t ha^{-1} of four date palm residue BCs produced with different pyrolysis temperatures (BC300, BC400, BC500, and BC600) averaged across the time of incubation (**A**) and incubation time averaged across the BC products (**B**). Bars sharing the same letter among treatments are not significantly different according to the SNK test ($p \leq 0.05$). Errors bars represent the standard error of the mean.

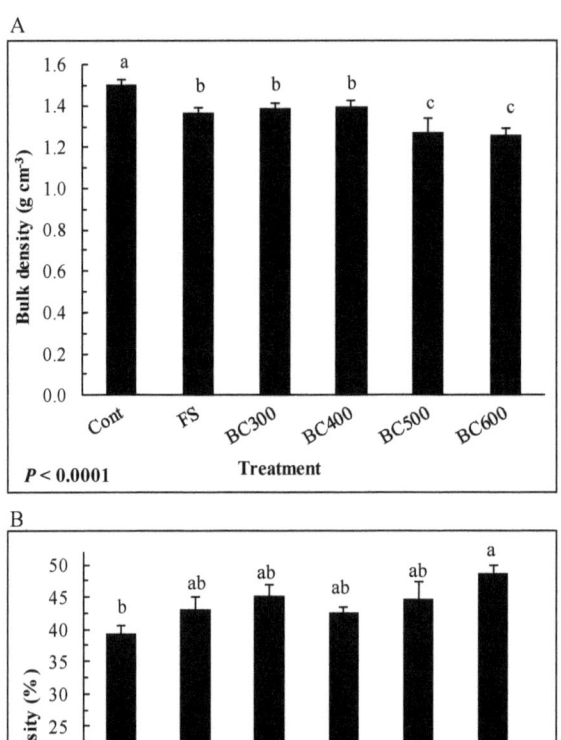

Figure 6. Effects of 8 t ha^{-1} of four date palm residues BCs (BC300, BC400, BC500, and BC600) and the FS on soil bulk density (**A**) and total porosity (**B**) after eight months of incubation. Bars sharing the same letter among treatments are not significantly different according to the SNK test ($p \leq 0.05$). Errors bars represent the standard error of the mean ($n = 4$).

3.4. Impact of Date Palm Residues Biochars on Selected Soil Chemical Properties

The soil pH, EC, OM, and CEC (Table 2) were significantly impacted by amendment treatment. The greatest decrease in soil pH was observed with the FS treatment followed by that with BC300 treatment, both of which were significantly different from each other and significantly lower than any other treatment (Table 2). None of the other treatments differed from the control. A significant increase in soil EC comparison to the control was observed, with the greatest increase in EC found in BC600 and FS treatments followed by that with BC500 or BC300 treatment (Table 2). Only the BC300 treatment resulted in higher soil OM and CEC in comparison to all the other treatments (Table 2).

Table 2. Effects of experimental treatments on the selected soil chemical properties.

Treatment	pH	EC	OM	CEC
		dS m^{-1}	%	meq/100
Control	8.25 ± 0.01 [a]	0.48 ± 0.03 [c]	0.62 ± 0.00 [b]	6.86 ± 0.42 [b]
FS	8.07 ± 0.02 [c]	1.51 ± 0.06 [a]	0.69 ± 0.04 [b]	8.24 ± 0.48 [b]
BC300	8.14 ± 0.04 [b]	1.22 ± 0.18 [a,b]	1.18 ± 0.12 [a]	9.86 ± 0.40 [a]
BC400	8.23 ± 0.02 [a]	1.09 ± 0.04 [b]	0.70 ± 0.07 [b]	7.09 ± 0.11 [b]
BC500	8.27 ± 0.01 [a]	1.22 ± 0.03 [a,b]	0.60 ± 0.04 [b]	7.92 ± 0.15 [b]
BC600	8.19 ± 0.01 [a,b]	1.54 ± 0.05 [a]	0.57 ± 0.02 [b]	7.72 ± 0.21 [b]
LSD (0.05)	0.06	0.26	0.19	0.99

Means within a column sharing the same letter are not significantly different at $p = 0.05$.

4. Discussion

4.1. Wheat Growth Response to Date Palm Residue Biochar Addition

The BCs derived from date palm residues were more effective when combined with NPK fertilizer, and a better wheat response was evident with BCs produced at the lower temperatures of 300 and 400 °C compared to those at 500 and 600 °C. Similarly, greenwaste BC produced at 450 °C increased radish yield in presence of N fertilizer, but had no effect when applied alone [8]. Rice husk BC produced at 450 °C was also found to increase total biomass and grain yields of wheat and rice when applied with N, but no effect was found in absence of N [13]. Many other studies found BC to be the most effective in crop yield when applied with mineral fertilizers [8,14,15,17,18,50,51]. Alburquerque et al. (2013) [14] found decrease in or no effect on total plant biomass in response to wheat straw and olive tree pruning BCs produced at low temperature and applied alone under growth chamber conditions. Oat hulls BC produced at 450 °C showed also a limited effect on crop yield under field conditions [16]. In contrast to the results of the current study, other studies, however, reported increases in yield when BC was applied alone [9–12,15]. Decrease in N plant uptake following BC addition has been reported [14], related to the ability of BC to adsorb N and limit its availability. The BC surface area and microporosity increase with increasing pyrolysis temperature, resulting in BCs of higher physisorption capacity [52,53]. Thus, the BC produced here at higher temperatures (>400 °C) reduced yield and N and P uptake, and this could be attributed to their high adsorption/sorption capacity, restricting soil nutrient availability for plant uptake. Addition of fertilizer may also enhance microbial decomposition and reduce any phytotoxic effects of BC as appeared to be evident with the high-pyrolysis-temperature BC in the current study. This may also explain the decreased yield and N and P plant uptake in higher-pyrolysis-temperature BC treatments without a fertilizer added observed in the current study. It was previously reported that mineral N availability is essential in stimulating microbial decomposition of organic materials [54,55]. Under the conditions of the current study, it appears that date palm residues BC did not supply nutrient, and its effect on wheat yield and N and P uptake was related to other factors, mainly improvement of soil water holding capacity and physical properties.

4.2. Effect of Date Palm Residue Biochar Addition on Water Retention

Several studies that previously investigated soil water retention in sandy soils after BC addition showed variable results [20–22,24,26,56–60]. The results of the current study clearly indicated that the residence time of BC in soil and the pyrolysis temperature are significant factors in determining effect of BC on water retention. All the BCs used in the current study significantly increased soil water retention in the sandy desert soil, and the effect increased with the time of incubation. The low pyrolysis temperature BCs (BC300 and BC400) enhanced water retention the most when compared to the high-pyrolysis-temperature BCs. Some other findings have indicated BC hydrophobicity (water repellency) to decrease with increasing pyrolysis temperature, indicating potentially greater soil water retention enhancement with higher-pyrolysis-temperature BC [61–63]. Jeffery et al. (2015) [59] reported that the hydrophobicity of BC reduced water infiltration into the BC particles and this can limit water retention. The hydrophobic surface of low-temperature BC is created by formation of aliphatic compounds that get volatilized at high pyrolysis temperature [61,62,64]. However, low pyrolysis temperature BCs were also reported to show higher oxidation rates that could render BC surfaces less hydrophobic via increasing the amount of oxygen-containing functional groups [62,63,65]. This can explain the increase in soil water retention with time in the current study, especially with low-temperature BCs which are considered to be more labile, and with greater potential for oxidation changes in their surface functionalities and porosity. In agreement with the current findings, BC produced at 500 °C increased sandy soil water retention and the increase was more evident with time of incubation, attributed to a change of BC surfaces from hydrophobic to hydrophilic [56]. Similarly, clay soil amended with BC produced at 400 °C retained less water immediately after application, but after a 180-day incubation, its effect on water retention was comparable to those produced at 600 or 800 °C [66]. BC addition was also found to increase available water content in sandy loam and loamy soils in the first growing season, and this impact continued to increase in the second growing season, highlighting the importance of BC aging in soil [67].

4.3. Impact of Date Palm Residues Biochars on Selected Soil Physical and Chemical Properties

All BC treatments in the current study reduced soil BD. The observed decrease in BD and increase in TP in soil amended with BC in the current study are in agreement with previous studies [22,23,25,68]. However, the impact of BC on BD and TP varies according to the FS type, rate of application, soil texture, and pyrolysis temperature [25]. In the current study, the high-temperature BCs (BC500 and BC600) showed the greatest decrease in BD, and this was also associated with the highest increase in TP, especially in BC600 treatment, which was significantly higher than in the control. Increase in pyrolysis temperature yields BC with high surface area and porosity, the most important physical properties that would enhance BC capabilities of improving soil physical properties, such as adsorption capacity, BD, and porosity [62,69]. However, the lower soil BD and higher porosity observed with the high-temperature BC used here were not associated with higher water retention when compared to low-temperature BC. This can be in part attributed to changes in surface chemistry of BC during the incubation as explained earlier or effects on pore size distribution.

Generally, pyrolysis of most organic FSs results in BCs with alkaline pH; however, the effect varies according to pyrolysis temperature and FS type [3]. As a result, BC addition is frequently found to increase soil pH in acid soils [3,7,27–29]. Conversely, in the current study with a soil of alkaline pH, the BCs had little effect on soil pH, with the exception of FS and BC300 treatments that showed a minor but significant decrease in soil pH. Other studies in neutral to alkaline soils also showed a limited effect of BC amendment on soil pH [70]. A decrease in soil pH was reported in alkaline soils following BC application [36,71,72]. This is probably related to some chemical oxidation and microbial decomposition of BC in soil, resulting in acidic compounds being produced and therefore lowering soil pH. This may be more evident with low-temperature BC which is expected to go through microbial decomposition to some extent. The pronounced effect of BC300 treatment in this study can support this explanation. All treatments increased soil EC, and this can be attributed to the salts contained within these materials, which is expected to elevate soil EC. Increases in soil EC were also reported in other studies [28,71]. This impact needs to be further monitored if these materials are to be applied frequently to arid soils, as without removal by leaching, they may cause soil salinity. It was only the BC produced at 300 °C which significantly increased soil organic matter content. Previous findings also report that greater increase in soil organic carbon with BC produced at low temperature than with high temperature BC [72,73]. A positive impact of BC on CEC is commonly reported, particularly in course-textured soils [7,36,73]. However, the impact of BC treatments on CEC in the current study was minimal and may be related to the low rate of application (8 t ha^{-1}) compared to those in other studies, and only the BC300 treatment showed significantly higher CEC than any other treatment. The CEC increase in soil amended with BC produced at 300 °C is consistent with the positive impact of this treatment on the other crop and soil parameters tested in this study.

5. Conclusions

The effect of date palm residues BC addition on sandy desert soil was evaluated under controlled environment conditions. The BCs produced at low pyrolysis temperatures (300 and 400 °C) were the most effective in promoting wheat growth when applied with a mineral fertilizer, enhancing soil water retention, particularly with aging in soil, whereas high temperature BCs better improved the selected soil physical properties (BD and TP). Soil pH was slightly reduced whereas all treatments increased soil EC. Soil OM content and CEC were only increased in the BC300 treatment. It appeared the pyrolysis temperature and aging are key factors in determining the potential benefit of BC addition to sandy desert soil.

Author Contributions: K.D.A. designed and implemented the experiment, collected and analyzed the data, and wrote the manuscript. J.J.S. contributed significantly to the manuscript through interpretation of the results and reviewing and editing of the manuscript.

Funding: This research was funded by the Deanship of Scientific Research at King Saud University through the Research Project NO R5-16-03-04.

Acknowledgments: The authors extend their appreciation to the Deanship of Scientific Research at King Saud University for funding this work through the Research Project NO R5-16-03-04.

Conflicts of Interest: The authors declare no conflicts of interest.

References

1. Spokas, K.A.; Cantrell, K.B.; Novak, J.M.; Archer, D.W.; Ippolito, J.A.; Collins, H.P.; Boateng, A.A.; Lima, I.M.; Lamb, M.C.; McAloon, A.J.; et al. Biochar: A synthesis of its agronomic impact beyond carbon sequestration. *J. Environ. Qual.* **2012**, *41*, 973–989. [CrossRef] [PubMed]
2. Cayuela, M.L.; Van Zwieten, L.; Singh, B.P.; Jeffery, S.; Roig, A.; Sánchez-Monedero, M.A. Biochar's role in mitigating soil nitrous oxide emissions: A review and meta-analysis. *Agric. Ecosyst. Environ.* **2014**, *191*, 5–16. [CrossRef]
3. Gul, S.; Whalen, J.K.; Thomas, B.W.; Sachdeva, V.; Deng, H. Physico-chemical properties and microbial responses in biochar-amended soils: Mechanisms and future directions. *Agric. Ecosyst. Environ.* **2015**, *206*, 46–59. [CrossRef]
4. Xie, T.; Sadasivam, B.Y.; Reddy, K.R.; Wang, C.; Spokas, K. Review of the effects of biochar amendment on soil properties and carbon sequestration. *J. Hazard. Toxic Radioact. Waste* **2015**, *20*, 04015013. [CrossRef]
5. Ding, Y.; Liu, Y.; Liu, S.; Li, Z.; Tan, X.; Huang, X.; Zeng, G.; Zhou, L.; Zheng, B. Biochar to improve soil fertility. A review. *Agron. Sustain. Dev.* **2016**, *36*, 36. [CrossRef]
6. Hussain, M.; Farooq, M.; Nawaz, A.; Al-Sadi, A.M.; Solaiman, Z.M.; Alghamdi, S.S.; Ammara, U.; Ok, Y.S.; Siddique, K.H.M. Biochar for crop production: Potential benefits and risks. *J. Soils Sediments* **2016**, *17*, 685–716. [CrossRef]
7. El-Naggar, A.; Lee, S.S.; Rinklebe, J.; Farooq, M.; Song, H.; Sarmah, A.K.; Zimmerman, A.R.; Ahmad, M.; Shaheen, S.M.; Ok, Y.S. Biochar application to low fertility soils: A review of current status, and future prospects. *Geoderma* **2019**, *337*, 536–554. [CrossRef]
8. Chan, K.Y.; Van Zwieten, L.; Meszaros, I.; Downie, A.; Joseph, S. Agronomic values of greenwaste biochar as a soil amendment. *Aust. J. Soil Res.* **2007**, *45*, 629–634. [CrossRef]
9. Hossain, M.K.; Strezov, V.; Chan, K.Y.; Nelson, P.F. Agronomic properties of wastewater sludge biochar and bioavailability of metals in production of cherry tomato (Lycopersicon esculentum). *Chemosphere* **2010**, *78*, 1167–1171. [CrossRef]
10. Major, J.; Rondon, M.; Molina, D.; Riha, S.J.; Lehmann, J. Maize yield and nutrition during 4 years after biochar application to a Colombian savanna Oxisol. *Plant Soil* **2010**, *333*, 117–128. [CrossRef]

11. Deal, C.; Brewer, C.E.; Brown, R.C.; Okure, M.A.E.; Amoding, A. Comparison of kiln-derived and gasifier-derived biochars as soil amendments in the humid tropics. *Biomass Bioenergy* **2012**, *37*, 161–168. [CrossRef]
12. Solaiman, Z.M.; Murphy, D.V.; Abbott, L.K. Biochars influence seed germination and early growth of seedlings. *Plant Soil* **2012**, *353*, 273–287. [CrossRef]
13. Wang, J.; Pan, X.; Liu, Y.; Zhang, X.; Xiong, Z. Effects of biochar amendment in two soils on greenhouse gas emissions and crop production. *Plant Soil* **2012**, *360*, 287–298. [CrossRef]
14. Alburquerque, J.A.; Salazar, P.; Barrón, V.; Torrent, J.; del Campillo, M.D.C.; Gallardo, A.; Villar, R. Enhanced wheat yield by biochar addition under different mineral fertilization levels. *Agron. Sustain. Dev.* **2013**, *33*, 475–484. [CrossRef]
15. Abbasi, M.K.; Anwar, A.A. Ameliorating effects of biochar derived from poultry manure and white clover residues on soil nutrient status and plant growth promotion-greenhouse experiments. *PLoS ONE* **2015**, *10*, e0131592. [CrossRef] [PubMed]
16. Alotaibi, K.D.; Schoenau, J.J. Application of Two Bioenergy Byproducts with Contrasting Carbon Availability to a Prairie Soil: Three-Year Crop Response and Changes in Soil Biological and Chemical Properties. *Agronomy* **2016**, *6*, 13. [CrossRef]
17. Van Zwieten, L.; Kimber, S.; Morris, S.; Chan, K.Y.; Downie, A.; Rust, J.; Joseph, S.; Cowie, A. Effects of biochar from slow pyrolysis of papermill waste on agronomic performance and soil fertility. *Plant Soil* **2010**, *327*, 235–246. [CrossRef]
18. Li, S.; Shangguan, Z. Positive effects of apple branch biochar on wheat yield only appear at a low application rate, regardless of nitrogen and water conditions. *J. Soils Sediments* **2018**, *18*, 3235–3243. [CrossRef]
19. Atkinson, C.J.; Fitzgerald, J.D.; Hipps, N.A. Potential mechanisms for achieving agricultural benefits from biochar application to temperate soils: A review. *Plant Soil* **2011**, *337*, 1–18. [CrossRef]
20. Abel, S.; Peters, A.; Trinks, S.; Schonsky, H.; Facklam, M.; Wessoolek, G. Impact of biochar and hydrochar addition on water retention and water repellency of sandy soil. *Geoderma* **2013**, *202*, 183–191. [CrossRef]
21. Ulyett, J.; Sakrabani, R.; Kibbilewhite, M.; Hann, M. Impact of biochar addition on water retention, nitrification and carbon dioxide evolution from sandy loam soils. *Eur. J. Soil Sci.* **2014**, *65*, 96–104. [CrossRef]
22. Głąb, T.; Palmowska, J.; Zaleski, T.; Gondek, K. Effect of biochar application on soil hydrological properties and physical quality of sandy soil. *Geoderma* **2016**, *281*, 11–20. [CrossRef]
23. Omondi, M.O.; Xia, X.; Nahayo, A.; Liu, X.; Korai, P.K.; Pan, G. Quantification of biochar effects on soil hydrological properties using meta-analysis of literature data. *Geoderma* **2016**, *274*, 28–34. [CrossRef]
24. Aller, D.; Rathke, S.; Laird, D.; Cruse, R.; Hatfield, J. Impacts of fresh and aged biochars on plant available water and water use efficiency. *Geoderma* **2017**, *307*, 114–121. [CrossRef]
25. Blanco-Canqui, H. Biochar and soil physical properties. *Soil Sci. Soc. Am. J.* **2017**, *81*, 687–711. [CrossRef]
26. Baiamonte, G.; Crescimanno, G.; Parrino, F.; De Pasquale, C. Effect of biochar on the physical and structural properties of a desert sandy soil. *Catena* **2019**, *175*, 294–303. [CrossRef]
27. Stewart, C.E.; Zheng, J.; Botte, J.; Cotrufo, F. Co-generated fast pyrolysis biochar mitigates greenhouse gas emissions and increases carbon sequestration intemperate soils. *Glob. Chang. Biol. Bioenergy* **2013**, *5*, 153–164. [CrossRef]
28. Chintala, R.; Mollinedo, J.; Schumacher, T.E.; Malo, D.D.; Julson, J.L. Effect of biochar on chemical properties of acidic soil. *Arch. Agron. Soil Sci.* **2014**, *60*, 393–404. [CrossRef]
29. Xu, G.; Sun, J.; Shao, H.; Chang, S.X. Biochar had effects on phosphorus sorption and desorption in three soils with differing acidity. *Ecol. Eng.* **2014**, *62*, 54–60. [CrossRef]
30. Jeffery, S.; Abalos, D.; Prodana, M.; Bastos, A.C.; Van Groenigen, J.W.; Hungate, B.A.; Verheijen, F. Biochar boosts tropical but not temperate crop yields. *Environ. Res. Lett.* **2017**, *12*, 053001. [CrossRef]
31. Raboin, L.-M.; Razafmahafaly, A.H.D.; Rabenjarisoa, M.B.; Rabary, B.; Dusserre, J.; Becquer, T. Improving the fertility of tropical acid soils: Liming versus biochar application? A long term comparison in the highlands of Madagascar. *Field Crops Res.* **2016**, *199*, 99–108. [CrossRef]
32. Liu, X.; Zhang, A.; Ji, C.; Joseph, S.; Bian, R.; Li, L.; Pan, G.; Paz-Ferreiro, J. Biochar's effect on crop productivity and the dependence on experimental conditions—A meta-analysis of literature data. *Plant Soil* **2013**, *373*, 583–594. [CrossRef]
33. Uzoma, K.C.; Inoue, M.; Andry, H.; Fujimaki, H.; Zahoor, A.; Nishihara, E. Effect of cow manure biochar on maize productivity under sandy soil condition. *Soil Use Manag.* **2011**, *27*, 205–212. [CrossRef]

34. Foster, E.J.; Hansen, N.; Wallenstein, M.; Cotrufo, M.F. Biochar and manure amendments impact soil nutrients and microbial enzymatic activities in a semi-arid irrigated maize cropping system. *Agric. Ecosyst. Environ.* **2016**, *233*, 404–414. [CrossRef]
35. Mulcahy, D.N.; Mulcahy, D.L.; Dietz, D. Biochar soil amendment increases tomato seedling resistance to drought in sandy soils. *J. Arid Environ.* **2013**, *88*, 222–225. [CrossRef]
36. Laghari, M.; Mirjat, M.S.; Hu, Z.; Fazal, S.; Xiao, B.; Hu, M.; Chen, Z.; Guo, D. Effects of biochar application rate on sandy desert soil properties and sorghum growth. *Catena* **2015**, *135*, 313–320. [CrossRef]
37. Ibrahim, H.M.; Al-Wabel, M.I.; Usman, A.R.; Al-Omran, A. Effect of Conocarpus biochar application on the hydraulic properties of a sandy loam soil. *Soil Sci.* **2013**, *178*, 165–173. [CrossRef]
38. Khalifa, N.; Yousef, L.F. A short report on changes of quality indicators for a sandy textured soil after treatment with biochar produced from fronds of date palm. *Energy Procedia* **2015**, *74*, 960–965. [CrossRef]
39. El-Naggar, A.H.; Usman, A.R.; Al-Omran, A.; Ok, Y.S.; Ahmad, M.; Al-Wabel, M.I. Carbon mineralization and nutrient availability in calcareous sandy soils amended with woody waste biochar. *Chemosphere* **2015**, *138*, 67–73. [CrossRef]
40. Usman, A.R.A.; Al-Wabel, M.I.; Ok, Y.S.; Al-Harbi, A.; Wahb-Allah, M.; El-Naggar, A.H.; Ahmad, M.; Al-Faraj, A.; Al-Omran, A. Conocarpus biochar induces changes in soil nutrient availability and tomato growth under saline irrigation. *Pedosphere* **2016**, *26*, 27–38. [CrossRef]
41. Bashour, I.I.; Al-Mashhady, A.S.; Prasad, J.D.; Miller, T.; Mazroa, M. Morphology and composition of some soils under cultivation in Saudi Arabia. *Geoderma* **1983**, *29*, 327–340. [CrossRef]
42. Nelson, D.W.; Sommers, L.E. Total carbon, organic carbon, and organic matter. In *Methods of Soil Analysis. Part 3. Chemical Methods*; SSSA Book Series; Sparks, D.L., Ed.; SSSA and ASA: Madison, WI, USA, 1996; No. 5; pp. 961–1010.
43. Loeppert, R.H.; Suarez, D. Carbonate and gypsum. In *Methods of Soil Analysis. Part 3. Chemical Methods*; Sparks, D.L., Bigham, J.M., Eds.; SSSA and ASA: Madison, WI, USA, 1996; pp. 437–474.
44. Gee, G.W.; Bauder, J.W. Particle-size analysis. In *Methods of Soil Analysis. Part 1. Physical and Mineralogical Methods*, 2nd ed.; Klute, A., Ed.; SSSA and ASA: Madison, WI, USA, 1986; pp. 383–411.
45. Usman, A.R.; Abduljabbar, A.; Vithanage, M.; Ok, Y.S.; Ahmad, M.; Ahmad, M.; Elfaki, J.; Abdulazeem, S.S.; Al-Wabel, M.I. Biochar production from date palm waste: Charring temperature induced changes in composition and surface chemistry. *J. Anal. Appl. Pyrolysis* **2015**, *115*, 392–400. [CrossRef]
46. Thomas, R.L.; Sheard, R.W.; Moyer, J.R. Comparision of conventional and automated procedures for nitrogen, phosphorus and potassium analysis of plant material using a single digestion. *Agron. J.* **1967**, *59*, 240–243. [CrossRef]
47. Novak, J.M.; Lima, I.; Xing, B.; Gaskin, J.W.; Steiner, C.; Das, K.C.; Ahmendna, M.; Rehrah, D.; Watts, D.W.; Busscher, W.J.; et al. Characterization of designer biochar produced at different temperatures and their effects on a loamy sand. *Ann. Environ. Sci.* **2009**, *3*, 195–206.
48. Blake, G.R.; Hartge, K.H. Bulk density. In *Methods of Soil Analysis. Part 1*, 2nd ed.; Klute, A., Ed.; SSSA: Madison, WI, USA, 1986; pp. 363–376.
49. Rhoades, J.D. Cation exchange capacity. In *Methods of Soil Analysis. Part 2: Chemical and Mineralogical Properties*; Page, A.L., Ed.; American Society of Agronomy: Madison, WI, USA, 1982; No. 9; pp. 149–157.
50. Asai, H.; Samson, B.K.; Stephan, H.M.; Songyikhangsuthor, K.; Homma, K.; Kiyono, Y.; Inoue, Y.; Shiraiwa, T.; Horie, T. Biochar amendment techniques for upland rice production in Northern Laos: 1. Soil physical properties, leaf SPAD and grain yield. *Field Crops Res.* **2009**, *111*, 81–84. [CrossRef]
51. Schulz, H.; Glaser, B. Effects of biochar compared to organic and inorganic fertilizers on soil quality and plant growth in a greenhouse experiment. *J. Plant Nutr. Soil Sci.* **2012**, *175*, 410–422. [CrossRef]
52. Kookana, R.S.; Sarmah, A.K.; Van Zwieten, L.; Krull, E.; Singh, B. Biochar application to soil: Agronomic and environmental benefits and unintended consequences. *Adv. Agron.* **2011**, *112*, 103–143.
53. Nguyen, T.T.N.; Xu, C.Y.; Tahmasbian, I.; Che, R.; Xu, Z.; Zhou, X.; Wallace, H.M.; Bai, S.H. Effects of biochar on soil available inorganic nitrogen: A review and meta-analysis. *Geoderma* **2017**, *288*, 79–96. [CrossRef]
54. Recous, S.; Robi, D.; Darwis, D.; Mary, B. Soil inorganic N availability: Effect on maize residue decomposition. *Soil Biol. Biochem.* **1995**, *27*, 1529–1538. [CrossRef]
55. Sakala, W.D.; Cadisch, G.; Giller, K.E. Interactions between residues of maize and pigeonpea and mineral N fertilizers during decomposition and N mineralization. *Soil Biol. Biochem.* **2000**, *32*, 679–688. [CrossRef]

56. Basso, A.S.; Miguez, F.E.; Laird, D.A.; Horton, R.; Westgate, M. Assessing potential of biochar for increasing water-holding capacity of sandy soils. *GCB Bioenergy* **2013**, *5*, 132–143. [CrossRef]
57. Sun, Z.; Moldrup, P.; Elsgaard, L.; Artur, E.; Bruun, E.; Hauggaard-Nielsen, H.; de Jonge, L.W. Direct and indirect short-term effects of biochar on physical characteristics of an arable sandy loam. *Soil Sci.* **2013**, *178*, 465–473. [CrossRef]
58. Hardie, M.; Clothier, B.; Bound, S.; Oliver, G.; Close, D. Does biochar influence soil physical properties and soil water availability? *Plant Soil* **2014**, *376*, 347–361. [CrossRef]
59. Jeffery, S.; Meinders, M.B.J.; Stoof, C.R.; Bezemer, T.M.; van de Voorde, T.F.J.; Mommer, L.; van Groenigen, J.W. Biochar application does not improve the soil hydrological function of a sandy soil. *Geoderma* **2015**, *251*, 47–54. [CrossRef]
60. Kinney, T.J.; Masiello, C.A.; Dugan, B.; Hockaday, W.C.; Dean, M.R.; Zygourakis, K.; Barnes, R.T. Hydrologic properties of biochars produced at different temperatures. *Biomass Bioenergy* **2012**, *41*, 34–43. [CrossRef]
61. Gray, M.; Johnson, M.G.; Dragila, M.I.; Kleber, M. Water uptake in biochars: The roles of porosity and hydrophobicity. *Biomass Bioenergy* **2014**, *61*, 196–205. [CrossRef]
62. Suliman, W.; Harsh, J.B.; Abu-Lail, N.I.; Fortuna, A.M.; Dallmeyer, I.; Garcia-Pérez, M. The role of biochar porosity and surface functionality in augmenting hydrologic properties of a sandy soil. *Sci. Total Environ.* **2017**, *574*, 139–147. [CrossRef] [PubMed]
63. Chen, B.; Zhou, D.; Zhu, L. Transitional adsorption and partition of nonpolar and polar aromatic contaminants by biochars of pine needles with different pyrolytic temperatures. *Environ. Sci. Technol.* **2008**, *42*, 5137–5143. [CrossRef] [PubMed]
64. Zimmerman, A.R. Abiotic and microbial oxidation of laboratory-produced black carbon (biochar). *Environ. Sci. Technol.* **2010**, *44*, 1295–1301. [CrossRef] [PubMed]
65. Kameyama, K.; Miyamoto, T.; Iwata, Y.; Shiono, T. Effects of biochar produced from sugarcane bagasse at different pyrolysis temperatures on water retention of a calcaric dark red soil. *Soil Sci.* **2016**, *181*, 20–28. [CrossRef]
66. Günal, E.; Erdem, H.; Çelik, İ. Effects of three different biochars amendment on water retention of silty loam and loamy soils. *Agric. Water Manag.* **2018**, *208*, 232–244. [CrossRef]
67. Lei, O.; Zhang, R.D. Effects of biochars derived from different feedstocks and pyrolysis temperatures on soil physical and hydraulic properties. *J. Soils Sediments* **2013**, *13*, 1561–1572. [CrossRef]
68. Jindo, K.; Mizumoto, H.; Sawada, Y.; Sanchez-Monedero, M.A.; Sonoki, T. Physical and chemical characterization of biochars derived from different agricultural residues. *Biogeosciences* **2014**, *11*, 6613–6621. [CrossRef]
69. Ahmed, H.P.; Schoenau, J.J. Effects of biochar on yield, nutrient recovery, and soil properties in a canola (*Brassica napus* L)-wheat (*Triticum aestivum* L) rotation grown under controlled environmental conditions. *Bioenergy Res.* **2015**, *8*, 1183–1196. [CrossRef]
70. Liu, X.H.; Zhang, X.C. Effect of Biochar on pH of Alkaline Soils in the Loess Plateau: Results from Incubation Experiments. *Int. J. Agric. Biol.* **2012**, *14*, 745–750.
71. Al-Wabel, M.I.; Usman, A.R.; Al-Farraj, A.S.; Ok, Y.S.; Abduljabbar, A.; Al-Faraj, A.I.; Sallam, A.S. Date palm waste biochars alter a soil respiration, microbial biomass carbon, and heavy metal mobility in contaminated mined soil. *Environ. Geochem. Health* **2017**, 1–18. [CrossRef] [PubMed]
72. Ouyang, L.; Yu, L.; Zhang, R. Effects of amendment of different biochars on soil carbon mineralization and sequestration. *Soil Res.* **2014**, *52*, 46–54. [CrossRef]
73. El-Naggar, A.; Lee, S.S.; Awad, Y.M.; Yang, X.; Ryu, C.; Rizwan, M.; Rinklebe, J.; Tsang, D.C.; Ok, Y.S. Influence of soil properties and feedstocks on biochar potential for carbon mineralization and improvement of infertile soils. *Geoderma* **2018**, *332*, 100–108. [CrossRef]

© 2019 by the authors. Licensee MDPI, Basel, Switzerland. This article is an open access article distributed under the terms and conditions of the Creative Commons Attribution (CC BY) license (http://creativecommons.org/licenses/by/4.0/).

Article

The Influence of Biochar and Solid Digestate on Rose-Scented Geranium (*Pelargonium graveolens* L'Hér.) Productivity and Essential Oil Quality

Alessandro Calamai [1,*], Enrico Palchetti [1], Alberto Masoni [1], Lorenzo Marini [1], David Chiaramonti [2], Camilla Dibari [1] and Lorenzo Brilli [3]

1 DAGRI, University of Florence, Piazzale delle Cascine 18, 50144 Firenze, Italy; enrico.palchetti@unifi.it (E.P.); alberto.masoni@unifi.it (A.M.); lo.marini@unifi.it (L.M.); camilla.dibari@unifi.it (C.D.)
2 RE-CORD and CREAR, Department of Industrial Engineering, University of Florence, Viale Morgagni 40, 50134 Florence, Italy; david.chiaramonti@re-cord.org
3 IBIMET-CNR, Via G. Caproni 8, 50145 Firenze, Italy; l.brilli@ibimet.cnr.it
* Correspondence: alessandro.calamai@unifi.it; Tel.: +39-055-2755800

Received: 17 April 2019; Accepted: 20 May 2019; Published: 22 May 2019

Abstract: In recent years, biochar has generated global interest in the areas of sustainable agriculture and climate adaptation. The main positive effects of biochar were observed to be the most remarkable when nutrient-rich feedstock was used as the initial pyrolysis material (i.e., anaerobic digestate). In this study, the influence of solid anaerobic digestate and biochar that was produced by the slow pyrolysis of solid digestate was evaluated by comparing the differences in the crop growth performances of *Pelargonium graveolens*. The experiment was conducted in a greenhouse while using three different growth media (i.e., solid digestate, biochar, and vermiculite). The results indicated that: (i) the pyrolysis of solid digestate caused a reduction in the bulk density (−52%) and an increase in the pH (+16%) and electrical conductivity (+9.5%) in the derived biochar; (ii) the best crop performances (number of leaves, number of total branches, and plant dry weight) were found using biochar, particularly for plant dry weight (+11.4%) and essential oil content (+9.4%); (iii) the essential oil quality was slightly affected by the growth media; however, the main chemical components were found within the acceptable range that was set by international standard trade; and, iv) biochar induced the presence of leaf chlorosis in *Pelargonium graveolens*.

Keywords: biochar; solid digestate; *Pelargonium graveolens*; leaf chlorosis; essential oil quality

1. Introduction

In recent decades, biochar, a carbon-rich product that is generated by pyrolysis of biomass under anaerobic conditions [1], has garnered much interest as a soil amendment. Nevertheless, several studies report contrasting results when biochar is incorporated in the potting substrates [2,3] or in open soil systems [4].

Among the numerous positive effects of biochar, the improvement of the soil water-holding capacity, increased aeration and cation exchange capacity (CEC), liming the potential, and increased nutrient availability are among the most relevant ones [5–8]. Furthermore, biochar contributes to carbon sequestration, thanks to its high stability and resistance to biological degradation, thus contributing to climate change mitigation, i.e., reducing agricultural greenhouse gas emissions [1,9]. Nonetheless, despite the aforementioned positive properties, biochar application in agricultural soils could potentially generate risks of increased concentrations of pollutants and toxic compounds, such as heavy metals, polyhydroxyalkanoates (PHAs), etc., thus promoting their incorporation in the soil [10]. In fact, as biochar is obtained by heating the biomass in the complete absence of oxygen, tars are generated, and could condense onto the biochar; moreover, if biomass contains heavy metals, these will be retained in the solid matter [11]. The method through which biochar is produced and extracted by the plant, together with the type of feedstock, will determine the different characteristics of the carbonized products.

The chemical and physical biochar characteristics depend on the initial structure of the lignocellulosic biomass (i.e., woods and barks, agricultural wastes, green-waste, and animal manures) and process conditions (i.e., type of pyrolysis, heating rate, maximum temperature and holding time at maximum temperature, biochar extraction, and condensation of volatiles), which thus result in mixed and contrasting outcomes regarding the promotion of soil fertility [12]. Generally, biochar that is produced from nutrient-rich feedstock, such as crop residues, manures, and anaerobic digestion residues (i.e., digestate) possesses higher nutrient contents than biochar that is generated from nutrient-poor feedstock (i.e., wood residues), which, in turn, determines lower soil benefits due to lower mineralization rates [6,13].

The anaerobic fermentation process is considered as being one of the most environmentally friendly methods to convert organic material, such as municipal solid wastes, manures, energy crops, and agricultural residues, into biogas, an energy-rich product. The degradation process takes place in biogas plants under optimal conditions (temperature, mixing, pH, etc.) and it is driven by a microbial consortium [14,15]. Digestate by-product, which is characterized by a nutrient-rich and homogeneous mixture of microbial biomass and undigested material, is usually used as fertilizer for supplying nutrients [15]. Moreover, digestate can be mechanically separated into liquid (80–90%) and solid fractions (10–20%), in order to facilitate the storage, handling, transport, and discharge [14]. As digestate is produced in large quantities throughout the year, it is often stored and distributed on soils only during dry days in order to avoid nutrient leaching [8]. The pyrolysis of digestate can generate biochar with a higher cation exchange capacity and higher phosphorous concentration when compared to the initial material and other feedstock [13–16].

The objectives of this study were: (i) the characterization of biochar as derived from solid digestate and the initial feedstock; (ii) the determination of the physical–chemical characteristics of the growth media after addition of soil amendments; and, (iii) evaluation of the growth-media-induced influence on yield, growth, and oil quality in rose-scented geranium (*Pelargonium graveolens* L'Hér.) cultivation.

The crop, which belongs to the *Geraniaceae* family, is an important aromatic plant that is cultivated in several countries (e.g., Algeria, Egypt, Morocco, Madagascar, France, China, and India) for the extraction of essential oil (EO), widely used in the perfumery, cosmetic, pharmaceutical, and food industries [17–19], and chosen because of its sensitivity to growth when it is cultivated using different substrates [20,21].

2. Materials and Methods

2.1. Description of Substrates and Their Components

The substrates were generated by mixing three different components: solid digestate, biochar, and vermiculite. (i) Solid digestate was obtained from thermophilic anaerobic digestion of corn silage, manure, and vegetable waste, in a biogas plant that is located in Ravenna (Italy). The samples were collected during spring 2016 and air-dried at 25 °C for two weeks before use. (ii) Biochar was obtained by slow pyrolysis in a pilot unit, fed with the digestate. The biomass in the pyrolyser was continuously mixed by rotating paddles. Carbonization was carried out at a temperature of 500 °C for 3 h, under a continuous flow of N_2, continuous extraction of pyrogases, and a heating rate of 25 °C/min. At the end of the set reaction time, the reactor was left to cool down to ambient temperature (continuing the extraction of pyrogases) overnight and then the biochar was extracted. The process yield (33.6%) was calculated from triplicate samples and expressed as weight percentages of dry weight biochar recovered to dry weight initial biomass, according to Vaughn et al. [22]. (iii) The vermiculite, which was used as a neutral filler component, was composed of particle sizes with diameters of 2–5 mm.

The three growth media were prepared by mixing all of the components. The products were differentiated, as follows: control (CS): vermiculite 100%; biochar substrate (BCS): vermiculite and biochar in a ratio 2:1 by volume; solid digestate substrate (SDS): vermiculite and solid digestate in a ratio of 2:1 (*v:v*).

In setting up the trial, growth media and primary components were chemically and physically characterized. The samples (three replicates) of each growth media were previously dried at 30–35 °C for six days, ground in a ceramic mortar, and sifted through a 2 mm sieve. The total C, H, N, and S contents were determined using a CHNS analyzer (LECO Corp. St. Joseph, MI, USA) following the American Society for Testing and Materials (ASTM) method D5373, while the elemental composition (As, Ba, Ca, Cd, Cr, Cu, Fe, K, Mg, Mn, Na, Ni, P, Pb, and Zn) was detected using an Inductively-Coupled Plasma Optical Emission Spectrometer (ICP, model IRIS Intrepid II XSP Radial, Thermo Fisher Scientific, Waltham, MA, USA), according to the EPA Method SW6020. Electrical conductivity (EC) and pH were measured following Ahmedna et al. [23] in water extracts (sample: deionized water ratio of 1:1, *w/w*) by employing a portable EC meter (Hanna Instruments HI 9313-6, Rhode Island, USA) and a pH meter (pHenomenal pH 1100L, VWR International, Leuven, Germany).

The physical properties (i.e., bulk density, total porosity, container capacity, and air space) were determined by using the ring knife method [24]. A ring knife with a volume of 200 cm^3 and weight W_0 was filled with the air-dried sample and weighed (W_1). Afterwards, the sample was saturated with distilled water for 24 h and weighed again (W_2). The ring knife with the saturated sample was opened from one side and then placed upside down on a holder with a leaky screen until the water stopped dripping from the bottom for 3 h before the ring knife and sample were weighed (W_3). Finally, the oven dried sample and the ring knife were kept at 65 °C until a constant weight was reached and weighed (W_4). The physical characteristics were calculated while using the following formulas: bulk density (g cm^{-3}) = ($W_4 - W_0$)/200; total porosity (%) = ($W_2 - W_4$)/200 × 100; air space (%) = ($W_2 - W_3$)/200 × 100; and, container capacity (%) = total porosity − air space.

2.2. Experimental Design

The experiment was carried out in a greenhouse that was located at the University of Florence (Italy). Plants of rose-scented geranium (*Pelargonium graveolens*) were cloned by first obtaining tissue culture from the same mother plant native to Madagascar using the methods reported by Grassi et al. [25]. After seven months of in-vitro cultivation, plantlets of about 10 cm in height, with seven leaves, and well-developed rooting systems were transplanted into 2 L plastic pots containing the experimental substrates.

A total of 21 pots for each growth medium were prepared and arranged in a randomized complete block design (RCBD) with three replicates. The pots were placed at 60 cm from each other between rows and at 30 cm from each other within a row, similar to the optimum plantation density in an open field [14]. The trial was conducted during the spring–autumn season (July–September), where the maximum and minimum temperatures were 29.8 °C and 17.0 °C, respectively. Within each treatment, five pots were randomly selected and equipped with a soil moisture sensor. An automated irrigation scheduling program was set up to maintain the soil moisture content at 60%, and thus avoid water stress.

2.3. Sampling and Plant Analysis

The biomass was harvested 70 days after transplanting, during the early flowering stage (i.e., balsamic time) in order to simulate the common cultivation practice. Cutting the flowers' stems and leaving 1–2 growing buds for the subsequent regeneration harvested the plants [18]. All of the vegetative parameters, such as plant height, number of leaves, number of principal branches, number of total branches, plant dry weight, and a soil plant analysis development (SPAD) chlorophyll meter were collected and recorded at the time of harvest. Three fully expanded and healthy leaves (top, middle, and bottom of the plant height) were collected for each plant at harvest time, dried in oven at 65 °C, ground in a ceramic mortar, and sifted through a 2 mm sieve in order to determine the nutrient concentration in biomass. The concentrations of macro and micro-nutrients were determined according to the CHNS and ICP methods.

2.4. Essential Oil Extraction and GC–MS Analysis

The EO extraction was performed while using the hydro-distillation method using a Clevenger apparatus (Ambala Cantt, Haryana, India) [26]. The distilled EO of each sample was filtered, dehydrated with anhydrous sodium sulfate, and stored at 0 °C until analyzed. The essential oil yield (EO%) of the distilled biomass was obtained as the percentage on a volume basis (ml oil extracted from 100 g dry plant herbage), while the essential oil content for each plant (EOpl) was calculated by multiplying the EO% for the biomass weight and expressed in g oil/plant. GC–MS analysis was performed on a PerkinElmer 8500 gas chromatograph (Perkin–Elmer, Norwalk, CT, USA), which was equipped with a fused silica column BP1 25 m, 0.5 mm, and ID 25 µm. The column was programmed from 60 to 250 °C at a rate of 5 °C min^{-1} and the maximum temperature was maintained for 5 min. The injection and detector temperatures were 250 °C and 300 °C, respectively. Nitrogen (1.1 mL min^{-1}) was used as the carrier gas and the split ratio was 1:50. Compound identification was performed by comparing the retention times of spectral peaks using the NIST08 library database, and then the corresponding Kovats index was calculated.

2.5. Statistical Analysis

Statistical analysis was performed using IBM SPSS Statistics v.21 software (IBM Corp., New-York, NY, USA). Analysis of variance (ANOVA) was used to detect the significance of the effects of the treatments on the vegetative parameters and nutrient content in plants. Data that did not fulfill ANOVA assumptions were the arcsin square root that was transformed before running the model. The most significant parameters that were identified by ANOVA were successively analyzed using a post hoc Tukey's test implemented using the same software.

3. Results

3.1. Characterization of Digestate and Biochar

Table 1 reports the chemical and physical properties of biochar and its initial feedstock (solid digestate). The solid digestate fraction had a pH of 7.1 and EC value of 6.21 mS cm^{-1}. The pyrolysis process resulted in an increase of about one pH unit (8.5) and a slight increase in the EC value, reaching 6.86 mS cm^{-1}. An increase in several of the macro and micro-element contents was observed in the biochar with respect to the solid digestate. The only exception was observed for S content, which was totally absent in biochar. Specifically, biochar showed an increase in N (+11%), P (+41%), and K (+41%) contents when compared to solid digestate, and a large amount of Fe, Mn, and Na (i.e., 1935.00, 454.60, and 1235.00 mg kg^{-1}), with respect to the initial material (i.e., 998.74, 168.80, and 827.00 mg kg^{-1}). The carbon content (C) increased, from 39.8% in the digestate to 55.83% in biochar. The pyrolysis process also affected the bulk density, which shifted from 0.29 kg m^{-3} in solid digestate to 0.14 kg m^{-3} in the pyrolyzed product.

3.2. Characterization of Growth Media

Biochar and solid digestate influenced the main chemical and physical properties of the growth media (Table 1). BCS showed the highest pH value (7.8), followed by SDS (6.7) and CS (5.6). Regarding EC data, the substrate added with biochar (BCS) exhibited a slightly higher value (2.11 mS cm^{-1}) than SDS (1.98 mS cm^{-1}) and CS (1.79 mS cm^{-1}). The bulk density and air space yielded the highest values in SDS (0.16 kg m^{-3} and 29.6%, respectively), followed by BCS (0.11 kg m^{-3} and 27.4%, respectively) and CS (0.09 kg m^{-3} and 23.3%, respectively). By contrast, CS recorded the highest values for the container capacity (65.5%) and total porosity (88.8%). The macro and microelement contents were highest in BCS, whilst lower values were found in SDS and CS.

Table 1. Physical–chemical properties of organic material and growth media used in *P. graveolens* cultivation. All values are means of three replicates ± standard deviation. Solid digestate; biochar; SDS (Solid Digestate Substrate); BCS (Biochar Substrate); CS (Control Substrate).

Parameters	Primary Components		Growth Media		
	Solid Digestate	Biochar	SDS	BCS	CS
pH	7.1 ± 0.12	8.5 ± 0.07	6.7 ±0.09	7.8 ± 0.08	5.6 ± 0.09
EC (mS cm^{-1})	6.21 ± 0.03	6.86 ± 0.09	1.98 ± 0.02	2.11 ± 0.03	1.79 ± 0.04
Bulk density (kg m^{-3})	0.29 ± 0.04	0.14 ± 0.10	0.16 ± 0.02	0.11 ± 0.06	0.09 ± 0.10
Air space (%)	NA	NA	29.60 ± 1.71	27.40 ± 2.11	23.30 ± 0.89
Container capacity (%)	NA	NA	50.20 ± 0.11	56.80 ± 1.01	65.50 ± 0.65
Total porosity (%)	NA	NA	79.80 ± 0.78	84.20 ± 0.29	88.80 ± 0.57
C (%)	39.80 ± 0.23	55.83 ± 0.64	16.11 ± 0.58	19.01 ± 0.03	13.91 ± 0.09
H (%)	5.68 ± 1.14	2.03 ± 0.99	0.91 ± 0.68	0.94± 0.12	0.83 ± 0.28
N (%)	1.23 ± 1.18	1.46 ± 0.97	0.72 ± 1.89	0.84 ± 0.95	0.05 ± 1.06
S (%)	0.29 ± 0.09	ND	ND	ND	ND
K (%)	1.41 ± 1.03	3.43 ± 0.41	1.15 ± 0.83	2.53 ± 0.67	0.65 ± 0.76
P (%)	0.79 ± 0.51	1.91 ± 0.09	0.47 ± 0.52	0.71 ± 0.29	0.15 ± 0.38
Ca (%)	2.00 ±0.39	3.93 ± 0.87	2.22 ± 0.74	2.91 ± 0.03	1.76 ± 0.26
Mg (%)	0.47 ± 0.21	0.95 ± 0.27	2.61 ± 0.63	2.69 ± 0.11	2.86 ± 0.18
As (mg kg^{-1})	4.56 ± 0.03	8.03 ± 0.29	8.67 ± 0.07	8.78 ± 0.37	8.99 ± 0.09
Ba (mg kg^{-1})	15.54 ± 0.19	35.53 ± 0.12	256.15 ± 0.16	262.90 ± 0.23	319.61 ± 0.31
Cd (mg kg^{-1})	0.05 ± 0.22	0.28 ± 0.7	0.43 ± 0.81	0.50 ± 0.29	0.43 ± 0.41
Cr (mg kg^{-1})	7.53 ± 1.14	14.15 ± 1.28	59.47 ± 0.03	83.66 ± 0.07	79.04 ± 0.56
Cu (mg kg^{-1})	30.37 ± 2.12	61.52 ± 3.47	69.13 ± 2.93	72.41 ± 4.29	89.12 ± 3.49
Fe (mg kg^{-1})	998.74 ± 0.04	1935.00 ± 0.02	680.21 ± 0.14	880.36 ± 0.75	553.02 ± 0.39
Mn (mg kg^{-1})	168.60 ± 0.17	454.60 ± 0.64	356.50 ± 0.21	429.60 ± 0.22	431.80 ± 0.28
Na (mg kg^{-1})	827.00 ± 0.56	1235.00 ± 0.49	1675.00 ± 0.79	1848.00 ± 0.11	1704.00 ± 0.23
Ni (mg kg^{-1})	3.94 ± 0.03	10.58 ± 0.49	40.35 ± 0.87	42.45 ± 0.76	58.35 ± 0.24
Pb (mg kg^{-1})	1.92 ± 0.98	6.07 ± 1.68	50.43 ± 3.21	56.86 ± 3.23	43.01 ± 2.47
Zn (mg kg^{-1})	170.70 ± 0.86	380.70 ± 0.91	166.80 ± 0.49	228.40 ± 0.73	85.40 ± 0.55

Note: NA, not available; ND, not detectable.

3.3. Effect on Vegetative Parameters and Foliar Nutrient Content

Table 2 displays the growth media effects on vegetative parameters. Overall, the number of total branches, plant dry weight, and SPAD showed high statistical significance ($p < 0.01$), whilst the number of leaves and number of principal branches were less influenced by the substrates ($p < 0.05$). No statistical significance ($p > 0.05$) was found for plant height.

Using BCS, all of the parameters were significantly higher than those that were found in SDS and CS, with the exception of the SPAD parameters. On average, plants that were cultivated under BCS had about 146 leaves, 2.5 principal branches, 16 total branches, a dry weight of 122.7 g, and a SPAD value of 17.89. Using SDS, the most important decrease was found in the plant dry weight (−11.4%), thus reflecting the lower total number of branches (−41.6%) and leaves (−13.3%) when compared to BCS. By contrast, an increase was observed in SPAD (+42.7%). The lowest performances were observed under CS. Considerable reductions as compared to BCS were found in the number of leaves (−28%) and plant dry weight (−67.5%). SPAD showed an increase of about 40%.

Leaf decoloration was also observed when BCS was used. This condition was indicated by SPAD (17.9), which was much lower when compared to SDS (31.2) and CS (29.8).

Differences in growth media also significantly affected the accumulation of leaf nutrients (Table 3). In particular, plants that were cultivated in BCS showed the highest content in C (41.9%), N (2.2%), P (0.4%), K (3%), Fe (184.7 mg kg^{-1}), and Mn (79.4 mg kg^{-1}). Conversely, considerable decreases in nutrient contents were observed both using SDS and CS, particularly for Fe (−49% using SDS) and Mn (−70% using CS).

Table 2. Effects of growth media on vegetative parameters and essential oil response in *P. graveolens* cultivation. All the values are means of three replicates ± standard deviation. SDS, solid digestate growth media; BCS, biochar growth media; CS, control growth media; SPAD, soil plant analysis development; EO, essential oil.

Vegetative Parameters	Growth Media		
	SDS	BCS	CS
Plant height (cm)	33.90 ± 2.91 a	34.91 ± 3.09 a	35.10 ± 3.20 a
Number of leaves	126.81 ± 9.31 ab	146.23 ± 14.32 a	105.62 ± 7.59 b
Number of principal branches	2.33 ± 1.02 a	2.55 ± 0.89 a	1.71 ± 0.64 b
Number of total branches	9.38 ± 3.76 b	16.05 ± 5.06 a	8.24 ± 2.96 b
Plant dry weight (g)	108.67 ± 5.74 ab	122.70 ± 6.31 a	82.81 ± 4.49 b
SPAD	31.22 ± 4.37 a	17.89 ± 2.68 b	29.80 ± 5.19 a

EO Parameters	Growth Media		
	SDS	BCS	CS
Oil yield (%)	0.126 ± 1.15 a	0.122 ± 0.40 a	0.133 ± 0.63 a
Oil content (g)	0.136 ± 0.98 ab	0.150 ± 0.56 a	0.110 ± 1.02 b

Note: means followed by common letters within the same row are not significantly different ($p > 0.05$).

Table 3. Foliar nutrient content in a *P. graveolens* plant grown in different growth media. All values are means of three replicates ± standard deviation. SDS, solid digestate growth media; BCS, biochar growth media; CS, control growth media.

Elements	Units	Growth Media		
		SDS	BCS	CS
C	(%)	40.83 ± 1.51 a	41.94 ± 1.16 a	36.46 ± 0.88 b
H	(%)	7.36 ± 1.71 a	7.74 ± 2.01 a	7.05 ± 1.46 a
N	(%)	1.75 ± 0.63 ab	2.16 ± 0.88 a	1.23 ± 1.37 b
S	(%)	0.07 ± 0.33 a	0.08 ± 0.10 a	0.05 ± 0.54 a
K	(%)	1.57 ± 2.11 b	3.01 ± 0.49 a	1.33 ± 0.69 b
P	(%)	0.21 ± 1.36 ab	0.38 ± 1.76 a	0.10 ± 0.86 b
Ca	(%)	2.97 ± 0.10 a	3.73 ± 0.44 a	2.41 ± 0.46 a
Mg	(%)	0.41 ± 0.17 a	0.54 ± 0.27 a	0.35 ± 0.19 a
As	(mg kg^{-1})	2.13 ± 0.92 a	1.52 ± 0.41 a	0.89 ± 0.25 a
Ba	(mg kg^{-1})	24.06 ± 2.09 a	23.05 ± 1.87 a	26.50 ± 1.21 a
Cd	(mg kg^{-1})	ND	ND	ND
Cr	(mg kg^{-1})	0.47 ± 0.32 a	0.94 ± 0.13 a	0.88 ± 0.22 a
Cu	(mg kg^{-1})	3.30 ± 0.74 a	4.19 ± 0.93 a	3.78 ± 0.59 a
Fe	(mg kg^{-1})	90.32 ± 1.06 b	184.74 ± 1.50 a	90.93 ± 0.98 b
Mn	(mg kg^{-1})	43.39 ± 1.49 b	79.35 ± 1.32 a	23.32 ± 1.39 b
Na	(mg kg^{-1})	238.78 ± 0.13 a	235.26 ± 0.37 a	190.36 ± 0.28 b
Ni	(mg kg^{-1})	1.08 ± 0.06 a	1.41 ± 0.08 a	2.40 ± 0.04 a
Pb	(mg kg^{-1})	0.11 ± 0.35 a	0.14 ± 0.16 a	0.35 ± 0.98 a
Zn	(mg kg^{-1})	42.70 ± 0.75 a	38.88 ± 0.51 a	29.17 ± 0.43 b

Note: ND, not detectable. Means followed by common letters within the same row are not significantly different ($p > 0.05$).

3.4. Effect on Essential Oil Yield and Quality

Essential oil yield (EO%) was similar in plants that were cultivated in CS (0.13%), SDS (0.13%), and BCS (0.12%) (Table 2), while the essential oil content (EOPl) was higher in BCS (0.15 g/plant) as compared to that found in SDS (0.14 g/plant) and in CS (0.11 g/plant).

The GC–MS analysis (Table 4) of essential oil showed that the five most represented chemical oil components were: citronellol (27.9–33.2%), citronellyl formate (15–17%), geraniol (13.8–15.6%), isomenthone (4.4–5.4%), and linalool (2.4–3.1%). Among these, only citronellol and geraniol appear to be influenced by the growth media, with CS possessing the highest content of geraniol (33.2%) and the lowest content of citronellol (13.8%). Finally, the values of the citronellol/geraniol ratio were 1.8, 2.1, and 2.4 using BCS, SDS, and CS, respectively.

Table 4. The effects of growth media on the main chemical components of essential oil in *P. graveolens* cultivation. All values are means of three replicates ± standard deviation. SDS, solid digestate growth media; BCS, biochar growth media; CS, control growth media.

Parameters	Units	Growth Media		
		SDS	BCS	CS
Linalool	(%)	3.04 ± 0.16 a	2.39 ± 0.09 a	3.07 ± 0.10 a
Isomenthone	(%)	4.38 ± 0.06 a	5.41 ± 0.02 a	5.36± 0.01 a
Citronellol	(%)	30.2 ± 0.56 ab	27.86 ± 0.19 b	33.18 ± 0.29 a
Geraniol	(%)	14.28 ± 0.88 ab	15.62 ± 0.37 a	13.76 ± 0.48 b
Citronellyl formate	(%)	15.87 ± 0.03 a	16.95 ± 0.34 a	14.99 ± 0.09 a
C/G ratio		2.11	1. 78	2.41

Note: means followed by common letters within the same column are not significantly different ($p > 0.05$).

4. Discussion

4.1. Characterization of Digestate and Biochar

The pH increases that were observed during the pyrolysis process are coherent with the findings of Enders et al. [27], who observed that biochar deriving from animal manures (e.g., dairy manure, poultry manure, digested dairy manure, composted dairy manure), annual crop residues (corn, grass clippings, leaves), and waste leads to pH values that are generally above 7.5. These increases were likely due to the rapid rate of carbonization and the concentration of basic inorganic compounds [28–30]. Stefaniuk and Oleszczuk [30] reported higher pH values (i.e., >10) in the solid digestate fraction that was processed at 500 °C with respect to what is reported in this study. This was likely due to a longer pyrolysis time (5 h) when compared to that adopted in this study (3 h). This finding proposes the application of biochar that is produced by prolonged pyrolysis processes as a smart solution for soil pH re-equilibration and for alleviating nutrient stress on plant growth in areas that are characterized by acidic soils [31,32].

The increases of EC occurred during the thermal conversion process, a phase where chemical compounds responsible for salinity (i.e., production of K, Na, Mg, and Ca) are accumulated [33]. The EC values that were found in the three growth media were within the range of 0.007 and 8.3 mS cm^{-1} [16,28], thus they were below the threshold that was considered to be limiting for plant growth [34]. By contrast, when the pyrolysis of unseparated digestate produces biochar with higher EC values (>19.00 mS cm^{-1}), the addition of biochar to the soil should be carefully done in order to reduce the risks of salinity stress, which may negatively impact plant growth, root development, and soil macro and microorganisms [30,35,36].

As observed using other nutrient-rich feedstocks (i.e., sewage sludge, raw food waste, and food waste), the bulk density of solid digestate decreases upon pyrolysis [37,38]. In this study, the observed biochar bulk density was in the range that was reported by literature (0.08 to 0.8 kg m^{-3}) [22,38,39]. The pyrolysis process decreases the bulk density of the feedstock and increases biochar porosity at high process temperatures [40]. Biochar can be used for increasing soil aeration, facilitating root growth, promoting microbial respiration activity, and enhancing soil aggregate formation [5], thereby improving crop performances.

The contents of chemical elements found in biochar were mainly related to the process temperature [16]. Pyrolysis reduces mass, which thus enriches the concentration of chemical elements [41]. In our study, where pyrolysis occurred at a temperature of 500 °C for 3 h (slow pyrolysis), the considerable C increase (+71%) was associated with the loss of the hydroxyl group by dehydration [13,16]. A similar C increase was also observed in biochar that was obtained from digested sugarcane bagasse, agricultural residues, and anaerobic digestate [8,13]. By contrast, Garlapalli et al. [29], while using digestate in a pyrolysis process characterized by a short reaction time and higher temperatures, reported higher elemental carbon content (89%). This different process, despite producing a lower biochar yield, allows for higher fixed carbon and lower ash contents to be obtained, thus improving its mitigative potential due to a higher resistance to biotic degradation [28,29,42].

The results also indicated an N-content increase from solid digestate to biochar. This pattern had already been observed by Pituello et al. [16]; they reported that using silage digestate and pruning residues at temperatures below 550 °C caused an increase in the N content due to the formation of aromatic and heterocyclic structures. However, the different feedstocks also determine the N content in biochar. For instance, Tian et al. [43] observed a decrease in the N content (NO_x and NH_3) using sewage sludge coupled with cracking reactions of nitriles, N-heterocyclic compounds, and polymerization of amine-N when the pyrolysis temperature was between 300 to 700 °C.

The concentration of other inorganics in biochar (P, K, Ca, Mg, Fe, Zn) was probably due to the low volatility of their oxides, leading to gradual volatilization of oxygen and hydrogen [44]. This phenomenon was also observed in other studies during the pyrolysis of feedstocks as wastewater sludge, manure, and algae [33,45,46]. Stefaniuk and Oleszczuk previously highlighted the absence of S in biochar [30], who observed a decomposition of the sulfur-containing compounds into volatile SO_2 during the thermal process [44]. The highest presence of individual chemical elements in biochar was likely due to the more nutrient-rich feedstock as compared with the raw material (i.e., wood residues) used for the common biochar production [6]. The range of these elements was within the range that was suggested by the European Biochar Certificate (EBC) [47].

4.2. Characterization of Growth Media

Changes in the growth media properties after the addition of biochar and solid digestate were in the range that was suggested for the amendment to soil use [48]. However, the main changes were observed in soil pH. In SDS and CS, the pH fell within the optimum range indicated for the *Geraniaceae* family (5.7–6.6) [49]. By contrast, in BCS, the pH exceeded the optimal range, which was likely due to the presence of alkaline inorganic components [8].

However, the analysis of plant performances did not show any reduction on the vegetative parameters (i.e., plant height, plant dry weight, etc.). Ram et al. [50], who reported a similar herbage yield in *P. graveolens* cultivation, even when the pH was outside the optimal range (i.e., 4.9 and 8.4), also noted this. In a more recent study, Ram et al. [18] highlighted the lack of symptoms of nutrient deficiency or toxicity in growth media at a pH of 8.1. Other biochar properties that were not considered in this study (i.e., CEC, surface area, porosity) may have a hidden possible negative effect on the crop growth parameters.

4.3. Effect on Vegetative Parameters

The influence of growth media on vegetative parameters of *P. graveolens* showed that the highest performances were found using BCS. This result was partly expected, since BCS showed the highest nutrient availability, which favored biomass growth, herb yield, plant height, and leaf area in *P. graveolens* [5,18,20,21,51].

Under BCS, the high amount of different heavy metals, which are widely acknowledged as promoters of oxidative stress that is highly detrimental to cellular function and metabolism [52], did not negatively influence biomass or organ size. These results are in contrast with the findings of Patel and Patra [53], who, in *P. graveolens*, observed decreases in plant height and leaf area with concurrent increases in the content of heavy metals. Regarding the production of essential oil, the use of BCS increased the total biomass, particularly the size of the leaves and branches, which increases the yield of essential oil due to a higher presence of glandular trichomes [54].

4.4. Foliar Nutrient Composition

The foliar nutritional content was significantly influenced by the different growth media (Table 3). The high nutrients content (P, K, Ca, Mg, Na, Fe, Cu, Zn) that was found using biochar has also been reported by Awad et al. [51] for vegetables that were cultivated under a hydroponic system, and by Lehmann et al. [55] in cowpea and rice in Brazilian soils. However, when biochar exceeds the optimal rate, a reduction in the uptake of nutrients with a consequent decrease in biomass production can be observed. For instance, Conversa et al. [56] suggested that, in *P. zonale*, the application of biochar should not exceed 70:30 (v/v) in the peat/biochar substrate mixture. This is highly important, especially when considering the essential oil production, where a decrease in biomass may have negative economic consequences. Nevertheless, the processes governing biomass reduction are still unclear. Some studies [5,35] have suggested that this may be due to several interactions between biochar, plant roots, and microorganisms in the rhizosphere or complex reactions, depending on the properties of the soils and biochar involved. Despite awareness of these relationships, these mechanisms are still far from being completely understood and more studies are required.

Plants that are grown on BCS showed unexpected, based on the bibliographic database, leaf discoloration symptoms (Figure 1). This was likely due to leaf chlorosis, a physiopathy that is characterized by slight chlorotic speckling and marginal necrosis that initially spreads in older basal leaves and consequently to younger growths [57]. This physiopathy is particularly frequent in geranium due to the pH status of the growth media. The pH level can indeed affect the capability of plants to uptake nutrients, thus leading to nutrient deficiency or toxicity [57–62]. An analysis of the nutrient contents in the leaves of the symptomatic plants (Table 3) indicated a high concentration of Fe and Mn, as compared to those that are found in the leaves cultivated using SDS and CS. This condition can result in leaf discoloration that, by reducing the photosynthetic efficiency, indirectly affects crop growth [63]. Several studies reported that the toxicity can be due to an overload of Fe in different forms: ferrous sulfate, ferrous ammonium sulfate, ferric glucoheptonate, and ferric citrate are recognized to be moderately toxic to zonal geraniums at high application rates [61], while Fe chelates showed high toxicity when applied in liquid fertilizers [59]. Monteiro and Winterbourn [64] indicated that an excess of iron absorption might induce the production of free radicals that can oxidize the chlorophyll, which thus causes a decrease in the chlorophyll content. Arunachalam et al. [65], who reported decreases in the chlorophyll content as a consequence of increasing Fe concentrations, also confirmed this process. Conversely, Lee et al. [60] reported no visible toxicity symptoms regarding zonal geranium growth while using ferrous sulfate.

Figure 1. A non-symptomatic leaf (center) and leaf discoloration in symptomatic plants (lateral).

4.5. Essential Oil Yield and Quality

The observed oil yield decreases using growth media rich in nutrients, as reported by several studies: Ram et al. [18] reported that, in India, an increase of paddy straw mulch and nitrogen rates led to a considerable decrease in the oil yield in *P. graveolens*. Ram and Kumar [66], observed the same phenomenon in menthol mint leaves, which suggested that high nutrient availability can produce an increase in the size of the cell, with a consequent dilution of essential oil. The oil quality in geranium is commercially determined by the citronellol and geraniol ratio (C/G). These two components are known for their existing interconversion, where geraniol is the precursor of citronellol [19]. The range of the C/G ratio between 1 and 3, where the highest quality is found when the ratio is closer to 1, which generally indicates a high-quality product that is acceptable for the perfume industry, whilst oils exceeding this range are considered of low quality and, therefore, are used for the production of creams, toiletry, soaps, etc. [19,67]. In this study, the essential oils that were obtained from all growth media showed a C/G ratio within the range of 1–3, which confirms that, in *P. graveolens*, the use of organic amendments does not change the overall quality [18].

5. Conclusions

Biochar has stimulated great interest in recent years, since it appears to be capable of enhancing sustainable agricultural production and contributing to climate mitigation. In this study, biochar and feedstock (solid digestate) were evaluated as a soil amendment in *P. graveolens*. The results showed that the pyrolysis of solid digestate caused a decrease in the bulk density (−52%) and an increase in both the soil pH (+16%) and electrical conductivity (+9.5%). The best crop performances (number of leaves, number of total branches, and plant dry weight) and the final essential oil content (EOpl) were found while using biochar with respect to solid digestate. On the contrary, the use of three different growth media only slightly influenced the oil quality, leaving their chemical characteristics within the range (C/G ratio of 1–3) that is recommended by the international standard trade for high-quality oils. Finally, most likely due to the high content of Fe and Mn, biochar may have induced the presence of chlorosis in the leaves of *P. graveolens*. This suggests a careful application of biochar to the soil is necessary in order to avoid the risks of nutrient deficiency or toxicity, which may reduce the plant biomass. This research demonstrates that biochar obtained from solid digestate, when applied in *P. graveolens* cultivation, may increase crop performances, in particular, the size of leaves, in terms of essential oil production. However, several conclusive remarks on crop performances in relation to biochar application and fertilization rates are still missing, which thus suggests the need for additional studies on CEC, surface area, and porosity. A more detailed analysis of the relationship between biochar nutrient uptake and the degree of physiopathy is also recommended.

Author Contributions: Conceptualization, A.C. and E.P.; Data curation, A.C., A.M. and L.M.; Formal analysis, A.C., A.M. and L.M.; Investigation, A.C. and A.M.; Methodology, E.P. and D.C.; Resources, D.C.; Supervision, E.P.; Validation, C.D.; Visualization, D.C.; Writing—Original Draft, A.C.; Writing—Review & Editing, C.D. and L.B.

Conflicts of Interest: The authors declare no conflict of interest.

References

1. Lehmann, J.; Gaunt, J.; Rondon, M. Bio-char Sequestration in Terrestrial Ecosystems—A Review. *Mitig. Adapt. Strateg. Glob. Chang.* **2006**, *11*, 403–427. [CrossRef]
2. Kadota, M.; Niimi, Y. Effects of charcoal with pyroligneous acid and barnyard manure on bedding plants. *Sci. Hortic.* **2004**, *101*, 327–332. [CrossRef]
3. Rutto, K.L.; Mizutani, F. Effect of Mycorrhizal Inoculation and Activated Charcoal on Growth and Nutrition in Peach (Prunus persica) Seedlings Treated with Peach Root-Bark Extracts. *J. Jpn. Soc. Hortic. Sci.* **2006**, *75*, 463–468. [CrossRef]
4. Gaskin, J.W.; Speir, R.A.; Harris, K.; Das, K.C.; Lee, R.D.; Morris, L.A.; Fisher, D.S. Effect of Peanut Hull and Pine Chip Biochar on Soil Nutrients, Corn Nutrient Status, and Yield. *Agron. J.* **2010**, *102*, 623. [CrossRef]

5. Deenik, J.L.; McClellan, T.; Uehara, G.; Antal, M.J.; Campbell, S. Charcoal Volatile Matter Content Influences Plant Growth and Soil Nitrogen Transformations. *Soil Sci. Soc. Am. J.* **2010**, *74*, 1259–1270. [CrossRef]
6. Spokas, K.A.; Cantrell, K.B.; Novak, J.M.; Archer, D.W.; Ippolito, J.A.; Collins, H.P.; Boateng, A.A.; Lima, I.M.; Lamb, M.C.; McAloon, A.J.; et al. Biochar: A Synthesis of Its Agronomic Impact beyond Carbon Sequestration. *J. Environ. Qual.* **2012**, *41*, 973. [CrossRef] [PubMed]
7. Xu, G.; Lv, Y.; Sun, J.; Shao, H.; Wei, L. Recent Advances in Biochar Applications in Agricultural Soils: Benefits and Environmental Implications. *Clean Soil Air Water* **2012**, *40*, 1093–1098. [CrossRef]
8. Monlau, F.; Sambusiti, C.; Ficara, E.; Aboulkas, A.; Barakat, A.; Carrère, H. New opportunities for agricultural digestate valorization: Current situation and perspectives. *Energy Environ. Sci.* **2015**, *8*, 2600–2621. [CrossRef]
9. Zhang, M.; Ok, Y.S. Biochar soil amendment for sustainable agriculture with carbon and contaminant sequestration. *Carbon Manag.* **2014**, *5*, 255–257. [CrossRef]
10. Rogovska, N.; Laird, D.; Cruse, R.M.; Trabue, S.; Heaton, E. Germination Tests for Assessing Biochar Quality. *J. Environ. Qual.* **2012**, *41*, 1014. [CrossRef] [PubMed]
11. Ogbonnaya, U.O.; Semple, K.T. Impact of Biochar on Organic Contaminants in Soil: A Tool for Mitigating Risk? *Agronomy* **2013**, *3*, 349–375. [CrossRef]
12. Jindo, K.; Mizumoto, H.; Sawada, Y.; Sanchez-Monedero, M.A.; Sonoki, T. Physical and chemical characterization of biochars derived from different agricultural residues. *Biogeosciences* **2014**, *11*, 6613–6621. [CrossRef]
13. Inyang, M.; Gao, B.; Ding, W.; Pullammanappallil, P.; Zimmerman, A.R.; Cao, X. Enhanced Lead Sorption by Biochar Derived from Anaerobically Digested Sugarcane Bagasse. *Sep. Sci. Technol.* **2011**, *46*, 1950–1956. [CrossRef]
14. Pages-Díaz, J.; Pereda Reyes, I.; Lundin, M.; Sárvári Horváth, I. Co-digestion of different waste mixtures from agro-industrial activities: Kinetic evaluation and synergetic effects. *Bioresour. Technol.* **2011**, *102*, 10834–10840. [CrossRef] [PubMed]
15. Seadi, T.; Rutz, D.; Prassl, H.; Köttner, M.; Finsterwalder, T.; Volk, S.; Janssen, R. *Biogas Handbook*; University of Southern Denmark Esbjerg: Esbjerg, Denmark, 2008.
16. Pituello, C.; Francioso, O.; Simonetti, G.; Pisi, A.; Torreggiani, A.; Berti, A.; Morari, F. Characterization of chemical–physical, structural and morphological properties of biochars from biowastes produced at different temperatures. *J. Soils Sediments* **2015**, *15*, 792–804. [CrossRef]
17. Rao, B.R.R. Biomass yield, essential oil yield and essential oil composition of rose-scented geranium (Pelargonium species) as influenced by row spacings and intercropping with cornmint (Menthaarvensis L.f. piperascens Malinv. ex Holmes). *Ind. Crops Prod.* **2002**, *16*, 133–144. [CrossRef]
18. Ram, M.; Ram, D.; Roy, S. Influence of an organic mulching on fertilizer nitrogen use efficiency and herb and essential oil yields in geranium (*Pelargonium graveolens*). *Bioresour. Technol.* **2003**, *87*, 273–278. [CrossRef]
19. Saxena, G.; Verma, P.C.; Banerjee, S.; Kumar, S. Field performance of somaclones of rose scented geranium (*Pelargonium graveolens* L'Her Ex Ait.) for evaluation of their essential oil yield and composition. *Ind. Crops Prod.* **2008**, *27*, 86–90. [CrossRef]
20. Rezaei, N.A.; Ismaili, A. Changes in growth, essential oil yield and composition of geranium (*Pelargonium graveolens* L.) as affected by growing media. *J. Sci. Food Agric.* **2014**, *94*, 905–910. [CrossRef] [PubMed]
21. Pandey, V.; Patra, D.D. Crop productivity, aroma profile and antioxidant activity in *Pelargonium graveolens* L'Hér. under integrated supply of various organic and chemical fertilizers. *Ind. Crops Prod.* **2015**, *67*, 257–263. [CrossRef]
22. Vaughn, S.F.; Eller, F.J.; Evangelista, R.L.; Moser, B.R.; Lee, E.; Wagner, R.E.; Peterson, S.C. Evaluation of biochar-anaerobic potato digestate mixtures as renewable components of horticultural potting media. *Ind. Crops Prod.* **2015**, *65*, 467–471. [CrossRef]
23. Ahmedna, M.; Marshall, W.E.; Rao, R.M. Production of granular activated carbon from select agricultural by-products and evaluation of their physical, chemical, and adsorption properties. *Bioresour. Technol.* **2000**, *71*, 113–123. [CrossRef]
24. Tian, Y.; Sun, X.; Li, S.; Wang, H.; Wang, L.; Cao, J.; Zhang, L. Biochar made from green waste as peat substitute in growth media for Calathea rotundifola cv. Fasciata. *Sci. Hortic.* **2012**, *143*, 15–18. [CrossRef]
25. Grassi, C.; Palchetti, E.; Andreinelli, L. Messa a Punto Di Un Protocollo per l'introduzione in Vitro Di Pelargonium Graveolens. In *Un'importante Specie Tropicale per La Produzione Di Olio Essenziale*; VITROSOI: Pistoia, Italy, 2017; pp. 35–38.

26. Boukhatem, M.N.; Kameli, A.; Saidi, F. Essential oil of Algerian rose-scented geranium (*Pelargonium graveolens*): Chemical composition and antimicrobial activity against food spoilage pathogens. *Food Control* **2013**, *34*, 208–213. [CrossRef]
27. Enders, A.; Hanley, K.; Whitman, T.; Joseph, S.; Lehmann, J. Characterization of biochars to evaluate recalcitrance and agronomic performance. *Bioresour. Technol.* **2012**, *114*, 644–653. [CrossRef] [PubMed]
28. Yargicoglu, E.N.; Sadasivam, B.Y.; Reddy, K.R.; Spokas, K. Physical and chemical characterization of waste wood derived biochars. *Waste Manag.* **2015**, *36*, 256–268. [CrossRef] [PubMed]
29. Garlapalli, R.K.; Wirth, B.; Reza, M.T. Pyrolysis of hydrochar from digestate: Effect of hydrothermal carbonization and pyrolysis temperatures on pyrochar formation. *Bioresour. Technol.* **2016**, *220*, 168–174. [CrossRef]
30. Stefaniuk, M.; Oleszczuk, P. Characterization of biochars produced from residues from biogas production. *J. Anal. Appl. Pyrolysis* **2015**, *115*, 157–165. [CrossRef]
31. Verheijen, F.; Jeffery, S.; Bastos, A.C.; van der Velde, M.; Diafas, I. *Biochar Application to Soils: A Critical Scientific Review of Effects on Soil Properties, Processes and Functions*; European Commission, Joint Research Centre, Institute for Environment and Sustainability, Publications Office: Varese, Italy, 2010.
32. Pandit, N.R.; Mulder, J.; Hale, S.E.; Martinsen, V.; Schmidt, H.P.; Cornelissen, G. Biochar improves maize growth by alleviation of nutrient stress in a moderately acidic low-input Nepalese soil. *Sci. Total Environ.* **2018**, *625*, 1380–1389. [CrossRef]
33. Cantrell, K.B.; Hunt, P.G.; Uchimiya, M.; Novak, J.M.; Ro, K.S. Impact of pyrolysis temperature and manure source on physicochemical characteristics of biochar. *Bioresour. Technol.* **2012**, *107*, 419–428. [CrossRef]
34. Chong, C.; Rinker, D.L. Use of Spent Mushroom Substrate for Growing Containerized Woody Ornamentals: An Overview. *Compost Sci. Util.* **1994**, *2*, 45–53. [CrossRef]
35. Lehmann, J.; Joseph, S. *Biochar for Environmental Management: Science and Technology*; Earthscan: Sterling, VA, USA, 2009; Chapter 9.
36. Yan, N.; Marschner, P.; Cao, W.; Zuo, C.; Qin, W. Influence of salinity and water content on soil microorganisms. *Int. Soil Water Conser. Res.* **2015**, *3*, 316–323. [CrossRef]
37. Opatokun, S.A.; Kan, T.; Al Shoaibi, A.; Srinivasakannan, C.; Strezov, V. Characterization of Food Waste and Its Digestate as Feedstock for Thermochemical Processing. *Energy Fuels* **2016**, *30*, 1589–1597. [CrossRef]
38. Khanmohammadi, Z.; Afyuni, M.; Mosaddeghi, M.R. Effect of pyrolysis temperature on chemical and physical properties of sewage sludge biochar. *Waste Manag. Res.* **2015**, *33*, 275–283. [CrossRef]
39. Gundale, M.J.; DeLuca, T.H. Temperature and source material influence ecological attributes of ponderosa pine and Douglas-fir charcoal. *For. Ecol. Manag.* **2006**, *231*, 86–93. [CrossRef]
40. Ding, Y.; Liu, Y.; Liu, S.; Li, Z.; Tan, X.; Huang, X.; Zeng, G.; Zhou, L.; Zheng, B. Biochar to improve soil fertility. A review. *Agron. Sustain. Dev.* **2016**, *36*, 36. [CrossRef]
41. Novak, J.M.; Lima, I.; Xing, B.; Gaskin, J.W.; Steiner, C.; Das, K.C.; Ahmedna, M.; Rehrah, D.; Watts, D.W.; Busscher, W.J.; et al. Characterization of Designer Biochar Produced at Different Temperatures and Their Effects on a Loamy Sand. *Ann. Environ. Sci.* **2009**, *3*, 195–206.
42. Brewer, C.E.; Unger, R.; Schmidt-Rohr, K.; Brown, R.C. Criteria to Select Biochars for Field Studies based on Biochar Chemical Properties. *Bioenergy Res.* **2011**, *4*, 312–323. [CrossRef]
43. Tian, Y.; Zhang, J.; Zuo, W.; Chen, L.; Cui, Y.; Tan, T. Nitrogen Conversion in Relation to NH_3 and HCN during Microwave Pyrolysis of Sewage Sludge. *Environ. Sci. Technol.* **2013**, *47*, 3498–3505. [CrossRef]
44. Al-Wabel, M.I.; Al-Omran, A.; El-Naggar, A.H.; Nadeem, M.; Usman, A.R.A. Pyrolysis temperature induced changes in characteristics and chemical composition of biochar produced from conocarpus wastes. *Bioresour. Technol.* **2013**, *131*, 374–379. [CrossRef]
45. Bird, M.I.; Wurster, C.M.; de Paula Silva, P.H.; Bass, A.M.; de Nys, R. Algal biochar–production and properties. *Bioresour. Technol.* **2011**, *102*, 1886–1891. [CrossRef]
46. Hossain, M.K.; Strezov, V.; Chan, K.Y.; Ziolkowski, A.; Nelson, P.F. Influence of pyrolysis temperature on production and nutrient properties of wastewater sludge biochar. *J. Environ. Manag.* **2011**, *92*, 223–228. [CrossRef]
47. EBC, European Biochar Certificate. *Guidelines for Biochar Production, v. 4.2, June 13th 2012*; Ithaka Institute: Arbaz, Switzerland, 2012.
48. ANPA, Agenzia Nazionale per la Protezione dell'Ambiente. *I Fertilizzanti Commerciali e Aspetti Normativi, v. 3*; ISPRA: Varese, Italy, 2001.

49. Andrews, P.H.; Hammer, P.A. Response of zonale and ivy geranium to root medium pH. *Hortscience* **2006**, *41*, 1351–1355. [CrossRef]
50. Ram, M.; Prasad, A.; Gupta, N.; Kumar, S. Effect of soil pH on the essential oil yield in the geranium *Pelargonium graveolens*. *J. Med. Aromat. Plant Sci.* **1997**, *19*, 406–407.
51. Awad, Y.M.; Lee, S.E.; Ahmed, M.B.M.; Vu, N.T.; Farooq, M.; Kim, I.S.; Kim, H.S.; Vithanage, M.; Usman, A.R.A.; Al-Wabel, M.; et al. Biochar, a potential hydroponic growth substrate, enhances the nutritional status and growth of leafy vegetables. *J. Clean. Prod.* **2017**, *156*, 581–588. [CrossRef]
52. Panda, S.K.; Choudhury, S.; Patra, H.K. Heavy-Metal-Induced Oxidative Stress in Plants: Physiological and Molecular Perspectives. *Abiotic Stress Response Plants* **2016**, 221–236. [CrossRef]
53. Patel, A.; Patra, D.D. Effect of tannery sludge amended soil on glutathione activity of four aromatic crops: Tagetes minuta, *Pelargonium graveolens*, Ocimum basilicum and Mentha spicata. *Ecol. Eng.* **2015**, *81*, 348–352. [CrossRef]
54. Boukhris, M.; Ben Nasri-Ayachi, M.; Mezghani, I.; Bouaziz, M.; Boukhris, M.; Sayadi, S. Trichomes morphology, structure and essential oils of *Pelargonium graveolens* L'Hér. (Geraniaceae). *Ind. Crops Prod.* **2013**, *50*, 604–610. [CrossRef]
55. Lehmann, J.; da Silva, J.P.; Steiner, C.; Nehls, T.; Zech, W.; Glaser, B. Nutrient availability and leaching in an archaeological Anthrosol and a Ferralsol of the Central Amazon basin: Fertilizer, manure and charcoal amendments. *Plant Soil* **2003**, *249*, 343–357. [CrossRef]
56. Conversa, G.; Bonasia, A.; Lazzizera, C.; Elia, A. Influence of biochar, mycorrhizal inoculation, and fertilizer rate on growth and flowering of Pelargonium (*Pelargonium zonale* L.) plants. *Front. Plant Sci.* **2015**, *6*, 429. [CrossRef]
57. Biernbaum, J.A.; Shoemaker, C.A.; Carlson, W.H. Iron and manganese toxicity of seedling geraniums. *HortScience* **1987**, *22*, 1094.
58. Foy, C.D.; Chaney, R.L.; White, M.C. The Physiology of Metal Toxicity in Plants. *Annu. Rev. Plant Physiol.* **1978**, *29*, 511–566. [CrossRef]
59. Bachman, G.R.; Miller, W.B. Iron chelate inducible iron/manganese toxicity in zonal geranium. *J. Plant Nutr.* **1995**, *18*, 1917–1929. [CrossRef]
60. Lee, C.W.; Choi, J.-M.; Pak, C.-H. Micronutrient Toxicity in Seed Geranium (Pelargonium × hortorum Bailey). *J. Am. Soc. Hort. Sci.* **1996**, *121*, 77–82. [CrossRef]
61. Broschat, T.K.; Moore, K.K. Phytotoxicity of Several Iron Fertilizers and Their Effects on Fe, Mn, Zn, Cu, and P content of African Marigolds and Zonal Geraniums. *Hortscience* **2004**, *9*, 595–598. [CrossRef]
62. Smith, B.R.; Fisher, P.R.; Argo, W.R. Water-Soluble Fertilizer Concentration and pH of a Peat-Based Substrate Affect Growth, Nutrient Uptake, and Chlorosis of Container-Grown Seed Geraniums. *J. Plant Nutr.* **2004**, *27*, 497–524. [CrossRef]
63. Rout, G.R.; Sahoo, S. Role of iron in plant growth and metabolism. *Rev. Agric. Sci.* **2015**, *3*, 1–24. [CrossRef]
64. Monteiro, H.P.; Winterbourn, C.C. The superoxide-dependent transfer of iron from ferritin to transferrin and lactoferrin. *Biochem. J.* **1988**, *256*, 923–928. [CrossRef]
65. Arunachala, R.; Paulkumar, K.; Ranjitsin, A.J.A.; Annadurai, G. Environmental Assessment due to Air Pollution near Iron Smelting Industry. *J. Environ. Sci. Technol.* **2009**, *2*, 179–186. [CrossRef]
66. Ram, M.; Kumar, S. Yield and Resource Use Optimization in Late Transplanted Mint (Mentha arvensis). under Subtropical Conditions. *J. Agron. Crop Sci.* **1998**, *180*, 109–112. [CrossRef]
67. Palchetti, E.; Calamai, A.; Valenzi, E.; Vecchio, V. Pelargonio (*Pelargonium graveolens*). In *Oli e Grassi*, 1st ed.; Edagricole New Business Media: Milano, Italy, 2019; ISBN 978-88-506-5564-9.

© 2019 by the authors. Licensee MDPI, Basel, Switzerland. This article is an open access article distributed under the terms and conditions of the Creative Commons Attribution (CC BY) license (http://creativecommons.org/licenses/by/4.0/).

Article

Use of Carbonized Fallen Leaves of *Jatropha Curcas* L. as a Soil Conditioner for Acidic and Undernourished Soil

Takafumi Konaka [1], Shin Yabuta [2], Charles Mazereku [3], Yoshinobu Kawamitsu [4], Hisashi Tsujimoto [1,5], Masami Ueno [4] and Kinya Akashi [1,5,6,*]

1. United Graduate School of Agricultural Sciences, Tottori University, Tottori 680-8553, Japan; takafumi.konaka@gmail.com (T.K.); tsujim@alrc.tottori-u.ac.jp (H.T.)
2. Faculty of Agriculture, Kagoshima University, Kagoshima 890-8580, Japan; syabuta@agri.kagoshima-u.ac.jp
3. Department of Agricultural Research, Ministry of Agriculture, Private Bag 0033 Gaborone, Botswana; charmazereku@gmail.com
4. Faculty of Agriculture, University of the Ryukyus, Okinawa 903-0213, Japan; kawamitu@agr.u-ryukyu.ac.jp (Y.K.); ruenom@agr.u-ryukyu.ac.jp (M.U.)
5. Arid Land Research Center, Tottori University, Tottori 680-0001, Japan
6. Faculty of Agriculture, Tottori University, Tottori 680-8553, Japan
* Correspondence: akashi.kinya@tottori-u.ac.jp; Tel.: +81-857-315352

Received: 24 April 2019; Accepted: 6 May 2019; Published: 9 May 2019

Abstract: Jatropha (*Jatropha curcas* L.) represents a renewable bioenergy source in arid regions, where it is used to produce not only biodiesel from the seed oil, but also various non-oil biomass products, such as fertilizer, from the seed cake following oil extraction from the seeds. Jatropha plants also generate large amounts of fallen leaves during the cold or drought season, but few studies have examined the utilization of this litter biomass. Therefore, in this study, we produced biochar from the fallen leaves of jatropha using a simple and economical carbonizer that was constructed from a standard 200 L oil drum, which would be suitable for use in rural communities, and evaluated the use of the generated biochar as a soil conditioner for the cultivation of Swiss chard (*Beta vulgaris* subsp. *cicla* "Fordhook Giant") as a model vegetable in an acidic and undernourished soil in Botswana. Biochar application improved several growth parameters of Swiss chard, such as the total leaf area. In addition, the dry weights of the harvested shoots were 1.57, 1.88, and 2.32 fold higher in plants grown in soils containing 3%, 5%, and 10% biochar, respectively, compared with non-applied soil, suggesting that the amount of biochar applied to the soil was positively correlated with yield. Together, these observations suggest that jatropha fallen leaf biochar could function as a soil conditioner to enhance crop productivity.

Keywords: jatropha; biochar; arid region; acidic undernourished soil; fallen leaves

1. Introduction

Jatropha (*Jatropha curcas* L., Euphorbiaceae) has non-edible oils in its seeds that serve as a feedstock for biodiesel production. Furthermore, since this drought-tolerant species can thrive under a wide range of rainfall regimes, from 200 to over 1500 mm per annum [1,2], and can grow in poor soils, on eroded land, and on wasteland [3–6], jatropha seed oil also represents a promising feedstock for renewable energy in arid lands [7–9].

The utilization of non-oil biomass products of jatropha has also been evaluated from the perspective of whole-crop biorefineries [10]. Several studies have reported on the production of an organic fertilizer from the seed residues following oil extraction (the seed cake) [4,11–13] and have investigated use of the seed cake for bioenergy production, e.g., bio-oil and solid fuel [14–16]. Furthermore, the utilization of jatropha wood biochar as an energy source has been described [17]. The effects of jatropha leaf extracts as a herbal ointment [18], mite [*Rhipicephalus (Boophilus) annulatus*] repellent [19], and therapeutic agent [20] have been studied, and the positive effects of a biochar amendment made from a mixture of jatropha stems and leaves on jatropha seedling growth have been demonstrated [21]. Recently, production of catalytic biocarbons from jatropha biomass has been demonstrated [22,23].

Jatropha plants growing in the temperate zone are deciduous, losing their leaves in the cold season, so the fallen leaf biomass of this species cannot be neglected when considering whole-crop biorefineries [5,24,25]. A cultivation trial in Patancheru, India, estimated that jatropha trees produce 550 g of fallen leaves per plant at one year old and 1450 g at three years old, while the total plant biomass for four-year-old trees was estimated at 6.14 kg plant^{-1} [26], suggesting that fallen leaves make up a major proportion of the biomass that is produced during jatropha cultivation.

Previous studies have shown that biochar can contribute to agricultural production through improvement of the soil physicochemical properties [21,27,28] and fertilizing effects [29–31], and that the application of biochar derived from wood or organic waste has positive impacts on the growth of many crops, such as rice (*Oryza sativa*) and maize (*Zea mays*) [32–34]. Biochar derived from fallen leaves has also been investigated for several species—for example, biochars made from the fallen leaves of ginkgo (*Ginkgo biloba*) and maple (*Acer* spp.) trees have been studied as materials for the stabilization of heavy metals in polluted soils [35,36], and the physicochemical properties of fallen leaf biochars made from maple [37] and *Eucalyptus saligna* [38] have been reported. However, to the best of our knowledge, the use of biochar made from jatropha fallen leaves in the production of edible crops and vegetables has been limited.

In this study, fallen leaves of jatropha derived from a cultivation trial in Botswana [24,25] were pyrolyzed and applied to an acidic and undernourished soil to evaluate the use of jatropha fallen leaves as a feedstock for producing a soil conditioner. In addition, we investigated the performance of a simplified and inexpensive oil-drum carbonizer for the production of jatropha fallen leaf biochar to facilitate the implementation of biochar production in resource-poor villages in rural areas [39–41]. We chose to use Swiss chard (*Beta vulgaris* subsp. *cicla*) as a model plant for evaluating the impact of jatropha fallen leaf biochar on crop production because this is commonly grown in Botswana.

2. Materials and Methods

2.1. Production and Analysis of Jatropha Fallen Leaf Biochar

Fallen leaves of jatropha were collected in July (winter) 2016 from a jatropha cultivation site [24] located approximately 4 km northeast of the Department of Agricultural Research station in Gaborone, Botswana. The cold-induced defoliation of jatropha in this climatic region has been described previously [24,25]. The fallen leaves were dried in the sun and subsequently used for biochar production.

A simplified carbonizer was constructed from a used oil drum of standard 200 L capacity (572 mm diameter, 851 mm high). The lid of the drum was removed and replaced with a custom-made steel lid that contained a funnel (15 cm high, 7 cm internal diameter) in the center (Figure 1A). Several holes were then made in the bottom of the drum at approximately 10–15 cm intervals for ventilation (Figure 1B) and the drum was placed on top of four bricks to ensure air flow from the bottom at the beginning of ignition.

Figure 1. Jatropha fallen leaf biochar production using an oil-drum carbonizer. (**A**) Side view of the carbonizer. An oil drum was placed on top of bricks and covered with a custom-made lid with a funnel. (**B**) Bottom view of the carbonizer. Several ventilation holes were made in the bottom of the oil drum.

Approximately 44 kg of the jatropha fallen leaves were tightly packed into the drum up to the height of the upper rim by stamping them down with the body weight of an adult. A pilot burner was then provided by placing a burning piece of paper on top of the piled leaves in the drum. After 30 min of burning, the lid was placed on top of the drum and 1 h later, the top funnel was sealed by a brick and the bottom ventilation was closed by filling the area between the bricks and the bottom of the drum with soil, to suppress combustion inside the drum by blocking the in-flow of the air. The drum was maintained for the next 46.5 h to promote carbonization, during which time the internal temperature of the carbonizer was continuously measured by three thermocouple sensors with a measurable temperature range of −50 to 800 °C (TL-13K; Sato Shouji, Tokyo, Japan), which were inserted laterally into the center of the drum through small holes at heights of 15, 45, and 75 cm from the bottom, and a data logger (Ondotori MCR-4; T&D Corporation, Nagano, Japan). Then, 48 h after ignition, the carbonization was stopped by opening the lid manually using refractory gloves, and immediately applying approximately 10 L of water to the top of the pile. The drum was then left for 6 h, after which time the resulting biochar was spread out on a steel sheet in the open air and left to dry for 10 days, and then mixed extensively with a shovel to ensure homogenization. The surface structure of the biochar was then observed under a VHX-D500 scanning electron microscope (Keyence, Osaka, Japan).

The pH of the biochar was measured using a Laqua D-51 pH meter (HORIBA Scientific, Kyoto, Japan) at a biochar to water ratio of 1:50 (w/v) and the electrical conductivity (EC) of the biochar was measured with a JENCO VisionPlus meter (JENCO, San Diego, CA, USA). The cation exchange capacity (CEC) of the biochar was measured as described previously [42] with the following modifications. Biochar (1 g) was placed in a centrifuge tube and 25 mL of 1 M ammonium acetate was added. The tube was then shaken for 1 h and centrifuged at 830× g for 5 min, following which the supernatant was removed and 80% ethanol was added. The tube was then further shaken for 5 min and recentrifuged as described above. This ethanol washing procedure was repeated three times. After the final centrifugation, the supernatant was removed and 25 mL of 10% sodium chloride was added. The tube was then shaken for 1 h and centrifuged at 830× g for 5 min, following which the supernatant was taken into an Erlenmeyer flask and 5 mL of 18% formaldehyde aqueous solution and 200 µL of 1% of thymol blue were added. The solution was then titrated with 0.1 N sodium hydroxide. The CEC value was calculated using a previously described formula [43].

2.2. Experimental Soil

To evaluate the effect of biochar on soils that would otherwise be unsuitable for vegetable cultivation, an experimental soil with acidic, Cu/Ni-rich, and nutrient-poor characteristics [44] was collected from a suburb of Selebi-Phikwe in the central district of Botswana. The organic carbon content of the soil was measured using the potassium dichromate method [45] with a UVD 2950 spectrophotometer (LABOMED, Los Angeles, CA, USA).

2.3. Plant Material and Monitoring of Growth

Seeds of Swiss chard 'Fordhook Giant' were purchased from Sakata Seed Southern Africa (Lindsay, South Africa). The seeds were initially germinated on a horticulture soil (Potting soil, New Frontiers, Lobatse, Botswana) in a plastic seed tray, where they were grown for 2 weeks until their second true leaves emerged. The seedlings were then transplanted into 2 L plastic pots filled with the experimental soil supplemented with 0% (control), 3%, 5%, or 10% (w/w) biochar. The plants were provided with 0.5 L of water every second day.

The length of the longest leaf on each plant was measured weekly using a steel measuring tape. Then, at 44 days after transplantation to the experimental soil, the number of expanded leaves on each plant was counted and the aerial and underground tissues were harvested. The total leaf area was measured with an area meter (LI-3100; LI-COR, Lincoln, NE, USA), following which the whole tissues were dried in an oven at 70 °C for 2 days and the dry biomass was weighed using an electric balance.

2.4. Mineral Nutrient Assays

The unamended soil was decomposed by concentrated sulfuric acid as described previously [46] and their mineral contents were measured by inductively coupled plasma atomic emission spectroscopy (ICP-AES) (SPECTRO CIROS CCD; SPECTRO Analytical Instruments GmbH, Nordrhein-Westfalen, Germany). The biochar and aerial parts of the harvested plants were decomposed as described previously [47] with the following modifications. Dry samples of the biochar (0.2 g) and aerial parts (0.1 g) were dissolved in 10 mL of concentrated nitric acid in a flask and digested on a hot plate for 1 h at each of 90 °C, 140 °C, and 190 °C. The solutions were then evaporated at 220 °C (biochar) or 240 °C (aerial parts) until the volume was reduced to approximately 1 mL and analyzed by ICP-AES as described above. The water-soluble ions NO_3^- and NH_4^+ were extracted from the soils by mixing 5 g of soil with 50 mL of distilled water for 1 h and were quantified using an RQflex 10 reflectometer (Merck, Darmstadt, Germany).

2.5. Evaluation of the Water-holding Capacities of the Soils

After harvest, the experimental soils were spread onto a sheet in a greenhouse with a daily maximum temperature that ranged from approximately 40 to 49 °C and were dried out for 1 month. Then, 1 kg of the soils were poured back into the original pots and 500 mL of water was added. The pots were maintained in the greenhouse and their weights were measured on a daily basis from the day after water addition to estimate the water-holding capacities of the amended and unamended soils.

3. Results and Discussion

3.1. Properties of the Jatropha Fallen Leaf Biochar

The 200 L oil-drum carbonizer that was used in this study (Figure 1A,B) was completely filled with dried jatropha leaves and ignited from the top of the piled leaves. Then, 1.5 h after ignition, both top and bottom ventilation holes were closed, and the biomass inside was subjected to pyrolysis in an oxygen-limited condition. The heating scheme employed in this study is based on the auto-thermal process, in which burning part of the raw biomass material with a controlled air inlet provides the energy necessary for the pyrolysis process [48].

During the heating process, the internal temperature of the drum was monitored at three different heights (Figure 2). In the upper part of the drum (75 cm from the bottom), the temperature increased immediately after ignition to reach a maximum temperature of 395.1 °C at 6 h after ignition (HAI), after which it dropped sharply to <100 °C at 10–15 HAI. In the middle of the drum (45 cm from the bottom), the temperature increased from around 1–2 HAI to reach a small peak of 156 °C at 7 HAI.

The temperature was then maintained between 120 and 200 °C at 8–42 HAI, after which it sharply increased to 370 °C at 48 HAI. In the lower part of the drum (15 cm from the bottom), the temperature increased sharply at 6–18 HAI to exceed 400 °C, after which it was maintained at 396–454 °C until the pyrolysis was extinguished. The maximum temperature in the bottom part was 454 °C at 46 HAI.

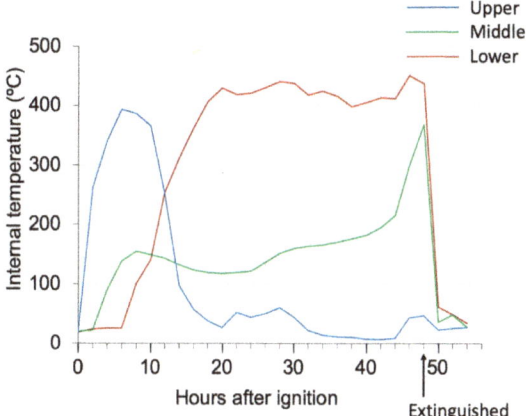

Figure 2. Temporal changes in the internal temperature of the oil-drum carbonizer at three different heights of 15 cm (lower), 45 cm (middle), and 75 cm (upper) from the bottom of the drum. The pyrolysis was terminated at 48 h after ignition (vertical arrow).

The timing of a sharp decrease in temperature from 365.3 to 26.3 °C in the upper part of the drum (at 10–20 HAI) appeared to be synchronized with a sharp increase in temperature from 25.6 to 429.5 °C in the bottom part of the drum (at 6–20 HAI), suggesting that the spatial location of the pyrolysis reaction migrated downward at this time. It is noteworthy that only a modest decreased in temperature from 137.7 to 118.5 °C was observed in the middle part of the drum at 6–18 HAI. The reason for this behavior is currently unknown, but one possibility is that the heat was transmitted downward via a peripheral route, bypassing the central axis of the drum where the mid-height sensor was located. It is also interesting to note that the temperature in the bottom part of the drum remained high for a prolonged period of time (18–48 HAI). The mechanism driving this phenomenon is also unclear, but it may have been related to a higher density of jatropha leaves occurring at the bottom of the drum due to compression under their own weight, allowing a high temperature to be sustained. The internal temperature profile of the drum suggested that the biomass was subjected to modestly high temperatures in the range of 100–450 °C. This temperature range was similar to those related to the degradation of polymers of hemicellulose, cellulose, and lignin during pyrolysis of plant biomass [22].

When the pyrolysis was terminated by applying water on top of the pile, soot and smoke at a height of approximately 50 cm evolved from the pile, but vapor explosion did not occur. Since the temperature profile suggested that the pyrolysis products in the drum were heterogeneous, the jatropha biochar was extensively mixed with a shovel for homogenization once the pyrolysis had been completely terminated. The resulting biochar had a mosaic appearance consisting of dark and light brownish fragments and grains of different sizes (Figure 3A), and scanning electron microscopy showed that the surface of subsets of these fragments had a porous structure (Figure 3B), which is a common hallmark of biochar [49].

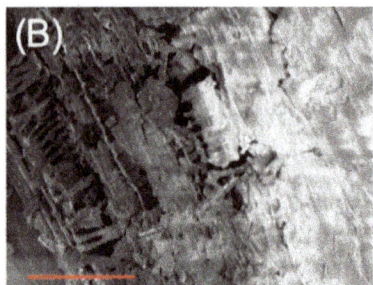

Figure 3. (**A**) Top view of the jatropha fallen leaf biochar that was produced by the carbonizer. Red scale bar represents 5 cm. (**B**) Close-up view of the surface of the jatropha fallen leaf biochar observed under an electron microscope. Red scale bar represents 100 μm.

The jatropha fallen leaf biochar had a pH of 9.84 ± 0.08 (Table 1), which was notably higher than that reported by Awasthi et al. [50] for jatropha leaves (pH 8.85). The EC of the biochar produced in this study was 575 ± 108 μS cm^{-1}, which was slightly lower than the reported value of 920 μS cm^{-1} for jatropha leaves [50], and the CEC of the biochar was 169 ± 19 mmol kg^{-1}, which was higher than the reported average values for wheat/barley biochar (103 mmol kg^{-1}) but lower than the average values for corn and rice straw/husk biochar (607 and 212 mmol kg^{-1}, respectively) [51].

Table 1. Properties of the jatropha fallen leaf biochar produced in this study.

Property	Value [1]
pH	9.84 ± 0.08
EC (μS cm^{-1})	575 ± 108
CEC (cmol kg^{-1})	16.9 ± 1.9
Element content (mg kg^{-1})	
Al	52,900 ± 700
C	227,000 ± 12,000
Ca	18,000 ± 1000
Cd	1.19 ± 0.03
Co	16.4 ± 0.2
Cr	29.8 ± 6.8
Cu	9.39 ± 0.66
Fe	5080 ± 340
K	6220 ± 530
Mg	741 ± 62
Mn	617 ± 46
N	15,000 ± 1000
Ni	43.4 ± 3.4
P	1450 ± 230
Pb	18.0 ± 0.5
Sn	<0.1
Ti	121 ± 3
Zn	80.1 ± 6.7

[1] Values are means ± standard deviations (n = 3). EC, electrical conductivity; CEC, cation exchange capacity.

The chemical constituents of biochars are influenced by the feedstock source and the pyrolysis temperature [29,33]. The jatropha fallen leaf biochar that was produced in the present study had P and K contents of 1450 ± 230 and 6220 ± 530 mg kg^{-1}, respectively (Table 1), which were higher than the reported average values for rice straw/husk biochar (1200 and 700 mg kg^{-1}, respectively) but lower than those for corn biochar (2350 and 19,000 mg kg^{-1}, respectively) [51].

High concentration of Al was observed in fallen leaf biochar (52,900 ± 700 mg kg^{-1}). Reasons for this high abundance is currently unknown; one possibility is that jatropha fallen leaves feedstock might be contaminated with Al-rich soil particles during harvest in the field. Alternatively, Al in fallen leaf biochar might be derived from intrinsic Al accumulated within jatropha leaves, although absorption and accumulation of Al in plant leaves are normally inefficient due to its phytotoxicity. The third possibility is that Al might be derived from leaching from the surface of the used oil drum. Although we extensively washed the inside of drum before usage, this possibility is not totally excluded. Origin of Al in the fallen leaf biochar should be examined in future studies.

3.2. Effects of Jatropha Fallen Leaf Biochar on Vegetable Growth in an Acidic and Undernourished Soil

The experimental soil that was collected from Selebi-Phikwe, Botswana, was highly acidic, with a pH of 3.39 ± 0.03 (Table 2), and it is well known that crop growth is generally retarded in acidic soils [52,53]. Furthermore, the experimental soil contained markedly lower contents of the major nutritious elements (e.g., P, 2.67 ± 0.91 mg kg^{-1}; K, 5.87 ± 2.76 mg kg^{-1}; Ca, 98.2 ± 8.5 mg kg^{-1}; and Mg, 20.7 ± 12.0 mg kg^{-1}; Table 2) than typical soils [54], suggesting that it was poor in nutrients. By contrast, the contents of Cu (772 ± 8 mg kg^{-1}) and Ni (249 ± 2 mg kg^{-1}) were particularly high in the experimental soil (Table 2).

Table 2. Properties of the experimental soil used in this study.

Property	Value [1]
pH	3.39 ± 0.03
Organic carbon (%)	0.12 ± 0.00
Nitrogen content (mg kg^{-1})	
NO$_3^-$	0.12 ± 0.04
NH$_4^+$	0.10 ± 0.01
Element content (mg kg^{-1})	
Al	69,400 ± 900
Ca	98.2 ± 8.5
Cu	772 ± 8
Fe	38,300 ± 100
K	5.87 ± 2.76
Mg	20.7 ± 12.0
Ni	249 ± 2
P	2.67 ± 0.91
Zn	40.4 ± 0.2

[1] Values are the means ± standard deviations of three samples for all parameters except pH and organic carbon, which were derived from two samples.

To examine the effect of jatropha fallen leaf biochar as a soil amendment, the biochar was applied to the experimental soil at rates of 3%, 5%, and 10% (w/w) and the growth of Swiss chard was examined. Swiss chard was chosen as a model crop because it is one of the major vegetables grown in Botswana. Plants that were grown in soil containing 5% and 10% biochar had significantly longer leaves than control plants at both 28 days after transplantation (DAT) (7.33 ± 0.58 cm and 9.67 ± 0.58 cm, respectively, versus 5.67 ± 0.58 cm, corresponding to 1.29 and 1.71 fold increases) and 44 DAT at harvest (9.33 ± 1.53 cm and 11.0 ± 1.00 cm, respectively, versus 6.00 ± 1.00 cm, corresponding to 1.56 and 1.83 fold increases) (Figure 4A).

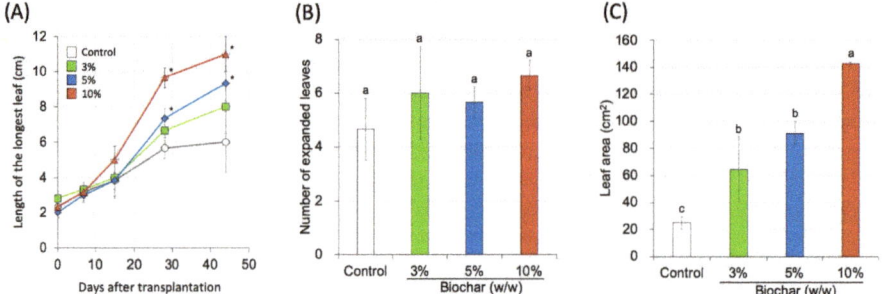

Figure 4. Effects of jatropha fallen leaf biochar application to the experimental soil on the growth of Swiss chard plants. Biochar was applied to the experimental soil at three levels (3%, 5%, and 10% (w/w)) and the growth of Swiss chard plants in the amended soils and unamended control soil was compared. (**A**) Temporal changes in the length of the longest leaf after transplantation in the biochar-amended soils. (**B**) Number of expanded leaves and (**C**) total leaf area per plant at harvest (44 days after transplantation). Values are means ± standard deviations ($n = 3$ plants). Significant differences are indicated by asterisks in (**A**) (*t*-test, $p < 0.05$) and different letters in (**B**,**C**) (Holm's test, $p < 0.05$).

The plants were harvested at 44 DAT and the number of leaves, total leaf area, and dried biomass weight were measured. Plants grown in soils supplemented with 3%, 5%, and 10% biochar had a similar total number of leaves to control plants (6.00 ± 1.73, 5.67 ± 0.58, and 6.67 ± 0.58 leaves, respectively, versus 4.67 ± 1.15; Figure 4B) but 2.58, 3.64, and 5.71 fold higher leaf areas, respectively, than the control plants (Figure 4C). Furthermore, although biochar treatment had no significant effect on the root dry biomass weight, the shoot and the total dry biomass weights were 1.57, 1.88, and 2.32 fold higher in the 3%, 5%, and 10% biochar-amended soils, respectively, compared with the unamended control soil (Figure 5). These observations indicate that the application of jatropha fallen leaf biochar to the acidic experimental soil improved the growth performance of stalk and leaves of Swiss chard.

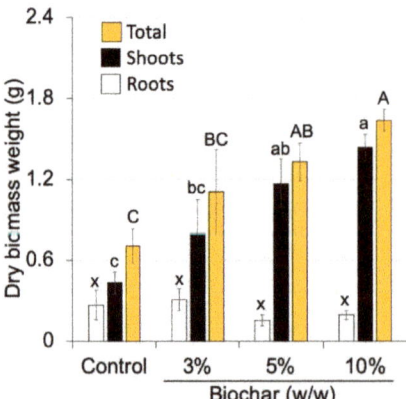

Figure 5. Effects of jatropha fallen leaf biochar application to the experimental soil on the dry biomass weight of Swiss chard plants at harvest (44 days after transplantation). Values are means ± standard deviations of three independent plants. Sets of bars with different letters are significantly different (Holm's test, $p < 0.05$).

The improved growth performance of Swiss chard in the biochar-amended experimental soils prompted us to examine the soil conditions and foliar mineral contents of the plants after harvest.

Biochar application caused the acidic experimental soil (pH 4.03 ± 0.21) to become more neutral in a concentration-dependent manner, reaching pH 6.62 ± 0.28 in the soil containing 10% biochar (Figure 6A). Furthermore, although the soil NH_4^+ levels were not affected by biochar application, the 5% and 10% biochar treatments resulted in a small but significant increase in the NO_3^- concentration (Figure 6B,C), suggesting that biochar application improved the nitrogen availability. Examination of the foliar mineral levels at harvest showed that the majority of minerals were not significantly affected by the biochar treatment, but potassium exhibited 2.17, 3.76, and 4.00 fold increases following the application of 3%, 5%, and 10% biochar, respectively (Table 3), which may reflect the relatively high potassium content of the jatropha fallen leaf biochar that was used in this study (6220 ± 530 mg kg^{-1}; Table 1).

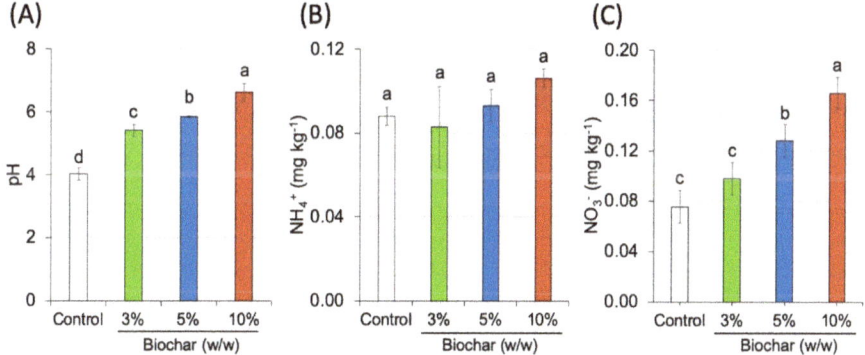

Figure 6. Soil pH and concentration of major inorganic nutrients in the jatropha biochar-amended soils at the time of harvest. The soil pH (**A**) and ammonium ion (**B**) and nitrate ion (**C**) contents are shown. Values are means ± standard deviations (n = 3 pots). Bars with different letters are significantly different (Holm's test, $p < 0.05$).

Table 3. Amount of minerals in the aerial parts of the Swiss chard plants.

Mineral	Content (mg plant^{-1})[1]			
	Control	3% Biochar	5% Biochar	10% Biochar
Al	5.35 ± 1.75 [a]	3.23 ± 1.41 [a]	3.72 ± 3.72 [a]	4.53 ± 2.82 [a]
Ca	7.79 ± 2.72 [a]	6.87 ± 1.96 [a]	5.89 ± 2.09 [a]	5.98 ± 1.47 [a]
Cu	0.01 ± 0.01 [b]	0.02 ± 0.00 [ab]	0.02 ± 0.00 [a]	0.01 ± 0.00 [ab]
Fe	0.41 ± 0.20 [a]	0.19 ± 0.10 [a]	0.22 ± 0.21 [a]	0.30 ± 0.18 [a]
K	22.0 ± 0.8 [c]	47.8 ± 14.5 [b]	82.7 ± 14.8 [a]	88.1 ± 9.7 [a]
Mg	0.34 ± 0.04 [b]	0.48 ± 0.10 [ab]	0.68 ± 0.08 [a]	0.68 ± 0.05 [ab]
Mn	0.38 ± 0.10 [b]	0.72 ± 0.24 [ab]	0.89 ± 0.16 [a]	0.78 ± 0.07 [ab]
Ni	0.04 ± 0.02 [b]	0.15 ± 0.06 [ab]	0.20 ± 0.06 [a]	0.04 ± 0.01 [b]
P	23.1 ± 1.9 [a]	19.9 ± 6.5 [a]	19.8 ± 5.7 [a]	21.8 ± 8.9 [a]
Ti	0.01 ± 0.01 [a]	0.00 ± 0.01 [a]	0.01 ± 0.01 [a]	0.01 ± 0.01 [a]
Zn	0.27 ± 0.07 [a]	0.28 ± 0.07 [a]	0.19 ± 0.03 [a]	0.15 ± 0.05 [a]

[1] Values are means ± standard deviations (n = 3 plants). Values with different letters within a row are significantly different (Holm's test, $p < 0.05$).

3.3. Effects of Jatropha Fallen Leaf Biochar on the Soil Moisture Content

The water-holding capacities of the experimental soils supplemented with different amounts of jatropha biochar were evaluated after harvest, by monitoring the pot weight after adding 500 mL of water to 1 kg of the dried soils. The weights of the pots containing biochar-applied soils increased in a concentration-dependent manner at each measurement time point (Figure 7). Furthermore, while the weights of the control pots had returned to their pre-watering levels at four days after water application, the pots that contained soils supplemented with 3%, 5%, and 10% biochar retained 3.3 ± 1.2 g, 16.3 ± 1.2 g, and 42.3 ± 2.1 g more water, respectively. These observations were consistent with those of previous studies [21,55,56] and suggest that the application of jatropha fallen leaf biochar improved the water-holding capacity of the experimental soil.

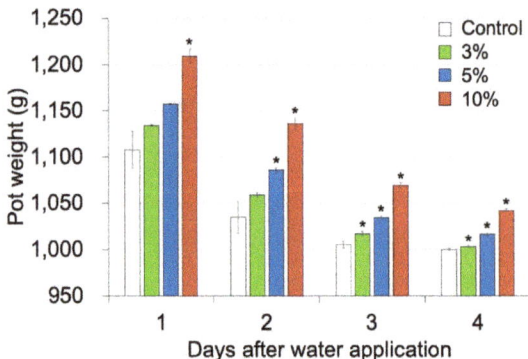

Figure 7. Effect of jatropha fallen leaf biochar application on the moisture contents of the experimental soils. The pot weights were measured daily to estimate the rates of water loss from the potted soils. Values are means ± standard deviations ($n = 3$ pots per biochar concentration). Asterisks indicate significant differences from the control (t-test, $p < 0.05$).

4. Conclusions

This study described the production of jatropha fallen leaves biochar using a simplified oil-drum carbonizer, based on the auto-thermal process in which partial combustion of the biomass provided the energy necessary for the pyrolysis of the remaining biomass. Internal temperature of the drum was maintained in the range of 100–450 °C, and the resultant products had porous surface structure characteristics for biochar. Application of the biochar to an acidic and undernourished soil significantly improved the growth performance of the model vegetable Swiss chard. These findings suggest that jatropha has potential applications not only for producing renewable energy and industrial products, but also as a feedstock for soil conditioner to improve agricultural production.

Author Contributions: T.K., M.U., and K.A. conceived, designed, and performed the experiments. S.Y. and Y.K. prepared experimental materials. M.C. and H.T. contributed to the laboratory works and editing of the manuscript. T.K. analyzed the data and wrote the paper with input from K.A.

Funding: Authors gratefully acknowledge funding of this co-research project under SATREPS program of Japan Science and Technology Agency, Japan International Cooperation Agency and the Government of Botswana, Joint Research Program and the Project Marginal Region Agriculture, the Arid Land Research Center, Tottori University, and the IPDRE Program, Tottori University.

Acknowledgments: The authors thank to the Tottori Agricultural Strategy Division of Tottori Prefecture in Japan for their useful advice on CEC analysis. The authors thank Enago (www.enago.jp) for the English language review. The authors thank for all the participants of the project "Information-Based Optimization of Jatropha Biomass Energy Production in the Frost- and Drought-prone Regions of Botswana" for their counsels.

Conflicts of Interest: The authors declare no conflict of interest. The funding sponsors had no role in the design of this study; in the collection, analyses, or interpretation of data; in the writing of the manuscript, or in the decision to publish the results.

References

1. Openshaw, K. A review of Jatropha curcas: an oil plant of unfulfilled promise. *Biomass Bioenerg.* **2000**, *19*, 1–15. [CrossRef]
2. Quinn, L.D.; Straker, K.C.; Guo, J.; Kim, S.; Thapa, S.; Kling, G.; Lee, D.K.; Voigt, T.B. Stress-tolerant feedstocks for sustainable bioenergy production on marginal land. *Bioenerg. Res.* **2015**, *8*, 1081–1100. [CrossRef]

3. Barua, P.K. Biodiesel from seeds of Jatropha found in Assam, India. *Int. J. Energy Inf. Commun.* **2011**, *2*, 53–65.
4. Montes, J.M.; Melchinger, A.E. Domestication and breeding of *Jatropha curcas* L. *Trends Plant Sci.* **2016**, *21*, 1045–1057. [CrossRef] [PubMed]
5. Pandey, K.K.; Pragya, N.; Sahoo, P.K. Life cycle assessment of small-scale high-input Jatropha biodiesel production in India. *Appl. Energy* **2011**, *88*, 4831–4839. [CrossRef]
6. Ye, M.; Li, C.; Francis, G.; Makkar, H.P.S. Current situation and prospects of *Jatropha curcas* as a multipurpose tree in China. *Agrofor. Syst.* **2009**, *76*, 487–497. [CrossRef]
7. Chauhan, B.S.; Kumar, N.; Cho, H.M. A study on the performance and emission of a diesel engine fueled with Jatropha biodiesel oil and its blends. *Energy* **2012**, *37*, 616–622. [CrossRef]
8. Jongschaap, R.E.E.; Corré, W.J.; Bindraban, P.S.; Brandenburg, W.A. *Claims and Facts on Jatropha Curcas L. Global Jatropha Curcas Evaluation, Breeding and Propagation Programme*; Report 158; Plant Research International. B.V.: Laren, The Netherlands, October 2007.
9. Parawira, W. Biodiesel production from *Jatropha curcas*: A review. *Sci. Res. Essays* **2010**, *5*, 1796–1808.
10. Navarro-Pineda, F.S.; Baz-Rodríguez, S.A.; Handler, R.; Sacramento-Rivero, J.C. Advances on the processing of *Jatropha curcas* towards a whole-crop biorefinery. *Renew. Sust. Energy Rev.* **2016**, *54*, 247–269. [CrossRef]
11. Abdul Khalil, H.P.S.; Sri Aprilia, N.A.; Bhat, A.H.; Jawaid, M.; Paridah, M.T.; Rudi, D. A Jatropha biomass as renewable materials for biocomposites and its application. *Renew. Sust. Energy Rev.* **2013**, *22*, 667–685. [CrossRef]
12. Srinophakun, P.; Titapiwatanakun, B.; Sooksathan, I.; Punsuvon, V. Prospect of deoiled *Jatropha curcas* seedcake as fertilizer for vegetables crops—A case study. *J. Agric. Sci.* **2012**, *4*, 211–226. [CrossRef]
13. Selanon, O.; Saetae, D.; Suntornsuk, W. Utilization of *Jatropha curcas* seed cake as a plant growth stimulant. *Biocatal. Agric. Biotechnol.* **2014**, *3*, 114–120. [CrossRef]
14. Biradar, C.H.; Subramanian, K.A.; Dastidar, M.G. Production and fuel quality upgradation of pyrolytic bio-oil from *Jatropha curcas* de-oiled seed cake. *Fuel* **2014**, *119*, 81–89. [CrossRef]
15. Chintala, V.; Kumar, S.; Pandey, J.K.; Sharma, A.K.; Kumar, S. Solar thermal pyrolysis of non-edible seeds to biofuels and their feasibility assessment. *Energy Convers. Manag.* **2017**, *153*, 482–494. [CrossRef]
16. Kongkasawan, J.; Nam, H.; Capareda, S.C. Jatropha waste meal as an alternative energy source via pressurized pyrolysis: A study on temperature effects. *Energy* **2016**, *113*, 631–642. [CrossRef]
17. Kumar, A.; Sharma, S. An evaluation of multipurpose oil seed crop for industrial uses (*Jatropha curcas* L.): A review. *Ind. Crops Prod.* **2008**, *28*, 1–10. [CrossRef]
18. Esimone, C.O.; Nworu, C.S.; Jackson, C.L. Cutaneous wound healing activity of herbal ointment containing the leaf extract of *Jatropha curcas* L. (Euphorbiaceae). *Int. J. Appl. Res. Nat. Prod.* **2008**, *1*, 1–4.
19. Juliet, S.; Ravindran, R.; Ramankutty, S.A.; Gopalan, A.K.K.; Nair, S.N.; Kavillimakkil, A.K.; Bandyopadhyay, A.; Rawat, A.K.S.; Ghosh, S. *Jatropha curcas* (Linn) leaf extract—A possible alternative for population control of *Rhipicephalus (Boophilus) annulatus*. *Asian Pac. J. Trop. Dis.* **2012**, *2*, 225–229. [CrossRef]
20. Oskoueian, E.; Abdullah, N.; Saad, W.Z.; Omar, A.R.; Ahmad, S.; Kuan, W.B.; Zolkifli, N.A.; Hendra, R.; Ho, Y.W. Antioxidant, anti-inflammatory and anticancer activities of methanolic extracts from *Jatropha curcas* Linn. *J. Med. Plants Res.* **2011**, *5*, 49–57.
21. Ogura, T.; Date, Y.; Maskujane, M.; Coetzee, T.; Akashi, K.; Kikuchi, J. Improvement of physical, chemical, and biological properties of aridisol from Botswana by the incorporation of torrefied biomass. *Sci. Rep.* **2016**, *6*, 28011. [CrossRef] [PubMed]
22. Álvarez-Mateos, P.; Alés-Álvarez, F.J.; García-Martín, J.F. Phytoremediation of highly contaminated mining soils by *Jatropha curcas* L. and production of catalytic carbons from the generated biomass. *J. Environ. Manag.* **2019**, *231*, 886–895. [CrossRef]
23. García-Martín, J.F.; Alés-Álvarez, F.J.; Torres-García, M.; Feng, C.H.; Álvarez-Mateos, P. Production of oxygenated fuel additives from residual glycerine using biocatalysts obtained from heavy-metal-contaminated *Jatropha curcas* L. roots. *Energies* **2019**, *12*, 740. [CrossRef]
24. Inafuku-Teramoto, S.; Mazereku, C.; Coetzee, T.; Gwafila, C.; Lekgari, L.A.; Ketumile, D.; Fukuzawa, Y.; Yabuta, S.; Masukujane, M.; George, D.G.M.; et al. Production approaches to establish effective cultivation methods for Jatropha (*Jatropha curcas* L.) under cold and semi-arid climate conditions. *Int. J. Agric. Plant Prod.* **2013**, *4*, 3804–3815.

25. Ishimoto, Y.; Kgokong, S.; Yabuta, S.; Tominaga, J.; Coetzee, T.; Konaka, T.; Mazereku, C.; Kawamitsu, Y.; Akashi, K. Flowering pattern of biodiesel plant Jatropha in frost- and drought-prone regions of Botswana. *Int. J. Green Energy* **2017**, *14*, 908–915. [CrossRef]
26. Wani, S.P.; Chander, G.C.; Sahrawat, K.L.; Rao, C.S.; Raghvendra, G.; Susanna, P.; Pavani, M. Carbon sequestration and land rehabilitation through *Jatropha curcas* (L.) plantation in degraded lands. *Agric. Ecosyst. Environ.* **2012**, *161*, 112–120. [CrossRef]
27. Anders, E.; Watzinger, A.; Rempt, F.; Kitzler, B.; Wimmer, B.; Zehetner, F.; Stahr, K.; Zechmeister-Boltenstern, S.; Soja, G. Biochar affects the structure rather than the total biomass of microbial communities in temperate soils. *Agric. Food Sci.* **2013**, *22*, 404–423. [CrossRef]
28. Sohi, S.P.; Krull, E.; Lopez-Capel, E.; Bol, R. A review of biochar and its use and function in soil. *Adv. Agron.* **2010**, *105*, 47–82.
29. Atkinson, C.J.; Fitzgerald, J.D.; Hipps, N.A. Potential mechanisms for achieving agricultural benefits from biochar application to temperate soils: A review. *Plant Soil* **2010**, *337*, 1–18. [CrossRef]
30. Glaser, B.; Lehmann, J.; Zech, W. Ameliorating physical and chemical properties of highly weathered soils in the tropics with charcoal—A review. *Biol. Fertil. Soils* **2002**, *35*, 219–230. [CrossRef]
31. Steiner, C.; Teixeira, W.G.; Lehmann, J.; Nehls, T.; Vasconcelos de Macêdo, J.L.; Blum, W.E.H.; Zech, W. Long term effects of manure, charcoal and mineral fertilization on crop production and fertility on a highly weathered Central Amazonian upland soil. *Plant Soil* **2007**, *291*, 275–290. [CrossRef]
32. Asai, H.; Samson, B.K.; Stephan, H.M.; Songyikhangsuthor, K.; Homma, K.; Kiyono, Y.; Inoue, Y.; Shiraiwa, T.; Horie, T. Biochar amendment techniques for upland rice production in Northern Laos 1. Soil physical properties, leaf SPAD and grain yield. *Field Crops Res.* **2009**, *111*, 81–84. [CrossRef]
33. Gaskin, J.W.; Steiner, C.; Harris, K.; Das, K.C.; Bibens, B. Effects of low-temperature pyrolysis conditions on biochar for agricultural use. *Trans. Am. Soc. Agric. Biol. Eng.* **2008**, *51*, 2061–2069.
34. Major, J.; Rondon, M.; Molina, D.; Riha, S.J.; Lehmann, J. Maize yield and nutrition during 4 years after biochar application to a Colombian savanna oxisol. *Plant Soil* **2010**, *333*, 117–128. [CrossRef]
35. Lee, M.E.; Park, J.H.; Chung, J.W. Adsorption of Pb(II) and Cu(II) by ginkgo-leaf-deriverd biochar produced under various carbonization temperatures and time. *Int. J. Environ. Res. Public Health* **2017**, *14*, 1528. [CrossRef]
36. Nejad, Z.D.; Kim, J.W.; Jung, M.C. Reclamation of arsenic contaminated soils around mining site using solidification/stabilization combined with revegetation. *Geosci. J.* **2017**, *21*, 285–396.
37. Mitchell, P.J.; Dalley, T.S.L.; Helleur, R.J. Preliminary laboratory production and characterization of biochars from lignocellulosic municipal waste. *J. Anal. Appl. Pyrolysis* **2013**, *99*, 71–78. [CrossRef]
38. Singh, B.; Singh, B.P.; Cowie, A.L. Characterisation and evaluation of biochars for their application as a soil amendment. *Aust. J. Soil Res.* **2010**, *48*, 516–525. [CrossRef]
39. Abdel-Fattah, T.M.; Mahmound, M.E.; Ahmed, S.B.; Huff, M.D.; Lee, J.W.; Kumar, S. Biochar from woody biomass for removing metal contaminants and carbon sequestration. *J. Ind. Eng. Chem.* **2015**, *22*, 103–109. [CrossRef]
40. Cobb, A.; Warms, M.; Maurer, E.P.; Chiesa, S. Low-tech coconut shell activated charcoal production. *Int. J. Serv. Lean. Eng.* **2012**, *7*, 93–104. [CrossRef]
41. Ishii, T.; Kadoya, K. Effects of charcoal as a soil conditioner on citrus growth and vesicular-arbuscular mycorrhizal development. *J. Jpn. Soc. Hort. Sci.* **1994**, *63*, 529–535. [CrossRef]
42. Schollenberger, C.J.; Simon, R.H. Determination of exchange capacity and exchangeable bases in soil – ammonium acetate method. *Soil Sci.* **1945**, *59*, 13–24. [CrossRef]
43. Borden, D.; Giese, R.F. Baseline studies of the clay minerals society source clays: cation exchange capacity measurements by the ammonia-electrode method. *Clay Miner.* **2001**, *49*, 444–445. [CrossRef]
44. Vurayai, R.; Nkoane, B.; Moseki, B.; Chaturvedi, P. Assessment of heavy metal pollution/ contamination in soils east and west Bamangwato Concession Ltd (BCL) Cu/Ni mine smelter in Selebi-Phikwe, Botswana. *J. Biol. Environ. Sci.* **2015**, *7*, 111–120.
45. Walkley, A.; Black, I.A. An examination of the Degtjareff method for determining soil organic matter, and a proposed modification of the chromic acid titration method. *Soil Sci.* **1934**, *37*, 29–38. [CrossRef]
46. Robinson, B.H.; Leblanc, M.; Petit, D.; Brooks, R.R.; Kirkman, J.H.; Gregg, P.E.H. The potential of *Thlaspi caerulescens* for phytoremediation of contaminated soils. *Plant Soil* **1998**, *203*, 47–56. [CrossRef]

47. Yamada, M.; Malambane, G.; Yamada, S.; Suharsono, S.; Tsujimoto, H.; Moseki, B.; Akashi, K. Differential physiological responses and tolerance to potentially toxic elements in biodiesel tree *Jatropha curcas*. *Sci. Rep.* **2018**, *8*, 1635. [CrossRef]
48. Boateng, A.A.; Garcia-Perez, M.; Mašek, O.; Brown, R.; del Campo, B. Biochar production technology. In *Biochar for Environmental Management*, 2nd ed.; Lehmann, J., Joseph, S., Eds.; Routledge: Oxon, UK; New York, NY, USA, 2015; pp. 63–87.
49. Chia, C.H.; Downie, A.; Munroe, P. Characteristics of biochar: Physical and structural properties. In *Biochar for Environmental Management*, 2nd ed.; Lehmann, J., Joseph, S., Eds.; Routledge: Oxon, UK; New York, NY, USA, 2015; pp. 89–109.
50. Awasthi, M.K.; Li, J.; Kumar, S.; Awasthi, S.K.; Wang, Q.; Chen, H.; Wang, M.; Ren, X.; Zhang, Z. Effects of biochar amendment on bacterial and fungal diversity for co-composting of gelatin industry sludge mixed with organic faction of municipal soil waste. *Bioresour. Technol.* **2017**, *246*, 214–223. [CrossRef]
51. Ippolito, J.A.; Spokas, K.A.; Novak, J.M.; Lentz, R.D.; Cantrell, K.B. Biochar elemental composition and factors influencing nutrient retention. In *Biochar for Environmental Management*, 2nd ed.; Lehmann, J., Joseph, S., Eds.; Routledge: Oxon, UK; New York, NY, USA, 2015; pp. 139–163.
52. Kochian, L.V.; Hoekenga, O.A.; Piñeros, M.A. How do crop plants tolerate acid soils? Mechanisms of aluminum tolerance and phosphorous efficiency. *Annu. Rev. Plant Biol.* **2004**, *55*, 459–493. [CrossRef]
53. Marschner, H. Mechanisms of adaptation of plants to acid soils. *Plant Soil* **1991**, *134*, 1–20. [CrossRef]
54. Bowen, H.J.M. *Environmental Chemistry of the Elements*; Academic Press: London, UK; New York, NY, USA, 1979; pp. 60–61.
55. Masiello, C.A.; Dugan, B.; Brewer, C.E.; Spokas, K.A.; Novak, J.M.; Liu, Z.; Sorrenti, G. Biochar effects on soil hydrology. In *Biochar for Environmental Management*, 2nd ed.; Lehmann, J., Joseph, S., Eds.; Routledge: Oxon, UK; New York, NY, USA; pp. 543–562.
56. Speratti, A.B.; Johnson, M.S.; Sousa, H.M.; Torres, G.N.; Couto, E.G. Impact of different agricultural waste biochars on maize biomass and soil water content in Brazilian Cerrado Arenosol. *Agronomy* **2017**, *7*, 49. [CrossRef]

© 2019 by the authors. Licensee MDPI, Basel, Switzerland. This article is an open access article distributed under the terms and conditions of the Creative Commons Attribution (CC BY) license (http://creativecommons.org/licenses/by/4.0/).

Article

Carbonaceous Greenhouse Gases and Microbial Abundance in Paddy Soil under Combined Biochar and Rice Straw Amendment

Supitrada Kumputa [1,2], Patma Vityakon [1,2], Patcharee Saenjan [1,2] and Phrueksa Lawongsa [1,2,*]

1. Department of Soil Science and Environment, Faculty of Agriculture, Khon Kaen University, Khon Kaen 40002, Thailand; nokie_msu@hotmail.com (S.K.); patma@kku.ac.th (P.V.); patsae1@kku.ac.th (P.S.)
2. Soil Organic Matter Management Research Group, Department of Soil Science and Environment, Faculty of Agriculture, Khon Kaen University, Khon Kaen 40002, Thailand
* Correspondence: phrula@kku.ac.th; Tel.: +66-43-364639

Received: 6 March 2019; Accepted: 30 April 2019; Published: 6 May 2019

Abstract: Little is known about the carbonaceous greenhouse gases and soil microbial community linked to the combination of biochar (BC) and rice straw (RS) in paddy soils. The objectives of this research were to evaluate the effects of combining BC and RS on (1) CH_4 and CO_2 production from paddy soil, (2) archaeal and bacterial abundance, and (3) rice grain yield. The experiments consisted of a pot trial and an incubation trial, which had a completely randomized design. The experiments included five treatments with three replications: (a) the control (without BC, RS, and chemical fertilizer (CF)); (b) CF; (c) BC 12.50 t ha^{-1}; (d) RS 12.50 t ha^{-1}; and (e) combined BC 6.25 t ha^{-1} + RS 6.25 t ha^{-1} + CF. In the sole RS treatment, CH_4 production (0.0347 mg m^{-2} season^{-1}) and the archaeal and bacterial abundance (5.81 × 10^8 and 4.94 × 10^{10} copies g^{-1} soil dry weight (DW)) were higher than outcomes in the sole BC treatment (i.e., 0.0233 mg m^{-2} season^{-1} for CH_4 production, and 8.51 × 10^7 and 1.76 × 10^{10} copies g^{-1} soil DW for archaeal and bacterial abundance, respectively). CH_4 production (0.0235 mg m^{-2} season^{-1}) decreased significantly in the combined BC + RS + CF treated soil compared to the soil treated with RS alone, indicating that BC lessened CH_4 production via CH_4 adsorption, methanogenic activity inhibition, and microbial CH_4 oxidation through bacterial methanotrophs. However, the archaeal abundance (3.79–5.81 × 10^8 copies g^{-1} soil DW) and bacterial abundance (4.94–5.82 × 10^{10} copies g^{-1} soil DW) in the combined BC+ RS + CF treated soil and the RS treated soil were found to increase relative to the treatments without RS. The increase was due to the easily decomposable RS and the volatile matter (VM) constituent of the BC. Nevertheless, the resultant CO_2 production was relatively similar amongst the BC, RS, and BC + RS treated soils, which was indicative of several processes, e.g., the CO_2 production and reduction that occurred simultaneously but in different directions. Moreover, the highest yield of rice grains was obtained from a combined BC + RS + CF treated soil and it was 53.47 g pot^{-1} (8.48 t ha^{-1}). Over time, the addition of BC to RS soil enhanced the archaeal and bacterial abundance, thereby improving yields and reducing CH_4 emissions.

Keywords: global warming; archaeal 16S rRNA gene; bacterial 16S rRNA gene; rice yields; qPCR; soil amendments

---

1. Introduction

To maintain the soil fertility and rice yield, the incorporation of rice straw (RS) into paddy soil has been widely practiced. However, in flooded soil conditions, the decomposition of RS results in high levels of CH_4 and CO_2 emissions from its high cellulose and hemicellulose content (at more than 50%

dry weight), both of which are easily decomposable C compounds [1]. In addition, some intermediate C products include dissolved organic carbon (DOC), which comprises low molecular-weight organic compounds such as acetates, formates, methylated compounds, primary and secondary alcohols, and some gases, e.g., CO_2 and H_2. All of these compounds are substrates for the methanogenic archaea which stimulate CH_4 production [2]. Concurrently, CH_4 oxidation mediated by methanotrophic microorganisms existing in the flooded soil system also occurs. The CH_4 oxidation results in the production of CO_2, as well as a decrease in CH_4 emissions into the atmosphere.

Contrary to the easily decomposable RS, biochar (BC), which is made from woody feedstock materials, has high contents of C resistant compounds, such as lignin [1]. Therefore, it is considered to be a resistant organic material. In particular, eucalyptus wood BC contains over 70% lignin (DW), which suppresses microbially mediated C mineralization [1,3]. The addition of BC creates a low available C condition, creating unsuitable circumstances for methanogenesis by archaea [4]. Although the BC incorporated into paddy soils suppresses CH_4 emissions, it also increases nutrient availability and rice yields [3]. When RS and BC were individually applied to paddy soils, contrasting effects on CH_4 production were found. Owing to contrasting chemical compositions, RS produced enhancing effects, whereas BC produced suppressing effects given its high content of fixed C, such as lignin [3], which are unfavorable to methanogenic activity [5]. Nevertheless, it is worth studying the incorporation of combined BC with RS, as well as the biological aspects of methanogen. In field situations, when BC is applied to paddy soil, it inevitably mixes with RS residues that remain after the paddy fields have been harvested. Therefore, it is imperative to investigate the effects of combining BC and RS on greenhouse gas production. An earlier study by Liu et al. [5] showed that when medium to high amounts of rice straw derived BC were mixed with RS, the CH_4 emissions from the incubated paddy soils declined by 21–35% compared to emissions without BC, citing the inhibiting effect of BC on methanogenic activity. However, the study did not investigate the microbial abundance. Findings on methanogen stimulation by BC were later reported by Feng et al. [6], who employed microbial gene abundance as the main indicator of microbial influence on CH_4 emissions in soils treated solely with BC. The research showed that corn stalk BC stimulated both the methanogenic archaea and the methanotrophic bacteria, as determined by their gene abundances. However, the CH_4 produced by the archaea could not meet the requirements of the bacteria. To our knowledge, there is no known work which has combined easily decomposable RS and resistant eucalyptus BC to test the effects of this mixture on the production of CH_4 and other carbonaceous greenhouse gases, using microbial gene abundance as a major indicator.

In this research, we addressed the hypothesis that adding the combined BC and RS to a paddy soil would reduce the soil's CH_4 production, raise its archaeal and bacterial abundance, and increase the rice grain yields. Therefore, the objectives of this research were as follows: (i) To evaluate the effects of the combined BC and RS on CH_4 production, CO_2 production, and the archaeal and bacterial abundance in paddy soils and (ii) to determine the effects that these conditions would have on the rice grain yields.

2. Materials and Methods

2.1. Organic Materials and Soil

BC was produced via pyrolysis at 350 °C under oxygen limited conditions in a traditional kiln commonly used in Northeastern Thailand. The feedstock consisted of the upper parts of the branches of 5 year-old eucalyptus trees (*Eucalyptus camaldulensis* Dehnh.). Meanwhile, the RS used was taken from a paddy field. The following chemical analyses of the BC and RS were conducted: (1) pH using a pH meter (BC or RS: water = 1:5); (2) total organic carbon content using a TOC Analyzer (multi EA 4000, Analytik Jena, Jena, Germany); (3) total nitrogen using the micro-Kjeldahl method [7]; (4) the content of cellulose, hemicellulose, and lignin in the RS and BC as described by Aravantinos-Zafiris et al. [8]; (5) the content of ash, VM, and fixed C in the BC, based on the American standard test method [9]; and the functional groups on the surface of the BC and RS were analyzed using Fourier transform infrared (FTIR) spectroscopy (TENSOR27, Bruker, Germany), at frequency ranges from 600 to 4000 cm^{-1}. The chemical characteristics of the BC and RS are shown in Table 1

Table 1. Characteristics of the BC and RS used in the experiments.

Organic Materials [1]	pH	OC [2]	TN [3]	C/N [4] Ratio	Cellulose	Hemicell [5]	Lignin	Fixed C	Ash	VM [6]
	(1:5)	%								
BC	6.32	60.2	0.56	101	1.24	1.65	75.69	61.72	3.3	34.97
RS	7.47	40.9	0.43	95	46.65	22.17	7.11	-	-	-

[1] BC = biochar, RS = rice straw; [2] OC = organic carbon; [3] TN = total nitrogen; [4] C/N = carbon/nitrogen; [5] Hemicell = hemicellulose; [6] VM = volatile matter.

The BC FTIR spectra contained the following peaks (Figure 1a): 3570–3200 cm^{-1} (hydroxy group); 2921 cm^{-1} (methylene C-H asymmetric); 1928–2113 cm^{-1} (aromatic combination bands) [10]; 1641–1737 cm^{-1} (C=O of aromatic group) [11]; 1373 and 1591 cm^{-1} (carboxylate); and 1205 cm^{-1} (phenol, C-O stretch) [10]. The RS spectra (Figure 1b) contained a 3570–3200 cm^{-1} (hydroxy group); 1637 cm^{-1} (carboxylate); and 1033 cm^{-1} (aromatic C-H in plane bend) [10].

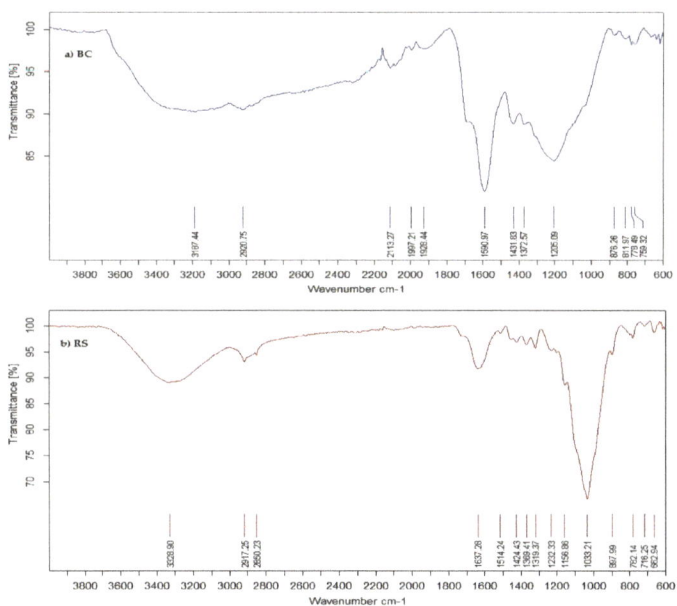

Figure 1. Fourier transform infrared (FTIR) spectra of the biochar (BC) (a) and rice straw (RS) (b) used in the experiments.

Paddy soil samples were randomly collected from the plow layer (0–15 cm) of an irrigated paddy field located in Ban Na Ngam in the Samran District of Khon Kaen, Thailand (N 16°32′45.9″, E 102°51′15.5″). The soil was classified as fine, mixed, and isohyperthermic Aeric Endoaquept. The soil was air-dried and then finely ground to be able to pass through a 2 mm sieve. The physical and chemical characteristics of the soil were analyzed for: (1) the soil texture using the hydrometer method [12]; (2) the soil organic carbon contents using wet digestion [13]; and (3) the total nitrogen using the micro-Kjeldahl method [7]. The soil showed the following physical-chemical properties: pH

(1:5) = 5.06; sandy loam texture with sand (65.8%); silt (21.9%); and clay (12.4%) with a soil organic carbon content of 0.83% and a nitrogen content of 0.08%.

2.2. Experiments

Two experiments (i.e., a pot and an incubation experiment) were conducted. The pot experiment was designed to evaluate the effects of the combined BC and RS on the production of carbonaceous gases, the microbial biomass, and the rice yields under non-leaching controlled conditions in the presence of rice plants. In contrast, the incubation experiment was designed to support the biological and biochemical data collected from homogeneous non-living root soils, to examine the effects of the combined BC and RS.

2.2.1. Pot Experiment

The pot experiment was performed from June to October 2015 in a greenhouse located at the Faculty of Agriculture at the Khon Kaen University in Khon Kaen, Thailand. Five treatments, performed in triplicate, were included as follows: (1) the control (without CF, BC, and RS amendments); (2) CF grade 16-16-8 [Urea (46% N), $(NH_4)_2HPO_4$ (18% N, 46%P_2O_5), KCl (60% K_2O)] at a rate of 0.188 t ha^{-1} as modified from the study by Thammasom et al. [3]; (3) BC 12.50 t ha^{-1}; (4) RS 12.50 t ha^{-1}; and (5) a mixture of BC:RS (1:1 w/w at a rate of 6.25 t ha^{-1} each) and CF. The experiment was arranged using a completely randomized design, wherein three kgs of sieved air-dried soil was placed in a pot (inner dimensions of 18 cm and a height of 23 cm, without a hole at the bottom). Based on the treatment parameters, the soil was then mixed with 2 mm sieved BC and/or RS (cut to a size of 2 cm in length). The soils in all the pots were submerged for 20 days before transplanting. This procedure was carried out to allow time for decomposition, so that the adverse effects from the toxic intermediate organic acid products of decomposition could be avoided. Then, three rice (*Oryza sativa* L.) seedlings (25 days old) of the Pitsanulok 2 (a photoperiod insensitive) varieties were transplanted to each pot. The CF was basally applied twice, that is, before transplanting and then 30 days after transplanting. Throughout the rice growing period, all the pots were maintained at a water level that was 5–7 cm above the soil surface without leaching, and the water was drained 10 days prior to the rice harvest.

2.2.2. Incubation Experiment

Treatments for the incubation trials were similar to the pot experiment treatments. Soil (2.5 g) was placed into a 60 mL glass bottle and then mixed with 2 mm BC and/or RS based on the treatments. Thereafter, 10 mL of the CF solution of the same strength as that used in the pot experiment was applied to the soil mixture. Calculations of the BC and RS weights were based on a soil bulk density of 1.39 g cm^{-3} and a soil weight of 2085 t ha^{-1}. The head space of the bottle was flushed with N_2 gas (99.99%), and then it was tightly closed using a septum and aluminum cap. The incubation of the soil was carried out at 28 °C for a period of 14 days under anaerobic conditions. The incubation period (14-days) was determined based on our previous study which found the highest CH_4 and CO_2 emissions after 14 days of incubation.

2.3. Data Collection

2.3.1. Rice Grain and Microbial Biomass C (MBC) in the Pot Experiment

After harvesting, rice grains collected from each pot were dried in an oven at 75 °C for 48 hours, and then weighed. A fresh soil sample was taken from each pot and analyzed for the MBC using the chloroform fumigation-extraction method described in Reference [14].

2.3.2. Gas Sampling, CH_4, and CO_2 Analysis in the Pot and Incubation Experiments

In the pot experiment, gas samples were collected using the closed chamber method. We used a transparent chamber made from acrylic, that was sized 21 × 21 × 100 cm (width × length × height). Gas sampling was performed once a week throughout the rice growing period. The process was carried out between 9.00 and 11.00 a.m., and a 1 ml insulin syringe was used to obtain the gas samples at 0, 10, and 20 min after the chamber cover had been placed over the potted soil as in Reference [15]. CH_4 and CO_2 concentrations were measured using a gas chromatograph (GC-2014, Shimadzu, Kyoto, Japan) equipped with a flame ionization detector (FID) as described in Reference [1]. The gas measurements were completed within 6 hours.

Under the incubation experiment, the 1 mL gas samples were collected 14 days after incubation from the head space of the glass bottles using a 1-mL insulin syringe. After collection, the CH_4 and CO_2 concentrations were immediately determined.

2.4. DOC Analysis and Determination of Archaeal and Bacterial Abundance in the Incubation Experiment

Extraction of DOC from the incubated soil was done by shaking the bottle for 30 min, followed by centrifuging at 4000 rpm for 15 min. The supernatant solution was filtered through a 0.45 µm syringe filter prior to the DOC analysis, using the TOC/TN_b analyzer (Multi N/C 2100s, Analytik Jena, Jena, Germany).

2.5. DNA Extraction and Quantitative Polymerase Chain Reaction (qPCR)

The total soil genomic DNA was extracted using a FastDNA™ SPIN Kit for Soil (MP Biomedicals, Santa Ana, CA, USA). The qPCR of the bacterial 16S rRNA gene and archaeal 16S rRNA gene were performed using a C1000 Touch™ thermal cycler combined with a CFX96™ detection module (BIO-RAD, Hercules, CA, USA). The primers and annealing conditions are listed in Table 2. The PCR mixtures (25 µL) contained 12.5 µL of EXPRESS SYBR® GreenER™ (Invitrogen, Carlsbad, CA, USA), 0.4 µM primer (each; final concentration), 1 µL of DNA template (10 ng µL^{-1}), and ultrapure water for the balance. Moreover, all the samples were analyzed in triplicate. Each reaction condition included an initial denaturing step of 10 min at 95 °C, followed by 40 cycles of 30 s of denaturing at 95 °C, 30 s of primer annealing (Table 2), and then 45 s of primer extension at 72 °C. The annealing temperatures were optimized for each primer pair. The abundances of bacteria and archaea determined using qPCR were reported as DNA copy numbers of 16S rRNA genes per g of dry soil.

Table 2. The qPCR primers and conditions used in this study.

Primers	Sequences (5' to 3')	Annealing Temps (°C)	Targeted Groups	References
Eub338	ACCTACGGGAGGCAGCAG	55	Bacteria	[16]
Eub518	ATTACCGCGGCTGCTGG	55	Bacteria	[17]
Ar109f	ACKGCTCAGTAACACGT	57.5	Archaea	[18]
Ar912r	CTCCCCCGCCAATTCCTTTA	57.5	Archaea	[18]

2.6. Statistical Analysis

One-way analysis of variance (ANOVA) was used to assess the treatment effects on various soil microbiological and biochemical properties, carbonaceous greenhouse gas emissions, and rice yields. Mean separation was performed using least significant difference (LSD) tests. We used the Statistix 10 software to carry out the statistical tests. To determine the correlation between the abundances of archaea and bacteria, the production of CH_4 and CO_2, and the DOC content in the incubated rice soil, the SigmaPlot 12.5 software program was used.

3. Results and Discussion

3.1. Carbonaceous Greenhouse Gases and Microbial Abundance in Paddy Soil as Affected by RS

Significant increases in the production of CH_4 were observed in the soil amended with RS alone, and in both pots (0.0347 mg m^{-2} season^{-1}) (Table 3) and incubation experiments (1379.3 mg kg^{-1}) (Table 4) relative to the other treatments. The increases in CH_4 production were due to the high contents of easily decomposed cellulose (46.65%) and hemicellulose (22.17%) in the RS (Table 1). When the RS was applied to the soil, it had a key role in stimulating the soil's microbial activity for C mineralization. This was indicated by a significantly higher DOC content (202.69 mg kg^{-1}), and a higher volume of archaeal abundance (5.81 × 10^8 copies g^{-1} soil DW) and bacterial abundance (4.94 × 10^{10} copies g^{-1} soil DW) in the RS compared to other treatments, with the exception of the mixed RS + BC treatment (Table 4). Dissolved organic C is a mixture of dissolved organic carbonaceous compounds with particle sizes that are smaller than 0.45 µm. It is derived from the degradation of organic materials and it contains carbohydrates, proteins, fats, hydrocarbons and their derivatives, and fractions of low molecular weight humic acids; as well as numerous simple organic compounds [19]. DOC is a crucial part of the organic labile pool which serves as substrates for soil microorganisms. Rice straw was found to generate a high content of low molecular weight DOC within two weeks after incorporation into the topsoil (0–15 cm) of a sandy soil from Northeastern Thailand [20]. During our 2-week incubation period, the soil treatments containing RS showed a higher abundance of archaea than the other treatments. Archaea was a dominant microbe that utilized the DOC, CO$_2$, and H$_2$ [21] from decomposing RS to produce CH$_4$. This revealed the archaea's function in methanogenesis, which involved CO$_2$ reduction.

Table 3. CH$_4$, CO$_2$ emissions, rice grains, and microbial biomass C (MBC) in the potted rice-soil treated with BC and RS.

Treatments [1]	CH$_4$ mg m^{-2}	CO$_2$ Season^{-1}	Rice Grains g pot^{-1}	MBC [2] mg kg^{-1}
Control	0.0298 ab	0.0018 b	41.52 b	79.81 c
CF	0.0263 b	0.0012 c	50.62 a	93.66 bc
BC 12.50 t ha^{-1}	0.0233 b	0.0013 c	35.55 b	224.08 a
RS 12.50 t ha^{-1}	0.0347 a	0.0021 a	35.43 b	129.84 bc
BC 6.25 t ha^{-1} + RS 6.25 t ha^{-1} + CF	0.0235 b	0.0012 c	53.47 a	167.94 ab
F-test	*	**	**	*
CV (%)	15.87	2.99	7.00	33.26

[1] CF = chemical fertilizer, BC = biochar, RS = rice straw, CV = coefficient of variation. [2] MBC = microbial biomass carbon. The different small letters in the columns indicate significant difference among treatments by LSD. *, ** = significant at $p \leq 0.05$ and $p \leq 0.01$, n = 3.

Table 4. The abundance of archaea and bacteria, CH$_4$ and CO$_2$ production, and DOC content in 14-day incubated soils treated with BC and RS.

Treatments [1]	Archaea Copies g^{-1}	Bacteria Soil DW	CH$_4$ mg kg^{-1}	CO$_2$ mg kg^{-1}	DOC [2] mg kg^{-1}
Control	1.21 × 10^8 b	1.88 × 10^{10} b	61.6 b	3459.7 a	98.00 b
CF	6.73 × 10^7 b	2.24 × 10^{10} b	32.6 b	1730.4 b	84.92 b
BC 12.50 t ha^{-1}	8.51 × 10^7 b	1.76 × 10^{10} b	45.7 b	229.3 c	78.88 b
RS 12.50 t ha^{-1}	5.81 × 10^8 a	4.94 × 10^{10} a	1379.3 a	507.1 c	202.69 a
BC 6.25 t ha^{-1} + RS 6.25 t ha^{-1} + CF	3.79 × 10^8 a	5.82 × 10^{10} a	4.5 b	557.5 c	186.63 a
F-test	**	*	**	**	**
CV (%)	36.16	45.70	20.15	29.96	15.75

[1] CF = chemical fertilizer, BC = biochar, RS = rice straw, CV = coefficient of variation. [2] DOC = dissolved organic carbon. The different small letters in the columns indicate significant difference among treatments by LSD. *, ** = significant at $p \leq 0.05$ and $p \leq 0.01$, n = 4.

In the soil treatments using RS alone, we observed high CH$_4$ production and high archaeal abundance. The DOC content was supported by the significantly high positive correlation between archaeal abundance and CH$_4$ production ($r = 0.799$ ***) and the DOC concentration ($r = 0.872$ ***). However, its moderately negative correlation with CO$_2$ production ($r = -0.403$) indicated that with an increasing abundance of archaea, the CO_2 had been consumed (Figure 2a–c). Therefore, the archaea had performed a crucial role in CH_4 production.

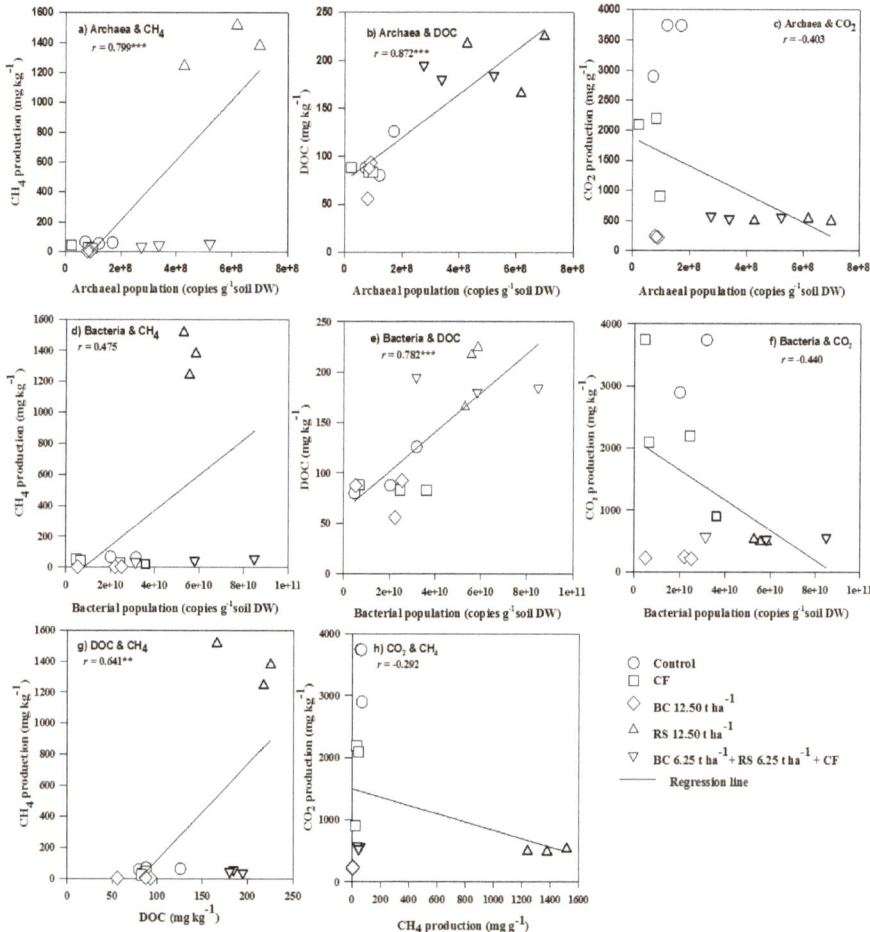

Figure 2. The correlation coefficients between Archaeal and CH_4 production (**a**), DOC content (**b**) and CO_2 production (**c**); between Bacterial abundance and CH_4 production (**d**), DOC content (**e**) and CO_2 production (**f**); between DOC content and CH_4 production (**g**); and between CH_4 and CO_2 production (**h**). ** Very significant at p-value < 0.01, *** extremely significant at p-value < 0.001, n = 15.

With respect to the bacteria, these appeared to be less effective in CH_4 production compared to the archaea. This was reflected in the moderately positive correlation between bacterial abundance and CH_4 production ($r = 0.475$) (Figure 2d), where the correlation between bacterial abundance and DOC concentration was significantly high ($r = 0.782$***) (Figure 2e). This indicated that the bacteria had played a crucial role in supplying DOC to the soil system, thereby supporting CH_4 production ($r = 0.641$**) (Figure 2g). However, a moderately negative correlation between the bacterial abundance and CO_2 production ($r = -0.440$) indicated that CO_2 had been consumed to form CH_4 with the increasing abundance of bacteria (Figure 2f). This assertion was supported by a weakly negative correlation coefficient of -0.292 between CO_2 and CH_4 production (Figure 2h).

With respect to archaea exerting its effects on CH_4 production, our results showed that compared to the bacteria, archaea was the more dominant microorganism. In the soil, especially in RS incorporated soil, *Methanomicrobia* was the main genus of archaea found, followed by *Methanobacteria* [22]. Moreover, in a previous pot experiment conducted with growing plants, the soil treated with RS showed a rapid decrease in soil redox potential to a range from −150 to −200 mV [1]. This was coupled with a rise in the soil pH to an optimal range of 7.5 to 8.5. This observed phenomenon resulted from electrons transferred from the RS, which had been utilized by microorganisms in the anaerobic respiration process [1]. This condition was assumed to be suitable for methanogenic archaea and bacteria. Moreover, these results were found to be concomitant with the experimental results reported by Yuan et al. [22].

3.2. Carbonaceous Greenhouse Gases and Microbial Abundance in Paddy Soil as Affected by BC

In contrast to RS, BC led to low levels of CH_4 production in soils from both the pot experiments (0.0233 mg m^{-2} season^{-1}) (Table 3) and the incubation experiments (45.7 mg kg^{-1}) (Table 4). This was because BC contained high levels of resistant C compounds, such as lignin (75.69%) and fixed C (61.72%) (Table 1). We discovered that the resistant constituents of BC had suppressed the soil's C mineralization. Compared to the RS treated soil, this suppression had led to a significantly lower DOC content (78.88 mg kg^{-1}), as well as lower archaeal abundance (8.51 × 10^7 copies g^{-1} soil DW) and bacterial abundance (1.76 × 10^{10} copies g^{-1} soil DW) in the BC treated soil (Table 4). This was consistent with a previous study by Liu et al, where it was found that paddy field soil amended with BC contributed to a low content of substrates [5]. In addition, the BC used in our experiments had a high content of VM (34.97%) (Table 1), which consisted of DOC, such as carboxylics and phenolics (as determined using FTIR, Figure 1), as well as aldehydes [23]. Not only could the DOC be consumed for CH_4 and CO_2 production, but it could be used by soil microorganisms for assimilation into the MBC. However, the archaeal and bacterial abundances were similar between the BC treated soil and the control soil (Table 4) because the amounts of DOC in both soil treatments were low. The archaeal and bacterial abundances in the BC treated soil (8.51 × 10^7 and 1.76 × 10^{10} copies g^{-1} soil DW, respectively) were significantly less than the abundances in RS treated soil (5.81 × 10^8 and 4.94 × 10^{10} copies g^{-1} soil DW, respectively). These results confirmed a lower CH_4 production in the BC treated soil (Table 4).

In terms of the microbial community, in the BC treated soil, a high MBC content of 224.08 mg kg^{-1} (Table 3) revealed the good biological quality of the soil containing archaea, bacteria (Table 4), and methanotrophs [6], which were involved in the CH_4 and CO_2 dynamics of such soil. Wang et al. [4] reported that soil amended with BC had significantly altered the composition of the soil's archaeal and bacterial communities. Furthermore, it was reported that the main constituents of the archaea communities included a miscellaneous Crenarchaeota group (MCG), Methanobacteria, and Thaumarchaeota archaea. In our study, the BC treatments were likely to be comprised of archaea similar to the composition reported by Wang et al. [4].

Moreover, in the BC alone and combined BC + RS + CF treated soils, the CH_4 production (0.0233 and 0.0235 mg m^{-2} season^{-1}, respectively) was found to be lower than in the soils treated with RS treatments (0.0347 mg m^{-2} season^{-1}). This may be attributed to the physical structure of BC given that it possessed several ≤ 2 mm micropores and had a large surface area [4,19], which could adsorb CH_4 gas [5] and serve as a CH_4-C substrate for the methanotrophs [6]. Biochar also supplied a habitat for the methanotrophs [6], where all these mechanisms had led to a reduction in CH_4 production in the BC amended soil (Table 3).

3.3. Carbonaceous Greenhouse Gases and Microbial Abundance in Paddy Soil as Affected by Combined BC and RS

In the incubation experiments using the combined BC + RS treatment (BC 6.25 t ha^{-1} + RS 6.25 t ha^{-1} + CF), there was an abundance of archaea and bacteria (3.79 × 10^8 and 5.82 × 10^{10} copies g^{-1} soil DW, respectively) (Table 4), which had proliferated at 6.25 t ha^{-1} RS to yield CH_4 of 4.5 mg kg^{-1} (Table 4). When the CH_4 results of the combined treatment (4.5 mg CH_4 kg^{-1}) were compared to the results of the RS alone (12.50 t ha^{-1} RS) (1379.3 mg CH_4 kg^{-1}), we found that the CH_4 had been drastically reduced in the combined treatment via the countering power of the BC, probably through the BC inhibition of methanogenic activity [5] and the adsorption process of BC. This result was despite the enhancing effects that RS had on CH_4 production in such anaerobic soil through methanogenic archaea and bacteria activities (Table 4). On the contrary, the amount of CO_2 production was found to be similar amongst the treatments of BC, RS, or their combination in the incubated soils. The non-different "net" CO_2 production was the net result of several microbial processes of microbial C mineralization (CO_2 production) and CO_2 reduction (CH_4 formation) that occurred simultaneously, but to different degrees and directions in the soils treated with these studied amendments. Biochar applied alone resulted in a low C mineralization, rendering a low content of CO_2 (Table 4). In contrast, RS alone favored C mineralization to form CO_2, which was furthered reduced to CH_4 and resulted in low CO_2 in the RS treated soil. In the combined BC + RS soil, the adverse effect of BC on RS resulted in a low CO_2 (Table 4). In the combined BC + RS + CF soil, CH_4 production in the potted soil (0.0235 mg m^{-2} season^{-1}) was significantly lower than in the RS alone (0.0347 mg m^{-2} season^{-1}) (Table 3). It appeared that the results for CH_4 production from the incubation experiments and rice pot experiments behaved in the same manner. However, conditions in the potted soil planted with rice differed from the conditions in the incubated soil, i.e., in the rice pot experiment, there were C substrates derived from rhizodeposition (root exudates, sloughed root, and dead root) [24], large areas of aerobic and anaerobic interface in the rice rhizosphere soil, and an oxidizing layer on the soil surface, whereas the incubated soil was presumably completely anaerobic. The BC-enhanced microbiological oxidation process was mediated by methanotrophs, which consumed CH_4 and transformed it into CO_2 at the aerobic–anaerobic interface [25] of the rice rhizosphere and the submerged soil. Therefore, as a consequence, CO_2 was released into the atmosphere (Table 3). This process is expected to exist within the rice rhizosphere at the aerobic–anaerobic interface, and it results in decreases in CH_4 and the release of CO_2 to the soil. More than 90% of the CH_4 production in soil is oxidized to CO_2 [6]. Our findings from the rice which had been planted in potted soil, enabled us to articulate the countering effects of the BC amendments on CH_4 production in soil via two possible mechanisms. These included the adsorption of CH_4 onto the BC surfaces and the oxidation of CH_4 to CO_2 by methanotrophs which utilize CH_4 as a source of C and energy [6].

The rice grain yields obtained from the combined treatment (BC 6.25 t ha^{-1} + RS 6.25 t ha^{-1} + CF) and individual CF treatment were 53.47 and 50.62 g pot^{-1}, respectively (Table 3), being the two highest yields in our pot experiments. In contrast, BC alone and RS alone depressed rice yields. It could be deduced that the enhanced yield under the combined BC + RS + CF treatment was due to the CF effect on grain yield. With CFs favorable supply of nutrients and the high nutrient adsorption characteristics of BC, the combined BC + RS + CF treatment could supply and retain sufficient plant nutrients for rice growth. In addition, chlorosis was observed in the rice plants treated with BC or RS alone. The studied soil had a low N content (0.08%) which caused soil N deficiency and led to a low rice yield. Therefore, CF is a necessary supplement for the amendment of organic materials in the soil.

4. Conclusions

Our results proved our hypothesis, that is, the incorporation of a combination of BC and RS in paddy soil reduces the soil's CH_4 production, raises archaeal and bacterial abundances, and increases the yield of rice grains compared to unamended soil. With such a combined BC + RS soil amendment, the RS component (a cellulose and hemicellulose-rich material) was able to rapidly decompose and enhance the archaeal and bacterial abundances relative to the without-RS amendments. Concurrently, in the combined BC + RS, there was a rapid production of the intermediate products of decomposition (i.e., DOC, CO_2, and H_2) which served as substrates for microbes to produce CH_4 in the methanogenesis process. Conversely, the recalcitrant lignin-rich BC component of the combined BC + RS amendment inhibited the activity of the archaea in methanogenesis resulting in lower CH_4 production than that of the RS alone. One of the proposed mechanisms for the suppression of methanogenesis was a high abundance of methanotrophic bacteria in the BC, which served as the bacteria's habitat. Methanotrophs performed the CH_4 oxidation process, which reduced the CH_4 content. Another mechanism was the adsorption of CH_4 onto the large and highly adsorptive surface area of the BC, which led to CH_4 reduction in the combined BC + RS treated soil. Compared to paddies receiving RS or BC applied individually, a further benefit was the high rice grain yield under the combined BC + RS treatment. The combined BC + RS material proved to be a more beneficial soil amendment than the RS or BC applied separately owing to the dual purposes of improving soil productivity and reducing greenhouse gas emissions.

Author Contributions: Conceptualization, S.K., P.S., and P.L.; Methodology, P.S. and P.L.; Formal analysis, S.K., P.S., and P.L.; Resources, P.S. and P.L.; Software, S.K.; Data curation, P.S. and P.L.; Writing—Original Draft Preparation, S.K., P.S., and P.L.; Writing—Review and Editing, S.K., P.S., P.V., and P.L.; Supervision, P.S. and P.L.; Validation, S.K. and P.L.; Project Administration, S.K., P.S., and P.L.; Funding Acquisition, P.S., P.V., and P.L.

Funding: This research was funded by the Soil Organic Matter Management Research Group of Khon Kaen University (KKU) and the Thesis Support Scholarship from the Graduate School of KKU.

Acknowledgments: Special thanks to Khon Kaen University for providing the facilities used to conduct the experiments. Acknowledgement is extended to the Soil Organic Matter Management Research Group of Khon Kaen University (KKU) for providing financial support, and the first author was provided with a Thesis Support Scholarship from the Graduate School of KKU.

Conflicts of Interest: The authors declare no conflict of interest. The funding sponsors had no role in the design of the study; in the collection, analyses, or interpretation of data; in the writing of the manuscript, and in the decision to publish the results.

References

1. Thammasom, N.; Vityakon, P.; Saenjan, P. Response of methane emissions, redox potential, and pH to eucalyptus biochar and rice straw addition in a paddy soil. *Songklanakarin J. Sci. Technol.* **2016**, *38*, 325–331.
2. Chowdhury, T.R.; Dick, R.P. Ecology of aerobic methanotrophs in controlling methane fluxes from wetlands. *Appl. Soil Ecol.* **2013**, *65*, 8–22. [CrossRef]
3. Thammasom, N.; Vityakon, P.; Lawongsa, P.; Saenjan, P. Biochar and rice straw have different effects on soil productivity, greenhouse gas emission and carbon sequestration in Northeast Thailand paddy soil. *Agric. Nat. Resour.* **2016**, *50*, 192–198. [CrossRef]
4. Wang, N.; Chang, Z.; Xue, X.; Yu, J.; Shi, X.; Ma, L.Q.; Li, H. Biochar decreases nitrogen oxide and enhances methane emissions via altering microbial community composition of anaerobic paddy soil. *Sci. Total Environ.* **2017**, 689–696. [CrossRef]
5. Liu, Y.; Yang, M.; Wu, Y.; Wang, H.; Chen, Y.; Wu, W. Reducing CH_4 and CO_2 emissions from waterlogged paddy soil with biochar. *J Soils Sediments.* **2011**, *11*, 930–939. [CrossRef]
6. Feng, Y.; Xu, Y.; Yu, Y.; Xie, Z.; Lin, X. Mechanisms of biochar decreasing methane emission from Chinese paddy soils. *Soil Biol. Biochem.* **2012**, *46*, 80–88. [CrossRef]
7. Bremner, J.M. Total nitrogen. In *Method of Soil Analysis. Part 2*; Black, C.A., Ed.; American Society of Agronomy: Madison, WI, USA, 1965; pp. 1149–1178.
8. Aravantinos-Zafiris, G.; Oreopoulou, V.; Tzia, C.; Thomopoulos, C.D. Fiber fraction from orange peel residues after pectin extraction. *LWT-Food Sci. Technol.* **1994**, *27*, 468–471. [CrossRef]
9. American Standard Test Method International. *Standard Test. Method for Chemical Analysis of Wood Charcoal*; ASTM International Destination: Pennsylvania, PA, USA, 2007; pp. D1762–D1784.
10. Coates, J. Interpretation of Infrared Spectra, A practical Approach. In *Encyclopedia of Analytical Chemistry*.; Meyers, R.A., Ed.; John Wiley & Sons Ltd: Chichester, UK, 2000; pp. 10815–10837.
11. Daffalla, S.B.; Mukhtar, H. and Shaharun, M.S. Characterization of Adsorbent Developed from Rice Husk: Effect of Surface Functional Group on Phenol Adsorption. *Appl. Sci.* **2010**, *10*, 1060–1067.
12. Bouyoucos, G.J. Directions for making mechanical analysis of soils by the hydrometer method. *Soil Sci.* **1936**, *4*, 225–228. [CrossRef]
13. Walkley, A.; Black, J.A. An examination of the dichormate method for determining soil organic matter and a proposed modification of the chromic acid titration method. *Soil Sci.* **1934**, *37*, 29–38. [CrossRef]

14. Vance, E.D.; Brookes, P.C.; Jenkinson, D.S. An extraction method for measuring microbial biomass C. *Soil Biol. Biochem.* **1987**, *22*, 703–707. [CrossRef]
15. Saenjan, P.; Tulaphitak, D.; Tulaphitak, T.; Soupachai, T.; Suwat, J. Methane emission from Thai farmers'paddy fields as a basis for appropiate mitigation technologies. In Proceedings of the 17th World Congress of Soil Science, Bangkok, Thailand, 14–21 August 2002.
16. Lane, D. 16S/23S rRNA sequencing. In *Nucleic Acid Techniques Systematics*; Stackebrandt, A., Goodfellow, M., Eds.; John Wiley: West Sussex, UK, 1991; pp. 115–175.
17. Muyzer, G.; De Waal, E.C.; Uitterlinden, A.G. Profiling of complex microbial populations by denaturing gradient gel electrophoresis analysis of polymerase chain reaction-amplified genes coding for 16S rRNA. *Appl. Environ. Microbiol.* **1993**, *59*, 695–700. [PubMed]
18. Lueders, T.; Friedrich, M. Archaeal population dynamics during sequential reduction processes in rice field soil. *Appl. Environ. Microbiol.* **2000**, *66*, 2732–2742. [CrossRef] [PubMed]
19. Gonetl, S.S.; Debska, B. Dissolved organic carbon and dissolved nitrogen in soil under different fertilization treatments. *Plant Soil Eviron.* **2006**, *52*, 55–63. [CrossRef]
20. Kunlanit, B. Decomposition of biochemistry contrasting organic residues regulating dissolved organic carbon dynamics and soil organic carbon composition in a sandy soil. Ph.D. Thesis, Khon Kaen University, Khon Kaen, Thailand, December 2014.
21. Lehmann, J.; Rilling, M.C.; Thies, J.; Masiello, A.C.; Hockaday, C.W.; Crowley, D. Biochar effects on soil biota—A review. *Soil Biol. Biochem.* **2011**, *43*, 1812–1836. [CrossRef]
22. Yuan, Q.; Hernandez, M.; Dumont, M.G.; Rui, J.; Fernandez Scavino, A.; Conrad, R. Soil bacterial community mediates the effect of plant material on methanogenic decomposition of soil organic matter. *Soil Biol. Biochem.* **2018**, *116*, 99–109. [CrossRef]
23. Butnan, S.; Deenik, J.L.; Toomsan, B.; Antal, M.J.; Vityakon, P. Biochar characteristics and application rates affecting corn growth and properties of soils contrasting in texture and mineralogy. *Geoderma.* **2015**, 105–116. [CrossRef]
24. Aulakh, M.S.; Wassmann, R.; Bueno, C.; Kreuzwieser, J.; Rennenberg, H. Characterization of Root Exudates at Different Growth Stages of Ten Rice (*Oryza sativa* L.) Cultivars. *Plant Biol.* **2001**, *3*, 139–148. [CrossRef]
25. Cao, M.; Gregson, K.; Marshall, S.; Dent, J.B.; Heal, O.W. Global methane emissions from rice paddies. *Chemosphere* **1996**, *33*, 879–897. [CrossRef]

© 2019 by the authors. Licensee MDPI, Basel, Switzerland. This article is an open access article distributed under the terms and conditions of the Creative Commons Attribution (CC BY) license (http://creativecommons.org/licenses/by/4.0/).

Article

Effect of Biochar Particle Size on Physical, Hydrological and Chemical Properties of Loamy and Sandy Tropical Soils

Sara de Jesus Duarte [1,*], Bruno Glaser [2] and Carlos Eduardo Pellegrino Cerri [1]

1. Soil and Plant Nutrition, Luiz de Queiroz College of Agriculture, University of São Paulo, Avenida Pádua Dias, 11, Agronomia, Piracicaba, SP 13418-900, Brazil; cepcerri@usp.br
2. Soil Biogeochemistry, Martin Luther University Halle-Wittenberg, Von-Seckendorff-Platz 3, 06120 Halle, Germany; bruno.glaser@landw.uni-halle.de
* Correspondence: saraduarte@usp.br; Tel.: +52-155-1884-9291

Received: 12 March 2019; Accepted: 24 March 2019; Published: 29 March 2019

Abstract: The application of biochar is promising for improving the physical, chemical and hydrological properties of soil. However, there are few studies regarding the influence of biochar particle size. This study was conducted to evaluate the effect of biochar size on the physical, chemical and hydrological properties in sandy and loamy tropical soils. For this purpose, an incubation experiment was conducted in the laboratory with eight treatments (control (only soil), two soils (loamy and sandy soil), and three biochar sizes (<0.15 mm; 0.15–2 mm and >2 mm)). Analyses of water content, bulk density, total porosity, pore size distribution, total carbon (TC) and total N (TN) were performed after 1 year of soil–biochar-interactions in the laboratory. The smaller particle size <0.15 mm increased water retention in both soils, particularly in the loamy soil. Bulk density slightly decreased, especially in the loamy soil when biochar > 2 mm and in the sandy soil with the addition of 0.15–2 mm biochar. Porosity increased in both soils with the addition of biochar in the range of 0.15–2 mm. Smaller biochar particles shifted pore size distribution to increased macro and mesoporosity in both soils. Total carbon content increased mainly in sandy soil compared to control treatment; the highest carbon amount was obtained in the biochar size 0.15–2 mm in loamy soil and <0.15 mm in sandy soil, while the TN content and C:N ratio increased slightly with a reduction of the biochar particle size in both soils. These results demonstrate that biochar particle size is crucial for water retention, water availability, pore size distribution, and C sequestration.

Keywords: biochar particle size; soil physics; soil chemistry; water retention

1. Introduction

The intensification of agricultural production on a global scale is necessary in order to secure the food supply for an increasing world population. However, in most tropical environments, sustainable agriculture faces large constraints due to low nutrient content and accelerated mineralization of soil organic matter (SOM) [1]. Therefore, the low cation exchange capacity (CEC) of the soils further decreases. Under such circumstances, the efficiency of applied mineral fertilizers is very low when the loss of mobile nutrients from the topsoil is enhanced by high rainfall [2]. Additionally, coarse-structured soils with low clay content are characterized by a lack of both water retention and nutrient-holding capacities that are necessary for plant growth [3]. Many farmers cannot afford the costs of regular applications of mineral fertilizers. Consequently, nutrient deficiency is prevalent in many crop production systems of the tropics [4].

In contrast to these deficient soils, the famous *Terra Preta* maintains its fertility, despite its 2000 years of age [5].This is partly due to the tremendous nutrient levels and SOM stocks that act as a long-term, slow-release fertilizer [5,6]. The physical and hydrological properties in this soil also contrast with adjacent soils. For instance, Glaser et al. [4] verified that the water retention of *Terra Preta* was 18% higher as compared to adjacent soils. The secret of the *Terra Preta* is in the biochar; this type of earth contains on average 50 Mg ha^{-1} of biochar per hectare in the upper 50 cm soil depth. Adjacent reference soils contain only 4 Mg biochar, which is about 10 times less than that in *Terra Preta* [5]. The existence of *Terra Preta* in Amazonia today proves that it is possible to convert infertile soils' insufficient physical and hydrological properties to sustainable, fertile soils with good physical and hydrological properties. It is evident that biochar is a key ingredient in making *Terra Preta* so special [6] and is the ingredient to improve the soil quality on intensive agriculture.

Biochar as key for *Terra Preta* formation can improve physical and hydrological properties such as water retention, water available content, bulk density, and porosity [7–9]. For example, the addition of 20 Mg ha^{-1} biochar to sandy soil in northeast Germany increased its water-holding capacity by 100% [10]. At the same time, the incorporation of biochar into soil has been shown to enhance soil capacity to retain plant nutrients, decrease nutrient losses from leaching, and increase soil water holding capacity, pH and SOM [11,12].

Many functions in one product are possible because the biochar is composed of condensed aromatic moieties that give biochar its black color and are responsible for its stability, which makes biochar an interesting compound for C sequestration [13,14]. In addition, biological degradation and consequently partial oxidation results in the formation of functional groups on the edges of biochar, causing reactivity in soil such as nutrient retention or organomineral stabilization [15,16]. The highly porous material leads to enhanced air and water storage in soil [6].

Because of these functions, biochar can be used as a soil amendment to improve the quality of agricultural soils [4]. The application of biochar to soil is considered as a win-win strategy to improve the soil physical conditions that influence soil hydraulic properties and water retention [2,17] and increase soil fertility [6,18].

The effects on the chemical and physical properties of soil are dependent on the biochar amount, pyrolysis temperature, biomass type, and biochar particle size [7,19,20]. However, few studies have focused comprehensively on the effects of biochar particle size on hydraulic, physical and chemical properties [7,21]. Understanding biochar particle size is important because it affects the interaction with the soil matrix. The greater and/or lesser interaction of biochar with the soil matrix may have a direct effect on its chemical, physical and hydrological properties [22]. This interaction is dependent of biochar particle size and therefore can influence the physical and hydraulic properties of soil [7].

Small biochar particles can more easily interact with soil particles to form aggregates than large biochar particles [23]. In addition, the greater specific surface area per unit of mass increases the water retention [7] and plant-available water [24]. In another study, Głąb et al. [25] found that bulk density decreased, total porosity increased, plant-available water content decreased, and water repellence decreased with an increase in the biochar size from 0.5 to 2 mm.

The role of biochar on temperate soils has been discussed in the literature [7,21,22]. However, for tropical soil conditions, there is a lack of information on this promising soil conditioner. The present study, as far as we understand, is the first research that presents data on the fate of biochar application on the physical, chemical and hydrological properties of tropical soil.

We hypothesize that the biochar application has a positive effect on the physical, chemical and hydrological properties of soil under tropical conditions. However, this effect is dependent on particle size; the reduction of particle size causes an increase in water retention and total porosity and a decrease in available water content and bulk density. Therefore, the objective of this study was to determine the effect of biochar particle size on the physical, hydrological and chemical properties of soil. The knowledge of the relationship between soil's physical, hydrological and chemical properties and biochar particle size is potentially useful in management applications, particularly those concerning irrigation and recovery of degraded areas.

2. Methodology

2.1. Biochar Production

In our study, we included biochar produced from biomass coming from agricultural residues (*Miscanthus giganteus*) (Table 1). This biochar is commercially produced by drying *Miscanthus* grass at 105 °C in a greenhouse, followed by 15 min of pyrolyzation in a second stationary metallic and cylindric 60 liters reactor at 450 °C. Both reactors were hermetically sealed. The pyrolysis process was performed by SPPT Research and Technology Company (Mogi Mirim, SP, Brazil) in a metal reactor, saturating the sample with N_2 and raising the temperature 10 °C every minute during the first 30 min and 20 °C per minute until reaching the desired temperature [26]. We used *Miscanthus giganteus*, because it is a residue of the growing biofuels industry and its high quantity of silicon can increase water retention in biochar, thus contributing to the high water storage capacity in the soil [27].

Table 1. Chemical characterization of the biochar.

Properties of Biochar	Unit	Value	Reference
pH in H_2O	-	5.9	[28]
Electric conductivity	$\mu S\ cm^{-1}$	605	[28]
Moisture	%	3.5	[29]
Cation Exchange Capacity	$mmol_c dm^{-3}$	33	[28]
Specific surface area	$m^2\ g^{-1}$	371	[30]
Labile C	%	2.7	[31]
Stable C	%	50.9	[31]

2.2. Biochar and Soil Characterization

2.2.1. Biochar Characterization

Analyses such as pH and electric conductivity were performed by Conz et al. [26] following the methods recommended by the International Biochar Initiative Guideline (IBI, 2015). The labile carbon, stable carbon and lability analyses were conducted following the method of [31]. Moisture was measured using the Standard Test Method for the Analysis (ASTM) method, specific surface area was measuring using the method reported by Cerato & Lutenegge [30] (Table 1).

2.2.2. Soil Characterization and Sampling

The soils used were Entisoland Oxisol, which were sampled in two different native forest areas located in Sao Paulo state, Brazil. The sampling sites were near Anhembi, Brazil (22°43′31.1″S and 48°1′20.2″W) and Piracicaba, Brazil (22°42′5.1″S and 47°37′45.2″W). Although these areas had never been cultivated, we identified the presence of alterations in the surrounding forest. However, the collection point was chosen in the middle of the forest in a location without alterations. The soils were sampled at the 0–20 cm layer, air-dried, homogenized, and sieved <2 mm.

The soil chemical characterization was performed by Feola Conz et al. [26], who followed the methodology proposed by Raij et al. [32] and determined the parameters such as pH in Calcium chloride 0.01 M ($CaCl_2$); P, K, Ca, Mg and K in the resin; Al extracted with potassium chloride (KCl, 1M); sulphate (S-SO_4) extracted with calcium phosphate ($Ca(H_2PO_4)_2$, 0.01M); percentage of saturation by bases (V%); percent saturation by aluminium (m%); and potential acidity (H + Al). The C and N contents were determined with elemental analyser (LECO CN-Truspec) (Table 2).

Table 2. Chemical and physical characterization of loamy and sandy soil.

Chemical Characteristics	Sandy Soil	Loamy Soil
pH (CaCl$_2$)	3.9	6.5
(%)		
Sand	90	41
Silt	2	27
Clay	8	32
Aluminium saturation	10	87
Base saturation	45	0
(mmol$_c$dm^{-3})		
Al	6	0
H + Al	62	18
SB	7	120
CEC	69	138

H + Al = potential acidity, SB = soma of bases (Ca, Mg, K), CEC = Cation exchange capacity (SB + Al + H), V = Base saturation (SB × 100/CTC); m = Aluminium saturation (100 × Al^{3+}/SB + Al^{3+}).

2.2.3. Experimental Setup and Sample Preparation

The design of the experiment was completely randomized with a 3 × 2 factorial, three biochar particle sizes (>2 mm, 2–0.15 mm and <0.15 mm) and two soil textures (sandy and loamy), with an additional treatment control (only soil), totaling eight different treatments with four replicates. We used 200 g of soil and 0.92 g of biochar (~25 Mg ha^{-1}) for biochar particle sizes of >2 mm; 2–0.15 mm and <0.15 mm. The biochar fractions were incorporated with soils in jars (500 mL volume) and vigorously mixed into the soil based on pre-calculated bulk densities, which included the contribution of the amendments to the final bulk density, resulting in 32 pots that were incubated for 1 year at a temperature of 30 °C under laboratory conditions. During that year, the moisture of the soil-biochar mixture was adjusted to 60% field capacity and maintained at that level throughout the experiment by weighing the jars three times per week and adding water if necessary. The biochar and soil were filled in the incubation, and after 12 months of interactions, the soil was sampled and physical and chemical analyses were performed.

To obtain the water retention curve, a volumetric ring of approximately 7 cm^3 was used to sample inside of the jar without disturbing the sample. First, the ring was inserted into the soil and the sample was removed with an aluminum device similar to a shovel. The soil remaining were used for chemical analysis.

2.2.4. Post-Incubation Analyses

C, N, and C:N Ratio

Carbon and nitrogen concentrations were determined using an elemental analyzer LECO TruSpec, (LECO Instruments ULC, Missossauga, Otario, Canada), and it was possible to obtain the atomic ratio (C:N).

Bulk Density and Particle Density

The bulk density was performed by the Black and Hartge method [33]. Deformed samples were used to determine the particle density (PD) in each soil and their different treatments. A helium gas pycnometer, model ACCUPYC 1330 (Micrometrics Instrument Corporation, Nacross, GA, USA) was used to determine the PD values. The principle of this method is based on the determination of the volume of solids by the variation of the pressure of one gas in a known volume chamber.

Porosity

The total porosity (*TP*) was calculated from the soil bulk density (*BD*) and the particle density (*PD*) [34] using the following Equation:

$$TP = \frac{PD - BD}{BD} \times 100 \quad (1)$$

The macro and microporosity were performed with a soil water retention curve using theoretical values for macroporosity superior to 50 µm and microporosity inferior to 15 µm [35].

Water Retention Curve

Soil water-holding capacity was measured by the moisture contents of the samples at different matric potentials (−15, −10, −3, −1, −0.3, −0.1, −0.06, −0.04, and −0.02 bar). The points −15, −10, −3, −1, −0.3, and −0.1 bar were used in the Richard chamber, and −0.06, −0.04, and −0.02 bar were used in Haines' apparatus. The moisture content was performed with gravimetric analysis. Using the multiplication of gravimetric moisture by bulk density, we converted the gravimetric moisture to volumetric moisture [36]. For the determination of curve parameters (θs, θs, α, n), we used a Van Genuchten (1980) type.

2.3. Data Analyses

Statistical analyses and the graphics were performed using R studio Version 1.1 [37] and Microsoft Excel. The data were checked for normality and homogeneity of variances to meet the assumptions of ANOVA and test of Tukey, with a probability threshold of 0.05.

3. Results and Discussion

3.1. Soil and Biocharcharacteristics

The *Miscanthus*-derived biochar had a higher quantity of C, N, Ca, Mg, Na, K, P, and S when compared to both soils (Table 1). Furthermore, the *Miscanthus*-derived biochar was characterized by hydrophobic groups such cyclic acid anhydrides (C-C and C-O); asymmetric carboxylates (CO_2), aromatic ketones (C=O); silicon (Si-O); and hydrophilic groups such carbonates (C-O) and silanol (Si-O-H). With these classifications, this biochar was determined to have a high hydrophobicity [38].

Sandy soil has a very low amount of K; a low amount of Mg and P; a base saturation of Ca; a medium amount of S, Al, CEC and m%; and a high content of sand at a low pH can contribute to the low fertility as compared to loamy soil (Table 3). The low pH (3.5) indicates the presence of exchangeable aluminum that can inhibit the root development and affect the availability of other nutrients and mineralization of organic matter. In addition, the values of Ca, Mg and K were very low, indicating that this soil is highly weathered and base saturation is very low, resulting in cation exchange sites occupied by components of the acidity H or Al [39]. Loamy soil has a low amount of Al, a medium amount of S and CEC, a high amount of V% and K, and a very high amount of Ca at a high pH; there is also no presence of Al, contributing to the low solubility of the Al^{+3}, rendering it harmless to roots and soil processes and increasing the availability of other nutrients (Table 2).

Comparing the biochar with sandy and loamy soil, the biochar had 7 and 2.5 times more N, 464 and 66 times more P, and 12.3 and 2 times more K in both soils, respectively (Table 3). This large difference can contribute to increasing the fertility in both of the soils. Glaser et al. [4] affirmed that the fertilization potential of the biochar is high, especially in tropical soils. This affirmation has been proven by Laird et al. [12] who observed a significant increase of P, K, Ca, and Mn after 500 days of biochar addition.

Table 3. Comparison between chemical properties of biochar and two soil textures: loamy and sandy soil.

Properties of Biochar	Biochar	Sandy Soil	Loamy Soil
	(mg kg^{-1})		
Ca	3361	5.3	97
Mg	2133	<1	20
K	8615	<0.7	4.1
P	1859	4	28
S	634	5.3	9.5
C	66.4	0.9	1.9
N	0.43	0.06	0.17
C:N	155	14	11

Adapted from Feola Conz et al. [26] C:N = ratio between Carbon and Nitrogen.

3.2. Biochar Effect on Physical Properties

3.2.1. Total Porosity

Naturally, the total porosity is higher in clay soil than in sandy soil. Though there is no significant difference between treatments ($p > 0.05$), the biochar addition (>2 mm) increased the total porosity in both soils. The increase in the particle size reduces the homogeneity of the pore distribution and increases the total porosity; this increase was more evident in loamy soil. Likewise, in a previous study, He et al. [8] reported an increase in the total porosity with the addition of biochar 4 mm.

The total porosity in sandy soil was higher when 0.15–2 mm biochar was added (Figure 1a), but this increase was not significant. Similar to our study, Głąb et al. [25] did not find a significant difference on total porosity in biochar *Miscanthus* with a different particle size (0.5–2 mm). However, they verified that the total porosity increased with biochar addition in loamy and sand soil, with an increase in biochar size from 0.5 mm to 2 mm. Similarly, in this research, we found that the size 0.15–2 mm sandy soil contributed to increasing the soil porosity.

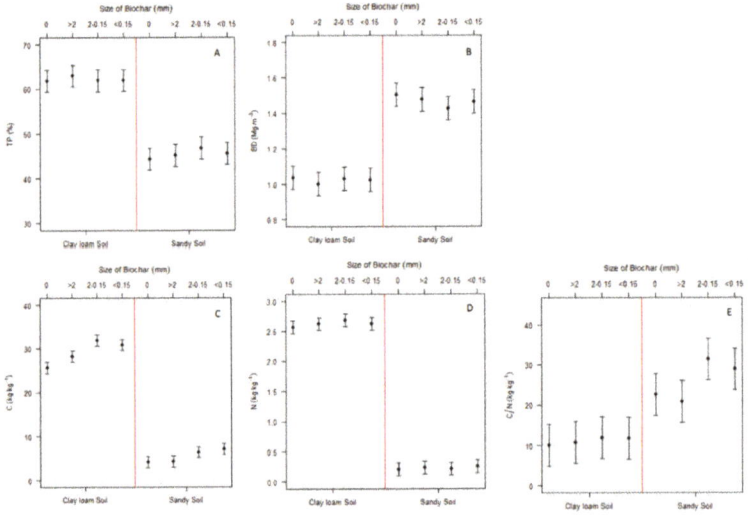

Figure 1. Effect of biochar size (0: Control (Only soil); >2 mm; 2–0.15 mm and <0.15 mm) on soil physical and chemical properties in loamy and sandy soil, where, TP: Total porosity; BD: Bulk density; C: Carbon; N: Nitrogen: C/N: Carbon and Nitrogen ratio.

The smaller fraction (<0.15 mm) slightly decreased the total porosity. The reduction of the porosity is proportional to the reduction of the particle size; this behaviour is attributed to biochar fragmentation; during the fragmentation process, most of the larger wood pores are destroyed in smaller pieces, thus increasing the number of micropores [40]. As we verify in this study, the reduction in the particle size to <0.15 mm increases the homogeneity in the pore distribution, increasing the micropore volume and reducing the total porosity, particularly in sandy soil [25,41]. In previous work, Glab et al. [25] found similar results for the particle size <0.3 mm.

The alteration in the biochar fraction and consequently in the porosity of the soil can influence the different impacts on soil physics and hydraulic properties. For example, wide spaces where water can freely move exist in larger biochar particles (>2 mm). Smaller fractions such as <0.15 mm have pores with the function of storing water [40]. Due to the fractioning process, these particles have a larger surface area and micropore volume that contributes to increasing the water retention. Therefore, the addition of smaller biochar particles could be beneficial to improving its ability to retain water. However, larger particles can increase aeration and drainage.

3.2.2. Bulk Density

The high bulk density in sandy soil and low bulk density in loamy soil can be attributed to the soil particle. For both soils, porosity and bulk density are inversely proportional to each other. There was no significant effect ($p > 0.05$) of biochar particle size on soil bulk density (Figure 1B). However, in sandy soil, there is a tendency for reduction of the bulk density with reduction of the particle size until the 2–0.15 mm fraction in loamy soil increases in the bulk density with the decrease of particle size.

The magnitude of the biochar effect on bulk density can be explained by simple dilution of the soil with the low bulk density of the biochar [12]. In clay loam soil and in the smaller fraction of sandy soil (<0.15 mm), the increase in the bulk density with a decrease in the biochar particle size can be associated with an arrangement of the particles of biochar in the volume of the soil, where the smaller particles can occupy the pores in the soil, this is not possible if the biochar particle is bigger. With this arrangement, more biochar is located inside the pores of the soil, contributing to a reduction of the total porosity and an increase in the soil bulk density. The reduction in the bulk density is so important for the increase in soil porosity because it contributes directly to root elongation and consequently plant development and production [42].

3.3. Biochar Effect on Chemical Properties

3.3.1. Total Carbon

The carbon amount in the loamy soil was higher than in sandy soil (Figure 1C). The addition of 25 Mg ha^{-1} biochar in the >2 mm fraction had a minor contribution to the increase in carbon in the loamy soil and especially sandy soils. In both soils, the carbon content increased significantly ($p < 0.05$) with a decrease in the biochar particle size (2–0.15 mm); the carbon amount for loamy soil increased by 6.6 kg kg^{-1} and in sandy soil by 4.2 kg kg^{-1} in the smaller biochar particle size (<0.15 mm) (Figure 1C).

The ability of biochar to improve the quantity of nutrients can be attributed to its large amount of carbon and its large specific surface area, porosity and amount of negative surface functional groups. All of these factors produce an enhanced soil cation exchange capacity [43] that can reduce nutrient leaching while increasing the quantity of the elements in the soil [44].

The increase of carbon content in loamy soil and sandy soil with biochar addition can contribute to an increase in aggregation stability, water retention, plant-available water content, and reduction of soil bulk density. These soil physical properties are essential for the soil physical quality and therefore for plant development [44].

3.3.2. Nitrogen

In loamy soil, there is a high amount of nitrogen as compared to sandy soil. The biochar addition contributed very little to an increase in the nitrogen amount. Although there is no significant difference on nitrogen amount with the different biochar particle size, we verified that in loamy soil, the 0.15–2 mm biochar size increased the nitrogen amount 0.12 kg kg^{-1} as compared to the control treatment. For sandy soil, the biochar size does not alter the nitrogen amount (Figure 1D). Similarly, Zhang et al. [45] verified that the difference was not significant in the first year of soil–biochar interaction. However, in loamy soil, we can see a little increase in the nitrogen amount in the particle size 0.15–2 mm and in sandy soil <0.15 um; this increase can be associated with the specific surface area that can contribute to the nitrogen amount in the soil.

3.3.3. C:N Ratio

In loamy and sandy soil, the C:N ratio was different. Though there is a tendency to increase the C:N ratio, for a small biochar particle size (2–0.15 mm), this increase was more evident in sandy soil. This study shows, that in loamy soil with the decrease of the biochar particle size (0.15–2 mm and <0.15 mm), the C:N ratio increased to 19 and 18, respectively, as compared to the control treatment. However, this increase was not significant ($p > 0.05$). In sandy soil, the difference in the biochar particle size 0.15–2 mm and <0.15 mm contributed to an increase in the C:N content of 88 and 64 kg kg^{-1}, respectively (Figure 1E). Zhang et al. [45]) verified an increase in the C:N ratio with biochar addition. The influence on the biochar particle size on the C:N ratio is associated with alteration of the carbon and nitrogen amount in the soil with biochar. As shown in this work, the C:N ratio is essential for the equilibrium of nutrients and their availability for plants and microorganisms survival. The only exception was in the sandy soil treatments of particle sizes of 0.15–2 mm and <0.15 mm, where the C:N ratio was inferior to 21 so there were no problems with decomposition and nitrogen immobilization [46].

3.4. Biochar Effect on Hydrological Properties

3.4.1. Water Retention Curve

According to a Shapiro–Wilk normality test at 5% of significance, there is no significant difference between the treatments (Figure 2) in clay loam and sandy soil. Comparing the addition of 25 Mg ha^{-1} of biochar with the control treatment (only soil), the biochar addition did not promote an increase in water retention in sandy soil. Although the effect of particle size on water retention properties was very low when the particle size decreased (<0.15 mm), the water retention increased in moderate (100 hPa) and in dry conditions (15000 hPa); the range of these conditions represents the variation between mesopores and micropores, respectively (Figure 2A). Similarly, in their experiment on sandy soil, Jeffery et al. [47] did not find significant effects of biochar application on soil water retention. Similar results were observed by Hardie et al. [48] with no improvement in soil moisture and water retention characteristics.

Biochar application improved water retention in wet and dry conditions in the loamy soil, as compared to the control. However, the scale of this effect is dependent on biochar particle size (Figure 2B). The biochar particle size (0.15–2 mm) contributed to an increase in water retention in wet (5 hPa) to dry (<10,000 hPa) conditions. The smaller particle size <0.15 mm held water more strongly than the larger particles.

In both soils (loamy and sandy soil), we verified that finer fractions increased water retention. For example, the particle size (0.15–2 mm) was responsible for an increase of 0.08 m^3; of water per m^3; of soil in both sandy and loamy soils [25].This increase occurred because small biochar particles often have more micropores than large biochar particles, holding more water than large particles [7]. For example, when considering biochar pyrolysis, the increase of pyrolysis temperature causes a decrease in the particle size. This size reduction contributes to the increase of microporosity and consequently an increase in the water retention in the permanent wilting point [49].

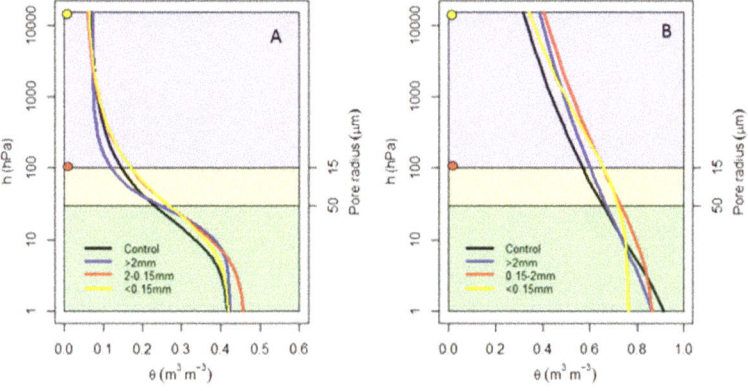

● Field capacity (FC); ○Wilting point (WP), FC-WP*BD= Water available content; ▨ Micropores (<15 um) -Water retention; ▨ Mesopores (15-50 um) - Conduction of the water after drainage; ▨ Macropore (>50 um) drainage and aeration.

Figure 2. Effect of biochar particle size on soil water retention in sandy (**A**) and loamy soil (**B**).

The increase in water retention with a decrease in particle size (especially in <0.15 mm) was also verified by Ibrahim et al. [50] in biochar with particle sizes ≤1 mm; this biochar contributed to an increase in the soil moisture content, especially when biochar was applied to the superficial layer (0–5 cm). Głąb et al. [25] verified that the increase in biochar size from 0.5 mm to 2 mm caused a decrease in the water retention in sandy soil; this result was found in this study with an effect that is more evident at a high matric potential. The increase of water retention at a high matric potential is attributed to water content that is affected more strongly by the organic carbon as compared to the low water potential [51].

The fact that smaller particles retain more water than larger particles is due to small biochar particles that can more easily mix or interact with soil particles to form aggregates than large biochar particles [23].These aggregates contribute to an increase in water retention. In addition, small biochar particles have a high specific surface area per unit of mass, and the water retention increases with an increase in the total specific surface area per unit of mass [7]; when the biochar with a high specific surface area is incorporated into the soil, it contributes to increasing the soil surface area [52]. That can result in the improvement of the soil water retention [4,52,53].

The increase in water retention with a reduction in particle size allows the soil to retain additional water, increasing the amount of available moisture in the root zone and permitting longer intervals between irrigations [10,54]. The biochar application improves soil water retention. From this point of view, biochar can be recommended as a valuable amendment that improves soil hydraulic properties. As the addition of biochar increases the volume of stored water in the soil, it may allow for a reduction in the frequency of irrigation [55]. The addition of biochar may have enhanced the effect on the soil water content, resulting in positive impacts on plant growth during periods of water deficit.

3.4.2. Plant-Available Water Content

A significant difference in the plant-available water content (AWC) was found for the soil textures. In all treatments, the loamy had more AWC than the sandy soil. For both soils, we found a significant effect of biochar addition on AWC ($p < 0.05$), and between biochar fractions the best treatment was 2–0.15 mm; this treatment differed from <0.15 mm and the control treatment. (Figure 3).

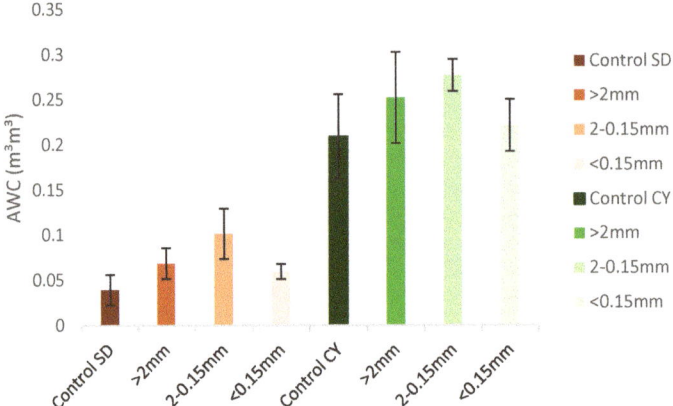

Figure 3. Effect of biochar size on plant-available water content in sandy soil (SD) and loamy soil (CY) AWC.

The plant-AWC is directly related to pore size distribution in different biochar-textured fractions [56]. Similar to our results, the increase in the plant-AWC in sandy and loamy soil with the decrease of the particle size was also verified by Głąb et al. [25] and Mukherjee & Lal [57].

The high porosity of biochar has a positive impact on soil water retention [58]. This high porosity is associated with a high specific surface area that increases with the decrease of the particle size, these are the essential factors that cause a rise in the soil available water content [59,60]. The sandy soil has a specific surface area <10 m^2 g^{-1} [61], while that of biochar *Miscanthus* can be as high as 371 m^2 g^{-1} in the fraction (>2 mm). This property of biochar is verified in the particle size (0.15–2 mm) and is therefore an important factor to increase the water-holding capacity of soil when mixed with biochar. However, when the biochar particle size is too small (<0.15 mm), the specific surface area is so large and retains the water so strongly that the available water content can be reduced. The application of the biochar in the soil with a smaller fraction (0.15–2mm) is advantageous in reducing the frequency of irrigation, especially where plants fully depend on irrigation [25].

3.4.3. Effect of Biochar Particle Size on Pore Size Distribution and its Relation with Water Retention Curve

Comparing only the biochar addition (25 Mg ha^{-1}) in the >2 mm fraction with the control treatment (only soil) in both soils (especially in loamy soil), the biochar addition increased the volume of micropores (<15 µm diameter) and decreased the volume of mesopores (15–50 µm of diameter) and macropores (>50 µm diameter). In the sandy soil, the biochar particle size affected the pore size distribution only slightly; only the fraction >2 mm had a larger increase of micropores (<15 µm) as compared to other treatments (Figure 4A). In loamy soil, comparing only the effect of particle size, smaller biochar particles (<0.15 mm) increased the volume of macropores (>50 µm) and mesopores (50–15 µm), yet reduced micropores (<15 µm). However, the treatments with biochar particle size 0.15–2 mm and >2 mm were similar. Similarity, Głąb et al. [25] found that smaller particles of biochar (0–500 µm in diameter) reduced the volume of small pores (0.5 µm) and fissures (500 µm) but increased the volume of pores in a diameter range from 0.5 to 500 µm.

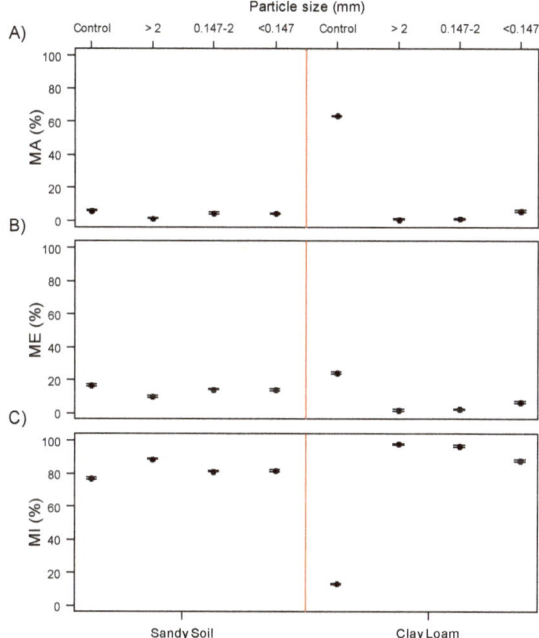

Figure 4. Effect of biochar doses on pore size distribution in Clay Loam and Sandy Soil. Macroporosity (**A**), mesoporosity (**B**) and microporosity (**C**).

The increase in the macroporosity and mesoporosity and the decrease in the microporosity under the addition of small biochar particle size (0.15–2 and <0.15 mm) can be associated with the biochar particles, which settle between the soil particle matrix without blocking the existing pores, thereby creating new pores to increase the macroporosity [62]. Moreover, the presence of the micropore was dominant in both soils (Figure 4), contributing to an increase in the soil water retention, as was previously reported by Tseng & Tseng [63] after 295 interactions of biochar and soil in incubation conditions.

In loamy and sandy soil, we found high water retention in smaller biochar fractions (0.15–2 and <0.15 mm (Figure 2). This increase can be associated with a high specific surface area that increases with the decrease of particle size. These particles can contribute to the creation of pores between biochar particles and soil particles (interpores). The pores inside biochar (intrapores) play fundamental roles in soil water retention [10].

Following soil pore classification, the α-type and β-type pores can be also indicated as transmission-like (≥50 µm), storage-like (0.5–50 µm) and residual-like (<0.5 µm) pores. The largest transmission-like pores are responsible for excess water drainage, thus permitting the aforementioned water. Storage-like pores retain water against gravity and release or diffusion within the biochar pores (from pores whose size belong to the top limit of the 0.5–50 µm interval). Finally, the smallest residual-like pores contain strongly bound water that cannot easily escape from the porous system [56].

The high quantity of micropores in all treatments and in both soils indicates that biochar may not sufficiently shift soil aeration conditions, as was previously reported [10]. However, the reduction in the biochar particle size contributed to an increase in the macroporosity and mesoporosity in sandy and loamy soils. The macropores and mesopores contribute by aeration and water conduction, and micropores contribute by water retention [23]. In loamy soil, the porous biochar particles can improve water flow. On fine-textured soils, biochar particle can also increase water infiltration and hydraulic conductivity by improving soil aggregation, thus increasing macroporosity. Therefore, biochar can be an important amendment to improve water movement in loamy soils [7]. Biochar could contribute to reducing the penetration resistance and increase the water holding capacity in soils, which would be beneficial for plant root elongation and available water [64].

4. Conclusions

The physical properties of soil, such as bulk density and total porosity, were dependent on the biochar size, especially in loamy soil. Small particles of biochar reduced the volume of soil pores (<0.15 mm diameter) but increased the volume of mesopores (0.15–0.50 mm diameter) and macropores (>0.50 mm diameter).

Biochar application improved the soil water characteristics by slightly increasing the plant-available water storage capacity, especially when the finest fraction was used in sandy soil. The biochar has a great potential to improve soil water retention in the finest fraction of loamy and sandy soils.

The benefits found in our research show that this material can be recommended for farmers as a soil amendment to improve the chemical, physical and hydrological quality of their soil.

For the farmers to obtain improvement in the chemical, physical and hydrological properties, the biochar can be used in the finest fraction <0.15 mm with rates of 25 Mg ha^{-1}. Moreover, due to the difficulty of applying a small particle size in agricultural soils, the biochar can be co-composted before it is applied to soil.

Further investigations are recommended to better understand the influence of biochar particle size on hydraulic conductivity, rates and time of interaction, as well as cost-to-benefit ratios in sandy and loamy soils.

Author Contributions: S.d.J.D. was responsible for conducting the experiment, writing the manuscript and contributing to the statistical analyses; B.G. contributed to the review and improvement of the paper and C.E.P.C. contributed to the experimental design, experimental conduction, and review of the paper.

Acknowledgments: This study was supported by CNPq (process number 404150/2013-6) and partially developed in University of Sao Paulo Brazil and in the Martin Luther University Halle (Saale), Germany and Fellowship were supported by Coordination for the Improvement of Higher Education Personnel (CAPES).

Conflicts of Interest: The authors declare no conflict of interest.

References

1. Tiessen, H.; Cuevas, E.; Chacon, P. The role of soil organic matter in sustaining soil fertility. *Nature* **1994**, *371*, 783–785. [CrossRef]
2. Cahn, M.; Extension, C. Cation and nitrate leaching in an oxisol of the Brazilian Amazon. *Agron. J.* **1993**, *85*, 334–340. [CrossRef]
3. Nasta, P.; Kamai, T.; Chirico, G.B.; Hopmans, J.W.; Romano, N. Scaling soil water retention functions using particle-size distribution. *J. Hydrol.* **2009**, *374*, 223–234. [CrossRef]
4. Glaser, B.; Lehmann, J.; Zech, W. Ameliorating physical and chemical properties of highly weathered soils in the tropics with charcoal—A review. *Biol. Fertil. Soils* **2002**, *35*, 219–230. [CrossRef]
5. Glaser, B.; Balashov, E.; Haumaier, L.; Guggenberger, G.; Zech, W. Black carbon in density fractions of anthropogenic soils of the Brazilian Amazon region. *Org. Geochem.* **2000**, *31*, 669–678. [CrossRef]
6. Glaser, B.; Birk, J.J. State of the scientific knowledge on properties and genesis of Anthropogenic Dark Earths in Central Amazonia (terra preta de I). *Geochim. Cosmochim. Acta* **2012**, *82*, 39–51. [CrossRef]
7. Blanco-Canqui, H. Biochar and Soil Physical Properties. *Soil Sci. Soc. Am. J.* **2017**, *84*, 687. [CrossRef]
8. He, X.; Yin, H.; Han, L.; Cui, R.; Fang, C.; Huang, G. Effects of biochar size and type on gaseous emissions during pig manure/wheat straw aerobic composting: Insights into multivariate-microscale characterization and microbial mechanism. *Bioresour. Technol.* **2019**, *271*, 375–382. [CrossRef]
9. Prendergast-Miller, M.T.; Duvall, M.; Sohi, S.P. Biochar-root interactions are mediated by biochar nutrient content and impacts on soil nutrient availability. *Eur. J. Soil Sci.* **2013**, *65*, 173–185. [CrossRef]

10. Liu, Q.; Liu, B.; Zhang, Y.; Lin, Z.; Zhu, T.; Sun, R.; Wang, X.; Ma, J.; Bei, Q.; Liu, G.; et al. Can biochar alleviate soil compaction stress on wheat growth and mitigate soil N_2O emissions? *Soil Biol. Biochem.* **2017**, *104*, 8–17. [CrossRef]
11. De Luca, T.H.; Mackenzie, M.D.; Gundale, M.J. Biochar effects on soil nutrient transformations. In *Biochar for Environmental Management: Science and Technology*; Earthscan: London, UK, 2009; pp. 251–270.
12. Laird, D.A.; Fleming, P.; Davis, D.D.; Horton, R.; Wang, B.; Karlen, D.L. Impact of biochar amendments on the quality of a typical Midwestern agricultural soil. *Geoderma* **2010**, *158*, 443–449. [CrossRef]
13. Tan, R.R. Data challenges in optimizing biochar-based carbon sequestration. *Renew. Sustain. Energy Rev.* **2019**, *104*, 174–177. [CrossRef]
14. Zhu, X.; Liu, Y.; Li, L.; Shi, Q.; Hou, J.; Zhang, R.; Zhang, S.; Chen, J. Nonthermal air plasma dehydration of hydrochar improves its carbon sequestration potential and dissolved organic matter molecular characteristics. *Sci. Total Environ.* **2019**, *659*, 655–663. [CrossRef]
15. Beusch, C.; Cierjacks, A.; Böhm, J.; Mertens, J.; Bischoff, W.-A.; de Araújo Filho, J.C.; Kaupenjohann, M. Biochar vs. clay: Comparison of their effects on nutrient retention of a tropical Arenosol. *Geoderma* **2019**, *337*, 524–535. [CrossRef]
16. Gronwald, M.; Don, A.; Tiemeyer, B.; Helfrich, M. Effects of fresh and aged chars from pyrolysis and hydrothermal carbonization on nutrient sorption in agricultural soils. *Soil* **2015**, *1*, 475–489. [CrossRef]
17. Kameyama, K.; Miyamoto, T.; Shinogi, Y. Increases in available water content of soils by applying bagasse-charcoals. In Proceedings of the 19th World Congress of Soil Science: Soil solutions for a changing world, Brisbane, Australia, 1–6 August 2010; pp. 105–108.
18. Dai, Z.; Zhang, X.; Tang, C.; Muhammad, N.; Wu, J.; Brookes, P.C.; Xu, J. Potential role of biochars in decreasing soil acidification-a critical review. *Sci. Total Environ.* **2017**. [CrossRef] [PubMed]
19. Liu, J.; Schulz, H.; Brandl, S.; Miehtke, H.; Huwe, B.; Glaser, B. Short-term effect of biochar and compost on soil fertility and water status of a Dystric Cambisol in NE Germany under field conditions. *J. Plant Nutr. Soil Sci.* **2012**, *175*, 698–707. [CrossRef]
20. Wang, B.; Gao, B.; Zimmerman, A.R.; Zheng, Y.; Lyu, H. Novel biochar-impregnated calcium alginate beads with improved water holding and nutrient retention properties. *J. Environ. Manag.* **2018**, *209*, 105–111. [CrossRef] [PubMed]
21. Novak, J.M.; Busscher, W.J.; Watts, D.W.; Amonette, J.E.; Ippolito, J.A.; Lima, I.M.; Gaskin, J.; Das, K.C.; Steiner, C.; Ahmedna, M.; et al. Biochars impact on soil-moisture storage in an ultisol and two aridisols. *Soil Sci.* **2012**, *177*, 310–320. [CrossRef]
22. Barnes, R.T.; Gallagher, M.E.; Masiello, C.A.; Liu, Z.; Dugan, B. Biochar-induced changes in soil hydraulic conductivity and dissolved nutrient fluxes constrained by laboratory experiments. *PLoS ONE* **2014**, *9*. [CrossRef]
23. Herath, H.M.S.K.; Camps-Arbestain, M.; Hedley, M. Effect of biochar on soil physical properties in two contrasting soils: An Alfisol and an Andisol. *Geoderma* **2013**, *209–210*, 188–197. [CrossRef]
24. Zhang, J.; Chen, Q.; You, C. Biochar Effect on Water Evaporation and Hydraulic Conductivity in Sandy Soil. *Pedosphere* **2016**, *26*, 265–272. [CrossRef]
25. Głąb, T.; Palmowska, J.; Zaleski, T.; Gondek, K. Effect of biochar application on soil hydrological properties and physical quality of sandy soil. *Geoderma* **2016**, *281*, 11–20. [CrossRef]
26. Conz, R.F.; Abbruzzini, T.F.; de Andrade, C.A.; Milori, D.M.; Cerri, C.E. Effect of Pyrolysis Temperature and Feedstock Type on Agricultural Properties and Stability of Biochars. *Agric. Sci.* **2017**, *8*, 914–933. [CrossRef]
27. Novak, J.M.; Cantrell, K.B.; Watts, D.W.; Busscher, W.J.; Johnson, M.G. Designing relevant biochars as soil amendments using lignocellulosic-based and manure-based feedstocks. *J. Soils Sediments* **2013**. [CrossRef]
28. International Biochar Initiative. Standardized Product Definition and Product Testing Guidelines for Biochar That is Used in Soil, International Biochar Initiative. 2015. Available online: http://www.biochar-international.org/sites/defaut/files/IBI_Biochar_Standards_V1.1.pdf (accessed on 2 December 2018).
29. ASTM. *Standard Test Method for the Analysis of Wood Charcoal*; ASTM D1762-84; ASTM International: West Conshohocken, PA, USA, 2007.
30. Cerato, A.B.; Lutenegger, A.J. Determination of surface area of fine-grained soils by the ethylene glycol monoethyl ether (EGME) method. *Geotech. Test. J.* **2002**, *25*, 315–321. [CrossRef]
31. Chan, K.Y.; Bowman, A.; Oates, A. Oxidizible organic carbon fractions and soil quality changes in an oxic paleustalf under different pasture leys. *Soil Sci.* **2001**, *166*, 61–67. [CrossRef]

32. Van Raij, B.; Andrade, J.C.; Cantarella, H.; Quaggio, J.A. *Análise Química para Avaliação da Fertilidade de Solos Tropicais*; Campinas Instituto Agronômico: Campinas, Brazil, 2001; 285p.
33. Blake, G.R.; Hartge, K.H. Bulk density. In *Methods of Soil Analysis, Part 1—Physical and Mineralogical Methods*, 2nd ed.; Agronomy Monograph 9; Klute, A., Ed.; American Society of Agronomy—Soil Science Society of America: Madison, WI, USA, 1986; pp. 363–382.
34. Flint, L.E.; Flint, A.L. Porosity. In *Methods in Soil Analysis, Part 4—Physical Methods*; Dane, J.H., Topp, G.C., Eds.; Soil Science Society of America: Madison, WI, USA, 2002.
35. Koorevaar, P.; Menelik, G.; Kirksen, C. *Elements of Soil Physics*; Developments in Soil Science, 13; Elsevier: Amsterdam, The Netherlands, 1983; 228p.
36. Klute, A.; Klute, A.; Dirksen, C. *Hydraulic Conductivity and Diffusivity: Laboratory Methods 9*; ASA: Madison, WI, USA, 1986.
37. RStudio Team. *RStudio: Integrated Development for R*; RStudio, Inc.: Boston, MA, USA, 2018; Available online: http://www.rstudio.com/ (accessed on 4 November 2018).
38. Duarte, S.J.; Glaser, B.; Lima, R.P.; Cerri, C.E.P. Chemical, physical and hydrological properties as affected by one year of *Miscanthus* biochar interaction with sandy and loamy tropical soils. *Soil Syst.* **2019**. submitted.
39. EMBRAPA. *Guia Prático para Interpretação de Resultados de Análises de Solo*; EMBRAPA: Aracaju, Brazil, 2015.
40. Conte, P.; Hanke, U.M.; Marsala, V.; Cimoò, G.; Alonzo, G.; Glaser, B. Mechanisms of water interaction with pore systems of hydrochar and pyrochar from poplar forestry waste. *J. Agric. Food Chem.* **2014**, *62*, 4917–4923. [CrossRef]
41. Lehmann, J.; Joseph, S. Biochar for Environmental Management: An Introduction. *Eartschcan* **2009**, *1*, 1–12.
42. Oussible, M.; Crookston, R.K.; Larson, W.E. Grain Yield of Wheat. *Agron. J.* **1992**, *38*, 34–38. [CrossRef]
43. Mukherjee, A. Physical and Chemical Properties of a Range of Laboratory-produced Fresh and Aged Biochars. Ph.D. Thesis, University of Florida, Gainesville, FL, USA, 2011.
44. Biederman, L.A.; Harpole, W.S. Biochar and its effect on plant productivity and nutrient cycling: A meta-analysis. *GCB Bioenergy* **2013**, *5*, 202–214. [CrossRef]
45. Zhang, A.; Bian, R.; Pan, G.; Cui, L.; Hussain, Q.; Li, L.; Zheng, J.; Zheng, J.; Zhang, X.; Han, X.; et al. Effects of biochar amendment on soil quality, crop yield and greenhouse gas emission in a Chinese rice paddy: A field study of 2 consecutive rice growing cycles. *Field Crop. Res.* **2012**, *127*, 153–160. [CrossRef]
46. Zimmerman, J.K.; Pulliam, W.M.; Lodge, D.J.; Quiñones-Orfila, V.; Fetcher, N.; Guzmán-Grajales, S.; Guzman-Grajales, S. Nitrogen Immobilization by Decomposing Woody Debris and the Recovery of Tropical Wet Forest from Hurricane Damage. *Oikos* **1995**, *72*, 314. [CrossRef]
47. Jeffery, S.; Bezemer, T.M.; Cornelissen, G.; Kuyper, T.W.; Lehmann, J.; Mommer, L.; Sohi, S.P.; Van De Voorde, T.F.; Wardle, D.A.; Van Groenigen, J.W. The way forward in biochar research: Targeting trade-offs between the potential wins. *GCB Bioenergy* **2015**, *7*, 1–13. [CrossRef]
48. Hardie, M.; Clothier, B.; Bound, S.; Oliver, G.; Close, D. Does biochar influence soil physical properties and soil water availability? *Plant Soil* **2014**, *376*, 347–361. [CrossRef]
49. Hina, K.; Bishop, P.; Arbestain, M.C.; Calvelo-Pereira, R.; Maciá-Agulló, J.A.; Hindmarsh, J.; Hanly, J.A.; Macìas, F.; Hedley, M.J. Producing biochars with enhanced surface activity through alkaline pretreatment of feedstocks. *Soil Res.* **2010**, *48*, 606–617. [CrossRef]
50. Ibrahim, A.; Usman, A.R.A.; Al-Wabel, M.I.; Nadeem, M.; Ok, Y.S.; Al-Omran, A. Effects of conocarpus biochar on hydraulic properties of calcareous sandy soil: Influence of particle size and application depth. *Arch. Agron. Soil Sci.* **2017**, *63*, 185–197. [CrossRef]
51. Rawls, W.J.; Pachepsky, Y.A.; Ritchie, J.C. Effect of soil organic carbon on soil water retention. *Geoderma* **2003**, *116*, 61–76. [CrossRef]
52. Downie, A.; Crosky, A.; Munroe, P. Physical properties of biochar. In *Biochar for Environmental Management, Science and Technology*; Lehmann, J.L., Joseph, J.S., Eds.; Earthscan Publishers Ltd.: London, UK, 2009; pp. 13–32.
53. Abel, S.; Peters, A.; Trinks, S.; Schonsky, H.; Facklam, M.; Wessolek, G. Geoderma Impact of biochar and hydrochar addition on water retention and water repellency of sandy soil. *Geoderma* **2013**, *202–203*, 183–191. [CrossRef]
54. Yin, S.H.; Chang, Z.H.; Han, L.B.; Lu, X.S. Study on optimization of moisture retention for golf green rootzone soil mixtures. *Acta Ecol. Sin.* **2012**, *32*, 26–32. [CrossRef]

55. Mccready, M.S.; Dukes, M.D. Landscape irrigation scheduling efficiency and adequacy by various control technologies. *Agric. Water Manag.* **2011**, *98*, 697–704. [CrossRef]
56. Conte, P.; Nestle, N. Water dynamics in different biochar fractions. *Magn. Reson. Chem.* **2015**, *53*, 726–734. [CrossRef]
57. Mukherjee, A.; Lal, R. Biochar Impacts on Soil Physical Properties and Greenhouse Gas Emissions. *Agronomy* **2013**, *3*, 313–339. [CrossRef]
58. Ogawa, M.; Okimori, Y.; Takahashi, F. Carbon sequestration by carbonization of biomass and forestation: Three case studies. *Mitig. Adapt. Stratag. Glob. Chang.* **2006**, *11*, 421–436. [CrossRef]
59. Crabbe, M.J.C. Modelling effects of geoengineering options in response to climate change and global warming: Implications for coral reefs. *Comput. Biol. Chem.* **2009**, *33*, 415–420. [CrossRef] [PubMed]
60. Uzoma, K.C.; Inoue, M.; Andry, H.; Zahoor, A.; Nishihara, E. Influence of biochar application on sandy soil hydraulic properties and nutrient retention. *J. Food. Agric. Environ.* **2011**, *9*, 1137–1143.
61. Herbrich, M.; Zönnchen, C.; Schaaf, W. Short-term effects of plant litter addition on mineral surface characteristics of young sandy soils. *Geoderma* **2015**, *239–240*, 206–212. [CrossRef]
62. Steiner, C.; Melear, N.; Harris, K.; Das, K.C. Biochar as bulking agent for poultry litter composting. *Carbon Manag.* **2011**, *2*, 227–230. [CrossRef]
63. Tseng, R.; Tseng, S. Pore structure and adsorption performance of the KOH-activated carbons prepared from corncob. *J. Colloid Interface Sci.* **2005**, *287*, 428–437. [CrossRef]
64. Andrenelli, M.C.; Maienza, A.; Genesio, L.; Miglietta, F.; Pellegrini, S.; Vaccari, F.P.; Vignozzi, N. Field application of pelletized biochar: Short term effect on the hydrological properties of a silty clay loam soil. *Agric. Water Manag.* **2016**, *163*, 190–196. [CrossRef]

© 2019 by the authors. Licensee MDPI, Basel, Switzerland. This article is an open access article distributed under the terms and conditions of the Creative Commons Attribution (CC BY) license (http://creativecommons.org/licenses/by/4.0/).

Article

Growth, Seed Yield, Mineral Nutrients and Soil Properties of Sesame (*Sesamum indicum* L.) as Influenced by Biochar Addition on Upland Field Converted from Paddy

Cosmas Wacal [1], Naoki Ogata [2], Daniel Basalirwa [1], Takuo Handa [1], Daisuke Sasagawa [1], Robert Acidri [1], Tadashi Ishigaki [1], Masako Kato [2], Tsugiyuki Masunaga [3], Sadahiro Yamamoto [4] and Eiji Nishihara [4],*

1. United Graduate School of Agricultural Sciences, Tottori University, 4-101 Koyama Minami, Tottori 680-8553, Japan; cwacal@gmail.com (C.W.); danielbasalirwa@gmail.com (D.B.); Takuo.Handa@me2.seikyou.ne.jp (T.H.); ssgw2653.02.06@gmail.com (D.S.); acidrirobert24@gmail.com (R.A.); tishigaki@adm.tottori-u.ac.jp (T.I.)
2. National Institute of Crop Science, National Agriculture and Food Research Organization, 2-1-2 Kannondai, Tsukuba, Ibaraki 305-8518, Japan; naokio@affrc.go.jp (N.O.); shunn@affrc.go.jp (M.K.)
3. Faculty of Life and Environmental Sciences, Shimane University, Matsue, Shimane 690-8504, Japan; masunaga@life.shimane-u.ac.jp
4. Faculty of Agriculture, Tottori University, 4-101 Koyama Minami, Tottori 680-8553, Japan; yamasada@tottori-u.ac.jp
* Correspondence: nishihar@tottori-u.ac.jp; Tel.: +81-(0)857-31-5385

Received: 12 December 2018; Accepted: 25 January 2019; Published: 27 January 2019

Abstract: Sesame is an important oilseed crop cultivated worldwide. However, research has focused on biochar effects on grain crops and vegetable and there is still a scarcity of information of biochar addition on sesame. This study was to assess the effect of biochar addition on sesame performance, with a specific emphasis on growth, yield, leaf nutrient concentration, seed mineral nutrients, and soil physicochemical properties. A field experiment was conducted on an upland field converted from paddy at Tottori Prefecture, Japan. Rice husk biochar was added to sesame cropping at rates of 0 (F), 20 (F+20B), 50 (F+50B) and 100 (F+100B) t ha^{-1} and combined with NPK fertilization in a first cropping and a second cropping field in 2017. Biochar addition increased plant height, yield and the total number of seeds per plant more in the first cropping than in the second cropping. The F+50B significantly increased seed yield by 35.0% in the first cropping whereas the F+20B non-significantly increased seed yield by 25.1% in the second cropping. At increasing biochar rates, plant K significantly increased while decreasing Mg whereas N and crude protein, P and Ca were non-significantly higher compared to the control. Soil porosity and bulk density improved with biochar addition while pH, exchangeable K, total N, C/N ratio and CEC significantly increased with biochar, but the effect faded in the second cropping. Conversely exchangeable Mg and its plant tissue concentration decreased due to competitive ion effect of high K from the biochar. Biochar addition is effective for increasing nutrient availability especially K for sesame while improving soil physicochemical properties to increase seed yield, growth and seed mineral quality.

Keywords: sesame; rice husk biochar; nutrient concentration; cropping

1. Introduction

Sesame (*Sesamum indicum* L.) is an important oilseed crop cultivated worldwide for its edible oil and food [1]. Sesame seeds contain high oil content (44–58%), proteins (18–25%), carbohydrates (13.5%), ash (5%) [2], and mineral components, such as K (815 mg/100 g), P (647 mg/100 g), Mg

(579 mg/100 g) and Ca (415 mg/100 g) [3]. This contributes to its health and nutritional benefits. Therefore, demand for sesame seeds is increasing due to the increasing knowledge on their dietary and health benefits, but there has been limited research on sesame evidenced by low yield in most growing areas hence hampering its adoption and expansion in the world [4]. Although sesame has been reconsidered a local specialty crop in Japan [5], the production of sesame is still low. For instance, the Food and Agriculture Organization (FAO) in 2016 estimated that 11 tons of sesame seeds were produced from an area of 21 hectares [6]. With the increase in abandoned paddy fields estimated at 360,000 ha by the year 2010, farmers were encouraged to convert such fields into cultivation of upland crops, such as wheat and soybeans [7,8], including sesame. However, crop yield on upland fields converted from paddy may decrease due to declining soil fertility status of the paddy soils that could require soil amendment with organic materials [9].

Biochar is a soil amendment produced from thermal decomposition of organic materials through pyrolysis and it has the potential to increase crop yields [10,11]. Earlier research has shown that biochar addition can improve plant growth and soil quality [12–14]. The positive responses in crop yield on biochar addition were attributed to improved soil properties, such as a decrease in soil bulk density, and subsequent increase in porosity and water holding capacity [15,16], increase in the cation exchange capacity (CEC) which enhances the retention of basic nutrients, [17], increased uptake of N and its availability in soil [18], adsorption of soil phytotoxins [19,20], liming effects [21], and increased plant nutrient concentration [22,23]. For instance, cultivation on sandy soils using biochar increased maize yield by 150% and 98% over the control at rates of 15 t ha^{-1} and 20 t ha^{-1} respectively [23]. It has also been reported that rice husk biochar addition at 41.3 t ha^{-1} significantly increased rice grain yield by 16–35% attributed to increased water holding capacity, available N and cation exchange capacity (CEC) [24]. Zhang et al. [25] also reported an increase in rice yields of 14% over the control in paddy soils with wheat straw biochar rate of 40 t ha^{-1}. Furthermore, it has been shown that rice husk biochar addition rates of up to 50 t ha^{-1} significantly increased maize seed protein by 27% compared to without biochar while increasing plant height by 23% compared with control [26]. The authors attributed these increases to the increase in soil fertility status improving nutrition required for maize grain quality improvement. There are also several reports on an increase in plant height, growth and grain quality with biochar application in crops [27–29], which indicate a positive effect of biochar on crops.

Biochar addition to soils is expected to promote sustainable crop production through a positive effect on yield, but these may depend on the cropping seasons. For instance, Cornelissen et al. [30] studied the effect of rice husk biochar and cacao shell biochar applied to Indonesia Utisol soil at rates of up to 15 t ha^{-1} and found that the maize yield with rice husk biochar become lower and faded in the second cropping while with cacao shell biochar was highest in the second cropping through third and fourth, but faded in the fifth cropping seasons. In addition, biochar applied to soil and tested over four cropping seasons on acidic soil in Brazil showed positive effects in the first cropping, but these faded in the following cropping [31]. Carter et al. [32] also reported a high yield of lettuce and cabbage in the first cropping, but yield decreased significantly in the third cropping with rice husk biochar rates of up to 167 t ha^{-1} field equivalent. Furthermore, rapeseed yield with wheat straw biochar faded after third cropping that suggested biochar needed to be applied after every three years to maintain positive effects on crop yield [33]. These lack of positive responses of crop yield could be attributed to the changes in biochar chemistry over time as it ages in the soil environment [34]. Hence, the properties of biochar responsible for crop improvement may be altered consequently leading to no effect on growth and yield.

Several pieces of research have focused on biochar effects on grain crops, such as rice, maize, wheat and vegetable of which plant growth responses to biochar addition varied [19,21,24,27,28,32]. However, there is still a scarcity of information of biochar addition on sesame [35,36], indicating a need to generate understanding of how biochar addition can effectively be used to increase sesame production. An earlier research has shown that rice and saw dust biochar addition at 10 t ha^{-1} significantly increased sesame seed yield by 55.5% compared to without biochar attributed to improved soil physico-chemical properties, such as bulk density and porosity, increased pH, total N, K, Mg and CEC after biochar addition on a highly leached ultisols with low base saturation and strongly acidic soils [36]. Although in a pot experiment, biochar addition to sesame has been shown to significantly increase plant height at increasing rates where the optimal rate of 11.21 g kg^{-1} (equivalent to 22.42 t ha^{-1}) was obtained beyond which biochar decreased pant height becoming harmful to sesame growth [35]. The authors attributed this negative effect on sesame growth to increase in pH due to biochar application on already neutral soil that had pH 6.4 before the experiment. Furthermore, coconut shell biochar addition at 10 t ha^{-1} on sandy coastal has been shown to increase sesame yield when grown on sandy coastal soil [37]. Therefore, biochar addition shows positive results on sesame. However, crop responses to biochar application depend on biochar type, application rates, soil properties and climatic conditions [38]. It important to explore the utilization of biochar in sesame to understand how seed yield and growth are influenced on upland fields converted paddy in Japan under different field climatic conditions, paddy soils with low pH and higher biochar rates in order to close existing gaps on biochar use for sesame.

In this study, we hypothesized that biochar addition would increase sesame growth, yield and nutrient availability with increasing rates of biochar in first and second cropping. To investigate the effect of biochar addition on sesame performance, we cultivated sesame on two fields of first and second cropping on upland field converted from paddy. The specific objectives this study were to determine the effect of biochar addition on the (a) growth and yield of sesame in the first and second cropping; (b) leaf tissue nutrient concentration and seed mineral nutrient contents; and (c) soil physicochemical properties of the upland field converted paddy under continuous cropping.

2. Materials and Methods

2.1. Experiment Site

This field experiment was conducted in 2017 at the Tottori Prefecture, Japan (35°29′14.85″ N, 134°07′47.01″ E). Most precipitation occurred between June to September during the cultivation period. The total monthly rainfall received in the region in 2017 were 66.5, 158.5, 161 and 224.5 mm in June, July, August and September respectively. The average daily maximum temperatures were 24.2 °C in June, 30.6 °C in July, 30.5 °C in August and 26.0 °C in September favorable for sesame growing. The region has primarily one sesame crop harvested per year. The dominant soil at this site was classified as Cambisols [39].

2.2. Soil and Biochar Properties

Analysis of the basic physicochemical properties of the soil samples (0–15 cm) taken from the experiment field before sowing in 2016 indicted that the topsoil had a pH (1:5 H$_2$O) of 5.39; electrical conductivity (EC) of 0.05 dSm^{-1}; bulk density, 1.27 g cm^{-3}; porosity, 49.09%; total C, 26.25 g kg^{-1}; total N, 2.29 g kg^{-1}; C/N ratio, 11.46; available P (Truog-P), 6.99 mg kg^{-1}; exchangeable K, 109.62 mg kg^{-1}; exchangeable Ca 1931.22 mg kg^{-1}; and exchangeable Mg 383.29 mg kg^{-1}. The commercial rice husk biochar added to the study soils was manufactured and bought from a local store in Tottori, Japan. The rice husk biochar was surface applied by hand in June 2016 in the old field and then in June 2017 in the newly opened field before sesame sowing and immediately incorporated into the soil to a depth of 0–15 cm with base fertilizer utilizing a rotatory power tiller. The rice husk biochar had a pH of 10.47 and EC of 1.66 dSm^{-1} determined from biochar suspension (1:5 w/v, biochar: water) mechanically shaken for 1 h and measured with a pH and EC meter (Horiba Aqua Cond Meter F-74). Total C and N were analyzed by dry combustion on the CN-corder (Macro corder JM 1000CN, J-Science Co., Ltd., Kyoto, Japan) and reported as C, N and C/N ratio of 39.76%, 0.51% and 78.26, respectively. The rice husk biochar had available P of 647.94 mg kg^{-1} determined according to Truog method [40]. Exchangeable K, Ca, and Mg in the rice husk biochar were K, 3640.73 mg kg^{-1}, 1207.78 mg kg^{-1}; and 369.26 mg kg^{-1} respectively, determined upon extraction of biochar with 1 N ammonium acetate (pH 7.1) and analysed by atomic absorption spectrophotometer (Model Z-2300, Hitachi Co., Tokyo, Japan). The biochar cation exchange capacity (CEC) was 7.53 cmol$_c$ kg^{-1}, measured by the 1N (pH 7.1) ammonium acetate (NH$_4$OAc) as described by Chapman [41]. The bulk density of the biochar was determined by measuring the weight of compacted biochar in 100 cm^3 steel cylinders and was found to be 0.29 g cm^{-3} whereas the ash content was measured by igniting the biochar sample at 550 °C for 5 h in a muffle furnace and found to be 38.04%.

2.3. Experimental Design and Treatments

In this study, we conducted the experiment on two different plots: One in which sesame was previously cultivated with rice husk biochar for one year (season) and a new field where sesame had not been cultivated before. Each of these fields measuring 10.5 m by 6.5 m were divided into micro plots measuring 2.4 m by 1.9 m (4.56 m^2) onto which biochar was incorporated. Each micro plot was separated by 0.4 m as buffer space. The one-year old sesame field had rice husk biochar rates of 0, 20, 50 and 100 t ha^{-1} applied already in the previous year's cultivation. This field was under cropping in 2016 with biochar and sesame, but due to typhoon winds that destroyed the sesame plants, we could not collect data. Therefore, in 2017, the new field opened adjacent to the old field received a similar amount of rice husk biochar treatments at the start of the experiment. The new field and old field are considered as first and second cropping respectively.

Prior to sowing, all fields were ploughed by a power tiller, harrowed to a fine tilth and basal inorganic fertilizer applied at a rate of N—P$_2$O$_5$—K$_2$O, 70:105:70 kg ha^{-1}, including dolomite (CaMg(CO$_3$)$_2$) at a rate of 1000 kg ha^{-1}. Sesame cultivar 'Gomazou' was planted on ridges of 75 cm wide separated by 40 cm and plant spacing of 45 cm between rows and 15 cm between plants. Sowing date was 11th July 2017 in which five sesame seeds were sown per hole, then thinned to two plants at 14 days after sowing and then one plant per hole at 21 days after sowing. The fields were kept without weeds by hand weeding whenever necessary until harvesting at the on 22nd September 2017. Growth was determined by measuring plant height, height of lowest capsule and number of branches while the seed yield and 1000-seed weight were determined after drying sesame seeds in a greenhouse. The growth and seed yield were determined by randomly selecting ten plants per replicates in each treatment whereas the total number of seeds per plant was calculated after obtaining a weight of 1000 seeds and total seed weights from 10 plants.

2.4. Sampling and Analyses

2.4.1. Plant Sampling

For leaf tissue nutrient concentration analysis, three representative plants were selected at the reproductive stage (50 Days After Sowing) whereas remaining plants were used for growth and yield determination at harvest. Mature leaves from the representative samples were separated from stem and roots and oven dried at 72 °C until a constant weight was attained (after a week). Leaf samples were then ground to fine powder and digested in a mixture of concentrated H$_2$SO$_4$ (98%) and H$_2$O$_2$ (30%) for P, K, Ca and Mg concentration. Plant P concentration was determined colorimetrically with a spectrophotometer at 420 nm (Model U-5100, Hitachi Co., Tokyo, Japan). Plant K, Ca, and Mg concentration was determined by using an atomic absorption spectrophotometer (Model Z-2300, Hitachi Co., Tokyo, Japan). The plant N was determined from the ground sample with the dry combustion method on a CN Corder machine (Macro corder JM 1000CN, J-Science Co., Ltd., Kyoto, Japan).

For the seed mineral nutrient, the analysis was conducted on sesame seeds after harvesting and drying. Seed mineral nutrient concentrations were determined by means of dry ashing as described by Estefan et al. [42]. Ground sesame seed samples (1.0 g) was placed in a crucible, ignited, and burnt to ash in a muffle furnace at 550 °C for 5 h. The ash was cooled and dissolved into 5 mL of 2 N HCl, diluted to 50 mL in a volumetric flask with reverse-osmosis water, and filtered through Advantec

110-mm filter paper. Phosphorous (P) concentration was determined from the filtrates by means of the ammonium vanadate-ammonium-molybdate yellow colorimetric method using a spectrophotometer (Model U-5100, Hitachi Co., Tokyo, Japan), with absorbance measured at 420 nm. The seed K, Ca and Mg contents were determined from the filtrates by means of atomic absorption spectrophotometry (Model Z-2300, Hitachi Co., Tokyo, Japan). The seed N concentration was measured by means of the dry combustion method using a CN-corder (Macro corder JM 1000CN, J-Science Co., Ltd., Kyoto, Japan) and the values of total N values (%) were multiplied by the N factor 6.25 to obtain crude protein values (N × 6.25%) [43].

2.4.2. Soil Sampling

To evaluate the effect of the rice husk biochar addition on soil physical properties, soil samples from fields were collected at 0–15 cm from the sesame fields after harvest. Three soil phases in each biochar treatment of the first and second cropping were calculated from two samples collected per replication at the top 10 cm layer by applying a soil three-phase meter (Model DIK-1130, Daiki Rika Kogyo Co., Ltd., Saitama, Japan) to a 100 cm^3 undisturbed soil core. The surface soil bulk density and porosity were determined on the undisturbed soil cores collected using sampling cores of 100 cm^3 after oven drying at 105 °C for 24 h.

For chemical analysis, at harvest, soil samples were collected with a trowel to a depth of 15 cm, air dried, and passed through a 2-mm sieve. Soil suspension (1:5 w/v soil: water) was used to measure pH and electrical conductivity with a pH meter and EC meter (Horiba Aqua Cond Meter F-74). Total C and N were analysed by dry combustion on the CN-corder (Macro corder JM 1000CN, J-Science Co., Ltd., Kyoto, Japan). Soil exchangeable K, Ca, and Mg were extracted in 1 N ammonium acetate (pH 7.1), and analysed by using an atomic absorption spectrophotometer (Model Z-2300, Hitachi Co., Tokyo, Japan). Cation exchange capacity (CEC) was measured by the 1N (pH 7.1) ammonium acetate (NH_4OAc) extraction methods in which the NH_4^+ saturated soil was equilibrated with 10% KCl and steam distilled by micro–kjeldhal distillation before titration with 0.1 N H_2SO_4 [41] and expressed to $cmol_c\ kg^{-1}$ soil. Soil available P was determined using 0.002 N H_2SO_4 (pH 3.0) in ammonium sulphate solution according to Troug method [40]. The P concentration in soil samples was measured by the ammonium molybdate–ascorbic acid method at an absorption wavelength of 710 nm on a spectrophotometer (Model U-5100, Hitachi Co., Tokyo, Japan).

2.5. Data Analyses

All results were the means of the three replicates. Data were analyzed using one-way analysis of variance (ANOVA) using SPSS version 22.0 (SPSS for windows Inc., Chicago, Illinois, USA) to evaluate the measured parameters as affected by the different rates of biochar addition. The pairs of means were also compared on significant ANOVA tests using Tukey's honestly significance difference (HSD) test ($p < 0.05$). A nonlinear regression analysis was utilized to investigate the relationship between sesame seed yield, plant height and the biochar rates. When considering the differences between the cropping fields, a two-way analysis of variance was used with the different biochar treatments and cropping as two fixed factors.

3. Results

3.1. Effect of Rice Husk Biochar on the Growth and Yield Components of Sesame

The plant height, height of the lowest capsule, number of branches per plant and 1000-seed weight as affected by varying rates of biochar addition in first and second cropping fields are shown in Table 1.

Table 1. The plant height, height of the lowest capsule, number of branches per plant and 1000-seed weight of sesame under the different biochar treatments in different cropping.

Cropping	Biochar Treatment	Plant Height (cm)	Height of First Capsule (cm)	Number of Branches/Plant	1000-Seeds Weight (g)
First cropping	F	140.60 ab	68.12 a	1.72 b	2.16 a
	F+20B	137.85 b	64.26 a	2.03 b	2.08 a
	F+50B	157.63 a	67.48 a	3.01 a	2.23 a
	F+100B	152.47 ab	70.37 a	2.07 ab	1.97 a
Second cropping	F	114.42 b	55.93 a	2.34 a	1.93 a
	F+20B	134.71 ab	57.50 a	2.29 a	2.13 a
	F+50B	124.30 ab	58.14 a	2.48 a	2.03 a
	F+100B	139.63 a	59.02 a	2.45 a	2.03 a
Source of variation					
Biochar (B)		**	ns	ns	ns
Cropping (C)		***	***	ns	ns
B × C		*	ns	ns	ns

Means followed by different lowercase letters within a column in the same cropping are significantly different $p < 0.05$ according to the Tukey test. *** Significant at $p < 0.001$; ** Significant at $p < 0.01$; * Significant at $p < 0.05$; ns, Non-significant.

The biochar rate showed a significant influence on the plant height with a significant interaction between biochar and cropping. However, there were no significant differences in the height of lowest capsule and 1000-seed weight and number of seeds per plant although a number of branches per plant and 1000-seed weight were higher compared to the control in the first and second cropping respectively. In comparison with the control (F), the plant height of the F+50B was non-significantly higher in the first cropping by 10.8% whereas the F+100B was significantly higher in the second cropping by 18.1%. There were no significant differences between control with F+20B and F+50B. The number of branches per plant in the first cropping were significantly increased by 42.7% in the F+50B treatment compared to the control that indicated biochar increased vegetative growth of the sesame. However, in the second cropping, this significant effect was not observed although the F+50B and F+100B treatments tended to have a greater number of branches per plant compared to the control and no significant interaction between biochar addition and cropping was observed.

The seed yield and a total number of seeds per plant of sesame affected by varying rates of biochar in the first and second cropping are shown in Figure 1.

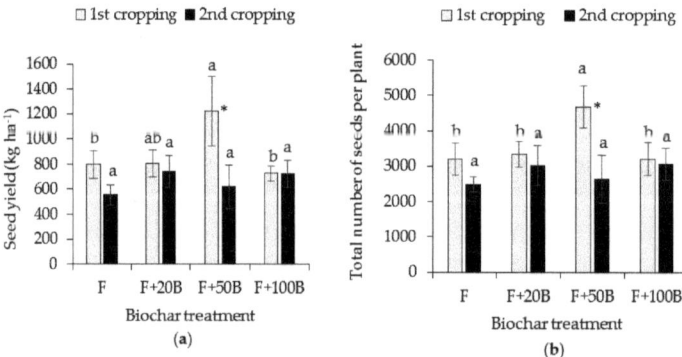

Figure 1. (a) The seed yield; (b) total number of seeds per plant as affected by rice husk biochar treatments. The bars represent the standard deviation of triplicates. Different lower case letters indicate significant difference ($p < 0.05$) among treatment means. * Significant difference ($p < 0.05$) between first and second cropping of the F+50B treatment.

The higher rate of biochar addition significantly improved the seed yield of sesame in the first cropping. In comparison with the control (797 kg ha^{-1}), the F+20B and F+50B treatments significantly increased seed yield by 0.9% and 35.0% (1226 kg ha^{-1}) whereas the F+100B significantly decreased seed yield by 9.4% (Figure 1a). However, there were no significant differences between control, F+20B and F+100B treatments. The increase in the seed yield was accompanied by an increase in the total number of seeds per plant (Figure 1b). The number of seeds per plant significantly increased by 31.7% in the F+50B treatments whereas there were no significant differences between the number of seeds per plant in the control, F+20B and F+100B although F+20B and F+100B increased by 2.4% and 0.2% respectively over the control. In the second cropping field, biochar addition did not significantly influence seed yield and the number of seeds per plant. However, the positive effects of biochar addition were observed. In comparison with the control, the seed yield of sesame in the second cropping increased in the F+20B, F+50B, F+100B treatments by 25.1%, 10.1% and 23.0% respectively. Similarly, the total number of seeds per plant non-significantly increased in the F+20B, F+50B, F+100B treatments by 17.6%, 5.5% and 19.1% respectively. In both the first and second cropping, the addition of biochar improved sesame yield with F+50B and F+20B showing higher seed yield compared to the control.

The analysis of variance indicated that the biochar rate did not exhibit statistically significant influences on both the seed yield and total number of seeds per plant while cropping exhibited a statistically significant ($p < 0.01$) influence and the interaction between the biochar rate and cropping was significant ($p < 0.01$). As the biochar addition increased, the sesame seed yield and number of seeds per plant increased and then decreased. A positive nonlinear relationship between the two cropping seed yields, plant height and biochar rates were observed (Figure 2). The determination coefficient (R^2) and the significance for the overall plant height were higher than that of seed yield indicating growth of sesame was significantly influenced more than seed yield.

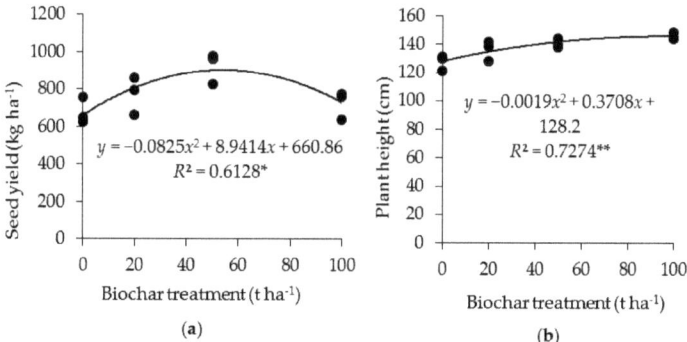

Figure 2. (a) The seed yield; (b) plant height response to rice husk biochar treatments. Relationships were fitted to first and second cropping data. * Significant at $p < 0.05$; ** Significant at $p < 0.01$.

The plant height showed a tendency to increase with the increasing rate of biochar without decreasing at high addition rates indicated by the nonlinear relationship. However, this increase in the plant height at the rate of biochar, above 100 t ha^{-1}, did not result into a significant increase in seed yield.

3.2. Effect of Rice Husk Biochar on the Leaf Tissue and Sesame Seed Nutrient Concentration

The leaf nutrient concentrations of sesame plants as affected by varying rates of biochar addition in the first and second cropping fields are shown in Table 2. The F+50B treatment significantly increased the leaf K concentration by 41.4% compared with the control in the first cropping and although there were no significant differences in leaf Mg concentration between control with biochar rates, the F+50B had a significantly higher leaf Mg than F+20B and F+100B in the first cropping. Biochar addition did not have a statistically significant effect on the leaf N, P and Ca concentrations in the first cropping. In the second cropping field, biochar addition did not have a statistically significant effect on any of the leaf nutrient concentrations. However, a non-significant increase in the N and K were observed compared to the control whereas leaf P increased in the F+20B, F+50B, but decreased in the F+100B. The leaf Ca and Mg concentrations were all non-significantly lower compared to the control.

Table 2. The leaf nutrient N, P, K, Ca and Mg of sesame under the different biochar treatments in different cropping.

Cropping	Biochar Rate	N (%)	K (%)	P (%)	Ca (%)	Mg (%)
First cropping	F	2.79 a	2.89 b	0.70 a	1.72 a	0.36 ab
	F+20B	2.99 a	3.10 ab	0.68 a	1.66 a	0.33 b
	F+50B	3.26 a	4.94 a	0.71 a	1.98 a	0.49 a
	F+100B	3.31 a	3.46 ab	0.67 a	1.58 a	0.32 b
Second cropping	F	2.91 a	2.41 a	0.44 a	1.53 a	0.35 a
	F+20B	3.04 a	2.69 a	0.49 a	1.47 a	0.34 a
	F+50B	3.15 a	2.75 a	0.51 a	1.46 a	0.32 a
	F+100B	3.11 a	2.49 a	0.44 a	1.33 a	0.30 a
Source of variation						
Biochar (B)		ns	**	ns	ns	*
Cropping (C)		ns	***	*	ns	*
B × C		ns	ns	ns	ns	*

Means followed by different lowercase letters within a column in the same cropping are significantly different $p < 0.05$ according to the Tukey test. *** Significant at $p < 0.001$; ** Significant at $p < 0.01$; * Significant at $p < 0.05$; ns, Non-significant.

The analysis of variance indicated that the biochar rate ($p < 0.01$) and cropping ($p < 0.001$) exhibited statistically significant influences on the concentration of leaf K, but no significant interaction whereas leaf Mg indicated significant ($p < 0.05$) interaction between biochar rates and cropping. The concentration of leaf P, Mg and K were significantly decreased by in the second cropping compared with first cropping irrespective of biochar addition.

The crude protein and mineral nutrient contents of sesame seed affected by varying rates of biochar addition in the first and second cropping fields are shown in Table 3.

The higher rates of biochar addition except in the F+100B, improved the crude protein, P, K, and Ca in the first cropping. However, there were no significant differences between biochar rates and control in the crude protein, P, Ca and Mg in first cropping. In comparison with control, the F+50B and F+100B treatments significantly increased the seed K content by 12.1% and 10.7% respectively whereas there was no significant difference between F+20B and control in the first cropping. However, no significant effect of biochar addition on seed K was observed in the second cropping. Biochar addition in the second cropping field significantly improved the crude protein content of sesame seeds. The crude protein of the F+50B and F+100B treatments were significantly higher than that of the control by 10.9% and 9.6%, respectively, whereas there were no significant differences between the F+20B treatment and control although an increase by 7.4% occurred in the F+20B compared with control. In the second cropping, biochar addition did not have a statistically significant effect on the seed P, K, Ca and Mg. The analysis of variance indicated that the biochar rate ($p < 0.01$) exhibited statistically significant influences on the seed K content, but no significant interaction between biochar rates and cropping. The content of seed crude protein, P, Ca and Mg were not significantly influenced by either biochar rates or cropping.

3.3. Effect of Biochar on Soil Physico-Chemical Properties in First and Second Cropping Fields

The soil physical properties of bulk density and porosity as affected by varying rates of biochar addition in the first and second cropping fields are shown in Figure 3. Biochar addition in the first and second cropping fields significantly decreased soil bulk density with increasing rates of biochar.

The F+100B treatments significantly decreased bulk density to 0.76 g cm^{-3} compared to the control (1.09 g cm^{-3}) in the first cropping and to 1.01 g cm^{-3} from 1.15 g cm^{-3} in the control of the second cropping field (Figure 3a).

Table 3. The seed macronutrient mineral nutrient contents of sesame seeds under the different biochar treatments in different cropping after harvesting.

Cropping	Biochar Rate	Crude Protein (%)	mg/100 g Seed			
			P	K	Ca	Mg
First cropping	F	20.0 a	612.6 a	744.7 b	1222.2 a	419.0 a
	F+20B	19.8 a	640.7 a	794.9 ab	1275.6 a	416.2 a
	F+50B	20.6 a	636.6 a	846.7 a	1321.5 a	409.6 a
	F+100B	19.1 a	617.6 a	833.8 a	1280.8 a	403.2 a
Second cropping	F	18.9 b	560.6 a	820.9 a	1252.2 a	396.9 a
	F+20B	20.4 ab	640.7 a	794.6 a	1419.7 a	398.7 a
	F+50B	21.2 a	617.3 a	857.6 a	1315.5 a	389.8 a
	F+100B	20.9 a	608.9 a	859.0 a	1382.3 a	422.4 a
Source of variation						
Biochar (B)		ns	ns	**	ns	ns
Cropping (C)		ns	ns	ns	ns	ns
B × C		ns	ns	ns	ns	ns

Means followed by different lowercase letters within a column in the same cropping are significantly different $p < 0.05$ according to the Tukey test. ** Significant at $p < 0.01$; ns, Non-significant.

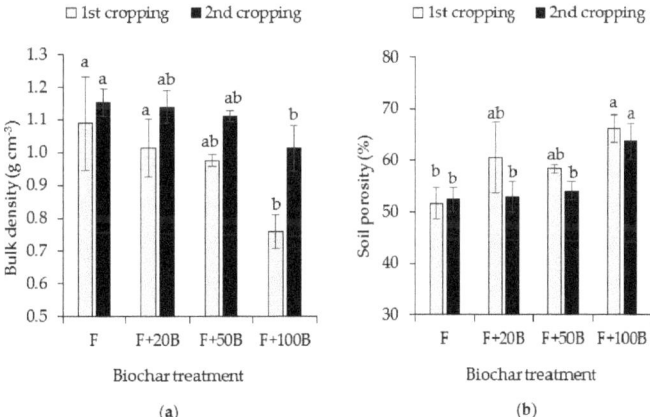

Figure 3. (a) The soil bulk density; (b) and soil porosity as affected by rice husk biochar treatments. The bars represent the standard deviation of triplicates. Different letters indicate significant difference ($p < 0.05$) among treatment means.

Similarly, the F+100B treatments significantly increased soil porosity to 66.1% compared to the control (51.6%) in the first cropping and to 63.7% from 52.4% in the control of the second cropping field (Figure 3b). In both the first and second cropping, the addition of biochar improved the physical property of the soil with higher rates of biochar addition. The analysis of variance indicated that the biochar rate exhibited statistically significant influences on both the soil bulk density ($p < 0.001$) and porosity ($p < 0.001$) (Table 4). Although the cropping had a significant influence on soil bulk density ($p < 0.001$) and porosity ($p < 0.05$), there were no significant interactions between the biochar rate and cropping.

Table 4. The soil chemical properties of sesame under the different biochar treatments in different cropping after harvesting.

Cropping	Biochar Rate	pH	EC (dSm^{-1})	TN (g kg^{-1})	C/N	P (mg kg^{-1})	CEC (cmol$_c$ kg^{-1})	Exchangeable Cations		
								K (mg kg^{-1})	Ca (mg kg^{-1})	Mg (mg kg^{-1})
First cropping	F	5.65 b	0.10 ab	2.40 a	10.1 c	13.3 a	9.9 b	190.7 c	3272.2 a	281.6 a
	F+20B	5.52 b	0.18 a	2.40 a	14.4 bc	46.4 a	12.3 ab	292.6 bc	3303.7 a	250.0 ab
	F+50B	5.97 ab	0.09 b	2.60 a	25.1 b	20.9 a	12.5 ab	408.0 b	2217.1 a	224.6 ab
	F+100B	6.38 a	0.12 ab	2.89 a	50.6 a	31.2 a	14.5 a	686.1 a	1683.5 a	198.1 b
Second cropping	F	5.54 a	0.08 b	2.09 b	10.3 c	40.2 a	11.5 b	179.5 b	3213.1 a	248.7 a
	F+20B	5.84 a	0.08 b	2.18 ab	15.6 bc	19.1 a	13.2 ab	158.8 b	1326.6 a	215.8 a
	F+50B	5.49 a	0.20 a	2.29 ab	19.1 b	36.9 a	12.9 ab	447.4 a	2575.9 a	244.7 a
	F+100B	5.68 a	0.09 b	2.53 a	31.4 a	33.5 a	14.3 a	386.6 a	1767.3 a	215.8 a
Source of variation										
Biochar (B)		ns	*	*	***	ns	***	***	ns	*
Cropping (C)		ns	ns	**	**	ns	ns	**	ns	ns
B × C		ns	***	ns	**	ns	ns	**	ns	ns

Means followed by different lowercase letters within a column in the same cropping are significantly different $p < 0.05$ according to the Tukey test. *** Significant at $p < 0.001$; ** Significant at $p < 0.01$; * Significant at $p < 0.05$; ns, Non-significant.

The soil chemical properties affected by varying rates of biochar addition in the first and second cropping fields are shown in Table 4.

Biochar addition in the first cropping significantly increased soil pH, EC, C/N ratio, exchangeable K and soil cation exchange capacity (CEC) compared to the control while the soil exchangeable Mg was significantly decreased. The F+100B treatment significantly increased soil pH by 0.73 units compared to the control (5.65) whereas the soil EC of the F+20B and F+50B were significantly different while of the control was not statically different from F+100B treatment. The C/N ratio significantly increased with increasing biochar rates by 4.3, 15.0 and 40.5 in the F+20B, F+50B and F+100B treatments respectively which indicated high carbon content in the soil with biochar treatments. The biochar addition significantly increased the soil exchangeable K with increasing rates of biochar treatments. The exchangeable K significantly increased by 101.9, 217.3 and 495.4 mg kg^{-1} in the F+20B, F+50B and F+100B respectively. Conversely, the exchangeable Mg significantly decreased by 31.55, 57.0 and 83.5 mg kg^{-1} in the F+20B, F+50B and F+100B respectively. However, there were no significant differences between the F+20B and F+50B treatments. The higher rates of biochar addition improved the soil CEC. The F+100B treatment significantly increased the CEC by 4.7 cmol$_c$ kg^{-1} compared to the control. In the first cropping, there were no significant effect of biochar addition on the available P and exchangeable Ca. However, these parameters showed higher values in the biochar addition treatments compared with the control, except exchangeable Ca which tended to decrease in the F+50B and F+100B treatments.

In the second cropping field, the addition of biochar significantly influenced soil EC, total N, C/N ratio, exchangeable K, and CEC whereas biochar did not show a significant effect on the soil pH, available P, exchangeable Ca or Mg. The F+50B treatment significantly increased soil EC by 0.12 dSm^{-1} compared to the control and no obvious significant differences were observed between the control, F+20B and F+100B treatments. The total N and C/N ratio significantly increased with increasing biochar rates and the in the F+100B treatments were significantly higher by 0.4 g kg^{-1} and 21.0 for total N and C/N ratio respectively when compared with control. The F+50B and F+100B treatments significantly increased exchangeable K by 267.9 and 207.1 mg kg^{-1} compared with the control. However, there was no significant difference between control and F+20B treatment. Similar to the first cropping, the higher rates of biochar addition improved the soil CEC. The F+100B treatment significantly increased the CEC by 0.23 cmol$_c$ kg^{-1} compared to the control.

In the second cropping, there was no obvious significant effect of biochar addition on the soil pH, available P and exchangeable Ca and Mg. However available P, exchangeable Ca and Mg were non-significantly decreased in the biochar addition treatments compared to the control. The analysis of variance indicated that the biochar rate exhibited statistically significant influences on EC, total N, C/N ratio, exchangeable K and Mg, and CEC whereas cropping significantly influenced total N, C/N ratio and exchangeable K (Table 4). There were also significant interactions between biochar rates and cropping for EC, C/N ratio, and exchangeable K that indicated an improvement of continuous cropping soil with biochar.

4. Discussion

4.1. Effect of Biochar on Soil Physico-Chemical Properties in First and Second Cropping Fields

Biochar additions significantly improved the soil physical properties of bulk density and porosity in both first and second cropping fields. A similar finding was reported in sesame that rice husk and saw dust biochar application at 10 t ha^{-1} significantly increased soil porosity and decreased the bulk density of a highly leached acidic Ultisol soil [36]. The increase in soil porosity and decrease in bulk density after biochar addition allow easy root penetration for water and nutrient absorption as a result of increased water holding capacity and reduced tensile strength of the soil [14–16]. This increase in soil porosity and decrease in bulk density is attributed to the low particle density and high porosity of the rice husk biochar compared with soil which enables the soil to hold more water and air [44].

In our study, the rice husk biochar used had a low bulk density of 0.29 g cm^{-3} which therefore greatly increased the soil porosity and decreased the soil bulk density after biochar addition on the upland field converted paddy soil. In addition, decreasing the bulk density and increasing soil porosity by biochar addition is important in the second cropping since continuous cropping results into a deterioration in the soil physical quality [45]. This suggests the integration of rice husk biochar in continuous cropping systems would improve soil conditions and overall crop performance.

Furthermore, biochar addition in the first cropping significantly increased soil pH with increasing rates of the rice husk biochar. Our result agrees with other findings that biochar addition increased soil pH [46,47]. Therefore, with rice husk biochar addition to slightly acidic upland field converted from paddy, the acidity would gradually be decreased due to the liming effect of the biochar. Rice husk biochar has been reported to increase soil pH of tea garden soil (acidic soil) from 3.33 to 3.63 [47]. In our study, increase in soil pH could be due to the alkaline nature of the biochar used since the rice husk biochar had a pH of 10.47 besides its high ash content (38.04%). The ash content of biochar plays important role in increasing soil pH and consequently determines the soil CEC [48]. This positive effect on soil pH was only observed in the first cropping. Our study showed that biochar in the second cropping did not have a significant effect on the soil pH. The pH of the control did not significantly differ from the biochar treatments in the second cropping. The addition of dolomite to increase soil pH and alleviate acidity in the second cropping possibly affected biochar liming effect. Whereas the biochar had a liming effect in the first cropping, this effect could have been offset by the addition of dolomite in the second cropping. This suggested that addition of both dolomite and biochar may not be good and could lead to loss of functional properties of biochar in the subsequent cropping although this hypothesis remains untested. However, Cornelissen et al. [30] reported that rice husk biochar addition on acidic Ultisol soil of Indonesia showed a less pronounced effect on the soil pH as the cropping season increased due to decreasing liming potential, which also agrees with our result. In addition to pH, biochar addition showed tendency to increase soil EC especially in the second cropping that could be attributed to the highly soluble minerals contained in the rice husk biochar that was readily added into the soil.

Although biochar addition significantly increased the soil total N and C/N ratio, possibly N immobilization occurred with increasing biochar rates. Similar findings have been reported that with biochar addition indicating an increase in the soil total C from 2.27% to 2.78% and total N from 0.24% to 0.25% [11]. In addition, research indicates that microbial immobilization of N with a C/N ratio of biochar above 16 occurs [46,49]. Furthermore, an increase in the soil total N, added from the rice husk biochar may not indicate N availability to plants and microbes [50]. Hence, the increase in the C/N ratio with increasing rates of biochar in both the first and second cropping could possibly be high enough to immobilize N thereby limiting N availability to sesame plant. The non-significant effect on soil available P and exchangeable Ca suggested the rice husk biochar had little ability to supply these nutrients. Possibly, the high Ca and P in the biochar was insoluble and remained in its micro-porous fabric of the biochar [51]. Limwikran et al. [51] reported that rice husk biochar could not readily release Ca into soil thus acted as a sink rather than a source for exchangeable Ca.

On the other hand, the biochar addition significantly increased the soil exchangeable K with increasing rates of biochar treatments whereas exchangeable Ca and Mg tended to decrease although not significantly different between biochar rates. This is contrary to Wang et al. [48] and Carter et al. [35] who reported the increase in the exchangeable cations Ca, Mg and K with rice husk biochar addition. In our study, only exchangeable K was significantly increased with increasing biochar rates. Wang et al. [47] reported that with biochar application, K levels in soil increased from 42 to 324 mg kg^{-1}. The rice husk biochar used had an ash content of 38.04% that was very high to increase soil K content in the soil. A similar increase in the exchangeable K was observed and attributed to high ash content in rice husk biochar and being a source of K itself (3640.73 mg kg^{-1} K in the rice husk biochar used) [12,52]. This increased exchangeable K with biochar addition could also be due to the increased soil pH which enhances release K into the soil [53]. Due to the increase in soil K, rice husk biochar could enhance crop growth and yield, including sesame on soils where K is a limiting factor.

Our study also showed that the higher rates of biochar addition improved the soil CEC which agrees with the findings of Ndor et al. [36] who reported increased CEC with rice husk and saw dust biochar in sesame cultivation. The biochar addition on a strongly acidic soil decreased in soil acidity through an increase in pH and the increasing the soil CEC thereby retaining nutrients into the soil for sesame [36]. In our study, the observed increase in the soil pH due to biochar addition and the rice husk biochar CEC (7.53 cmol$_c$ kg^{-1}) itself could also have increased the soil CEC. The increase in soil CEC levels is in agreements with findings of Laird et al. [54] who reported that biochar treatment increased soil CEC from by 4 to 30% more than the control while Jien and Wang [55] reported the CEC of a highly weathered soil was raised from 7.41 to 10.8 cmol$_c$ kg^{-1} after biochar application. In our study, the increase in the soil CEC is the indicator of nutrient retention suggesting that rice husk biochar could improve soil nutrient status in continuous cropping as observed in the second cropping. Similar results have shown that rice husk biochar addition to acid sulfate soils in Indonesia has been reported to increase CEC, in addition to improving soil porosity and exchangeable K [56]. Therefore, our result suggests that cultivation of sesame with rice husk biochar could improve not only soil physical properties promoting proper plant growth, but also hold sufficient nutrients through an increase in CEC upland fields converted paddy soils.

4.2. Effect of Rice Husk Biochar on the Leaf Tissue and Sesame Seed Nutrient Concentration

Biochar addition did not have a significant effect on leaf tissue N, P and Ca concentration and their contents in the sesame seeds. Overall, biochar addition had a significant influence on the leaf tissue K and Mg concentration of, and content of the seeds. Our result agrees with other findings, including a meta-analysis from different studies that biochar addition had no significant effect on plant tissue N and P concentrations whereas it increased the K tissue concentration [50,57]. Although the overall effect of biochar addition was not significant on crude protein, a significant increase in the crude protein content was observed in the second cropping suggesting an increase in protein synthesis in sesame. This could be attributed to the adequate supply of soil K from biochar allowing increased plant tissue K concentration. Potassium, K is an important nutrient that plays a significant role in protein synthesis in plants [58]. Hence, adequate supply of K from the rice husk biochar could enhance protein quality of sesame seeds. In addition, the increase in the crude protein content could be attributed to the slight increase in soil total N with biochar compared to the control in the second cropping. A similar finding shows that biochar addition (from acacia) of up to 50 t ha^{-1} in maize cropping system increased maize grain protein by 13% compared without biochar [26]. Therefore, it could be speculated that the protein content of sesame is increased with rice husk biochar addition although leaf tissue N concentration did not show statistical significance.

Biochar addition significantly increased leaf tissue K concentration and K content in the sesame seeds especially in the first cropping attributed to the high K content in the rice husk biochar reflected in the higher soil exchangeable K compared to the control. The soil K concentration in the 50 t ha^{-1} was twice the value in the control whereas it tripled in the 100 t ha^{-1} compared to control suggesting the rice husk biochar could possibly replace the 70 kg K$_2$O ha^{-1} inorganic fertilizer rate supplied to sesame in this study as a K source. A meta-analysis of several pieces of research shows that biochar addition treatments performed better than fertilizer at increasing plant tissue K concentration [50]. Hence, rice husk biochar with a high K content could be an alternative source of K fertilizers for sesame especially for resource poor farmers. Conversely, the leaf tissue Mg concentration tended to decrease compared to the control in the first cropping indicating that increasing biochar rates and cropping, the leaf tissue Mg concentration of sesame is significantly decreased that could influence seed yield at high rates of biochar addition. The content of Mg in the sesame seeds with biochar addition was also non-significantly lower than the control indicating the negative effect of high increasing rates of biochar addition on the sesame seed quality. Koyama and Hayashi [59] also found a similar decrease in the Mg concentrations in rice straw tissue with an increase in rice husk biochar addition rate up to 2 kg m^{-2} (equivalent to 20 t ha^{-1}). Our result also agrees Syuhada et al. [60] who found out that biochar addition at 10 and 15 g kg^{-1} from oil palm feedstock applied on to podsol soils significantly increased concentration and uptake of K while decreasing the Mg concentration in maize tissue. They attributed the low uptake of Mg to competitive ion effect between the uptakes of Mg and K. Although we did not measure the dry matter yield, the increased nutrient concentrations of K also implies high uptake of K since nutrient uptake is governed by the concentration of nutrients in plant tissues and the dry matter yield. Usually competitive ion effect occurs when there is a high concentration and uptake of K in plant tissue which decreases the concentration and uptake of Mg [61–63]. Major et al. [12] also found a similar decrease in the concentration of Mg in maize seeds in high wood biochar addition treatments (8 and 20 t ha^{-1}) attributed to declining stock of Mg in the soil due to this ion competition effect. In our study, the F+50B treatment significantly increased the leaf tissue K concentration by 41.4% compared with the control and increased seed K content by 12.1% in the first cropping, whereas Mg content in the seeds non-significantly decreased by 0.7%, 2.3% and 3.9% in the biochar addition of F+20B, F+50B and F+100B respectively. Therefore, the competitive ion effect is likely to occur when higher rates of the rice husk biochar with a high content of K is applied in sesame due to luxury consumption and decreased uptake of Mg. This could negatively affect yield and mineral quality of sesame.

4.3. Effect of Rice Husk Biochar on the Growth and Yield Components of Sesame

Biochar addition increased the overall sesame yield compared with the control that was consistent with Ndor et al. [36] who reported significant increase in the seed yield of sesame to 925 kg ha^{-1} in the 10 t ha^{-1} rice husk and sawdust biochar compared with the 595 kg ha^{-1} in the control in a field experiment. In particular, the yield increase with biochar addition, in the first cropping, suggested that sesame positively responds to biochar. In addition, 10 t ha^{-1} of coconut shell biochar added together with chicken manure to sesame resulted into higher seed weight per plant than the control on a sandy coastal soil [38]. In our study, the 50 t ha^{-1} (35.0% increase over control) in the first cropping significantly increased the sesame seed yield that could be attributed to increased number of seeds per plant rather than increase in the seed weight since there was no significant increase in the 1000-seed weight with biochar addition. The significant increase in the number of branches per plant with the biochar treatments in the first cropping could explain the increase in the total number of seeds per plant consequently increasing sesame yield. A similar increase in the number of branches per plant was reported that increased the seed yield of rapeseed (*Brassica napus* L.) with biochar addition [64]. In the second cropping, the F+20B (20 t ha^{-1}) non-significantly increased seed yield by 25.1% compared with control. The non-significant differences between the biochar treatments in the second cropping could be attributed to loss in the functional properties of the biochar although the seed yield increased in the 20 t ha^{-1}. This finding agrees with the recent report that rice husk biochar on acidic Ultisol soil of Indonesia at 15 t ha^{-1} significantly increased maize yield only in the first season, but the effect faded from the second season onwards [30].

The biochar addition also significantly increased sesame plant height in both the first and second cropping; and that could possibly explain the increase in the seed yield. It has been reported that poultry litter biochar at an optimal rate of 11.21 g kg^{-1} (equivalent to 22.4 t ha^{-1}) significantly increased sesame plant height [35], which is consistent with our results. The increase in the seed yield and plant height at 50 t ha^{-1} of rice husk biochar for sesame cropping on upland field converted paddy is consistent with other researchers who applied high biochar rates and obtained good crop performance [24,65]. Haefele et al. [24] reported that rice husk biochar applied at 4.13 kg m^{-2} (equivalent to 41.3 t ha^{-1}) in on Humic nitisols (pH 4.3) increased grain yield of rice by 16–35% at Sinilioan Phillipines whereas Schulz et al. [65] observed increased growth of oat plants with addition of 100 t ha^{-1} of composted biochar to sandy and loamy soil. Several pieces of field research indicated that biochar addition increased crop growth and yield [12,13]. For instance, cultivation on sandy soils using biochar increased maize yield by 150% and 98% over the control at rates of 15 t ha^{-1} and 20 t ha^{-1} respectively [23]. Zhang et al. [25] also reported an increase in rice yields of 14% over control in paddy soils with wheat straw biochar rate of 40 t ha^{-1}. However, the highest rate of 100 t ha^{-1} tended to have a negative effect on sesame in our study, which is consistent with the finding of Chan et al. [14]. The decrease in the seed yield in the 100 t ha^{-1} compared to 50 t ha^{-1} suggested the biochar addition rate had exceeded the beneficial amount. At this rate, the possible release of toxic substances like heavy metals and polycyclic aromatic hydrocarbons (PAHs) from biochar could have suppressed plant growth [66]. In addition, the low yield in the 100 t ha^{-1} is likely as a result of nutrient imbalances and N immobilization [67,68]. The decrease in the seed yield at 100 t ha^{-1} could be partly explained by the increased adsorption of available inorganic N at this high biochar addition rate. For instance, research shows that biochar addition improves the retention capacity of NH_4^+-N through enhanced CEC [69]. In our study, the increased CEC could have increased ammonium adsorption in the first cropping with 100 t ha^{-1} rice husk biochar rate. However, the adsorption of inorganic N onto biochar surfaces could decrease ammonia and nitrate losses from soil, but could as well potentially lead to the slow release of these nutrients to plants [70].

Although, the leaf tissue N concentration tended to increase non-significantly with biochar rates in the first cropping, the increase in growth and yield could be entirely attributed to an increase in the leaf tissue K concentration in sesame plant. Our study agrees with a meta-analysis of biochar research showing that biochar addition treatments performed better than fertilizer at increasing plant tissue K concentration and plant tissue N is unaffected by biochar addition thereby influencing yield [50]. On the other hand, our results indicated that leaf K concentration was significantly reduced by continuous cropping, but with the biochar addition, the concentration of K significantly increases. Therefore, the higher seed yield, number of seeds per plant and plant height compared to the control in the second cropping is attributed to this increased leaf K suggesting seed yield decline under continuous cropping could be recovered by adding more K fertilizer. This also suggests future research should consider comparing and contrasting biochar addition with more K fertilizer rates for sesame cropping. The increased seed yield was due to this increased K concentration due to the rice husk biochar addition. Moreover, the leaf tissue K concentration was above the adequate level of 2.4% required for sesame growth since K plays a significant role in increasing the internodes lengths consequently increase in sesame plant height [43]. However, the lower leaf tissue K concentrations in the second than first cropping accompanied by the low soil exchangeable K suggested a decrease in availability of K could have led to the lack of positive effect of rice husk biochar rates on sesame yield in the second cropping. Similar effects of lack of positive effect of biochar on crop yields have been attributed to decreasing nutrient contents in biochar addition after its addition [31]. This could suggest that the benefit of K addition from rice husk biochar could decrease over time affecting sesame yield. Therefore, the lack of non-significant effect on leaf tissue K concentration in the second cropping suggested K was the most determinant factor on growth and seed yield.

Furthermore, the non-significant effect on K concentration in sesame in combination with factors that affected the rice husk biochar properties in the second cropping could have contributed to low yields in the second cropping. For instance, the first cropping field had fresh biochar applied whereas the second cropping field had old biochar that could have influenced sesame yield due to biochar aging effect on the temperate soil. The meta-analyses by Biederman and Harpole [50] shows that the effect of biochar addition is less pronounced on temperate climate soils and freshly added biochar could perform better than the old or aged biochar. In our study, the seed yield did not show significant differences between biochar rates in the second cropping which also agrees with Persaud et al. [67] who reported no beneficial effect of rice husk biochar in on yield in the second cropping. The authors observed increase in above ground biomass of pak choy (*Brassica rapa* subsp. *chinensis*) by 32.81% in the 25 t ha^{-1} application rate (0, 5, 25 and 50 t ha^{-1}) to acidic Tabela sandy soil of Guyana in the first cropping compared with the control and attributed to increasing soil pH, exchangeable cations, CEC, and decrease in soil bulk density that is also consistent with our results. Therefore, improvement in the soil chemical properties and physical properties by the rice husk biochar enhanced sesame productivity especially in the first cropping. However, these benefits could deteriorate with an increase in the number of cropping as the biochar becomes old. For instance, changes in the physico-chemical properties as a result of aging when incorporated into the soil have been reported [71]. In our study, the porosity tended to decrease in the second cycle suggesting biochar particles had been crushed or broken down as a result of continuous cropping during tillage operation. The particles of biochar may also break and become smaller with time due to the physical interaction with drainage water [72]. With tillage, the particles could have been degraded and the ashes contents of the biochar increased leading to easy leaching in drainage water affecting soil pH that depends on the ash content. Thus, the lack of significant effect of the rice husk biochar in the second cropping could partly be due to leaching of the alkaline ashes [73,74]. Furthermore, the no effect on the second cycle could be attributed to the loss in the properties of biochar to adsorption and immobilization of heavy metals, polycyclic aromatic hydrocarbons (PAHs), phthalates etc. from the soil [75]. A study found that fresh rice husk biochar had a higher adsorption capacity for toxic compounds than aged one [76]. In addition, rice husk biochar addition up to 30 t ha^{-1} is reported to have a high affinity to removed cadmium from aqueous solutions when mixed in soil, attributed to high surface charge (net negative charge) of the bio-sorbents, thereby eliminating inhibition effect of cadmium (Cd^{2+}) on plant growth and improve yield [77]. Therefore, the fresh biochar as observed in the first cropping could have absorbed toxic heavy metals and other compounds that would hinder sesame growth and yield. The decrease in ability to increase pH, adsorb potential heavy metals, and overall changes in the physical properties of the rice husk biochar could be possible factors that led to the low yield in the second cropping.

Nonetheless, the rice husk biochar addition to sesame improved growth and yield at increasing rates; it could be recommended to apply rates not exceeding 50 t ha^{-1}. Although sesame is considered a high value oilseed crop, higher rates than 50 t ha^{-1} are not economically feasible under field conditions. The higher biochar rates are not economically feasible in most farming systems due to high costs [78,79]. Given the temperate climate where soils have favorable properties for plant production than tropical soils, large quantities of biochar may be required to achieve significant positive effects on yield as observed in this study [68]. However, further studies are needed to determine the optimal rice husk biochar rate while considering the cost and benefits of sesame cultivation on upland fields converted paddy.

5. Conclusions

The results demonstrated that sesame seed yield, plant growth and mineral content are improved with the biochar addition on upland field converted paddy. The biochar addition increased plant height with significant interaction in both cropping fields whereas the seed yield was only significantly influenced in the first cropping. The higher rate of biochar addition significantly improved the seed yield of sesame in the first cropping whereas in the second cropping field, biochar addition did not significantly influence seed yield and the number of seeds per plant. Among the seed mineral nutrients, K content was most increased by biochar. The overall improvement in the sesame growth, seed yield and mineral contents especially K was attributed to mainly increased K availability, soil pH, CEC, improved porosity and bulk density. Our study also suggests that rice husk biochar addition may not have a long lasting effect on sesame yield on upland field converted paddy since the positive effect of biochar tended to fade in the second cropping as its biochar aged suggesting one-time application would not be sufficient. However, further investigations are still required to clarify the non-significant influence of biochar addition on seed yield and growth of sesame in the second cropping when biochar had been incorporated in the first cropping to uncover the mechanisms underlying these processes with biochar addition in long-term field trials.

Author Contributions: C.W, N.O, M.K. and E.N. conceived and designed the experiment; C.W., D.B., D.S., T.H., T.I. and R.A. performed the experiments; C.W. and D.B. analyzed the data; T.M., S.Y. and E.N., T.M., S.Y. and E.N supervised the research and contributed in the discussion of the results; C.W. wrote the paper; E.N. revised the final draft manuscript.

Acknowledgments: We are grateful for the support from the Ministry of Education, Culture, Sports, Science and Technology, MEXT of Japan under the Japanese Government MEXT scholarship of the first author.

Conflicts of Interest: The authors declare no conflict of interest.

References

1. Ashri, A. Sesame. In *Oil Crops of the World*; Robbelen, G., Downey, R.K., Ashri, A., Eds.; McGraw Hill: New York, NY, USA, 1989; pp. 375–387.
2. Borchani, C.; Besbes, S.; Blecker, C.; Attia, H. Chemical Characteristics and Oxidative Stability of Sesame Seed, Sesame Paste, and Olive Oils. *J. Agric. Sci. Technol.* **2010**, *12*, 585–596.
3. Nzikou, J.M.; Matos, L.; Kalou, G.B.; Ndangui, C.B.; Tobi, N.P.G.P.; Kimbonguila, A.; Silou, T.; Linder, M.; Desobry, S. Chemical Composition on the Seeds and Oil of Sesame (*Sesamum indicum* L.) Grown in Congo-Brazzaville. *Adv. J. Food Sci. Technol.* **2009**, *1*, 6–11.
4. Dossa, K.; Diouf, D.; Wang, L.; Wei, X.; Zhang, Y.; Niang, M.; Fonceka, D.; Yu, J.; Mmadi, M.A.; Yehouessi, L.W.; et al. The Emerging Oilseed Crop Sesamum indicum Enters the "Omics" Era. *Front. Plant Sci.* **2017**, *8*, 1–16. [CrossRef] [PubMed]
5. Yasumoto, S.; Katsuta, M. Breeding a high-lignan-content sesame cultivar in the prospect of promoting metabolic functionality. *Japan Agric. Res. Q.* **2006**, *40*, 123–129. [CrossRef]
6. FAOSTAT. Food and Agriculture Statistical Database. 2016. Available online: http://www.fao.org/faostat/en/#data/QC/visualize (accessed on 20 July 2018).
7. Chono, S.; Maeda, S.; Kawachi, T.; Imagawa, C.; Buma, N. Optimization model for cropping-plan placement in paddy fields considering agricultural profit and nitrogen load management in Japan. *Paddy Water Environ.* **2012**, *10*, 113–120. [CrossRef]
8. MAFF. *FY2013 Annual Report on Food, Agriculture and Rural Areas in Japan Summary Ministry of Agriculture, Forestry and Fisheries*; MAFF: Tokyo, Japan, 2014.
9. Nishida, M. Decline in fertility of paddy soils induced by paddy rice and upland soybean rotation, and measures against the decline. *Japan Agric. Res. Q.* **2016**, *50*, 87–94. [CrossRef]
10. Lehmann, J.; Joseph, S. Biochar for environmental management: An introduction. *Biochar Environ. Manag. Sci. Technol.* **2009**, *1*, 1–12. [CrossRef]
11. Jones, D.L.; Rousk, J.; Edwards-Jones, G.; DeLuca, T.H.; Murphy, D.V. Biochar-mediated changes in soil quality and plant growth in a three year field trial. *Soil Biol. Biochem.* **2012**, *45*, 113–124. [CrossRef]
12. Major, J.; Rondon, M.; Molina, D.; Riha, S.J.; Lehmann, J. Maize yield and nutrition during 4 years after biochar application to a Colombian savanna oxisol. *Plant Soil* **2010**, *333*, 117–128. [CrossRef]
13. Lehmann, J.; Gaunt, J.; Rondon, M. Bio-char sequestration in terrestrial ecosystems—A review. *Mitig. Adapt. Strateg. Glob. Chang.* **2006**, *11*, 403–427. [CrossRef]
14. Chan, K.Y.; Van Zwieten, L.; Meszaros, I.; Downie, A.; Joseph, S.; Journal, A. Agronomic values of green waste biochar as a soil amendment. *Aust. J. Soil Res.* **2007**, *45*, 629–634. [CrossRef]
15. Nelissen, V.; Ruysschaert, G.; Manka'Abusi, D.; D'Hose, T.; De Beuf, K.; Al-Barri, B.; Cornelis, W.; Boeckx, P. Impact of a woody biochar on properties of a sandy loam soil and spring barley during a two-year field experiment. *Eur. J. Agron.* **2015**, *62*, 65–78. [CrossRef]
16. Lu, S.G.; Sun, F.F.; Zong, Y.T. Effect of rice husk biochar and coal fly ash on some physical properties of expansive clayey soil (Vertisol). *Catena* **2014**, *114*, 37–44. [CrossRef]
17. Lehmann, J.; Pereira da Silva, J.; Steiner, C.; Nehls, T.; Zech, W.; Glaser, B. Nutrient availability and leaching in an archaeological Anthrosol and a\rFerralsol of the Central Amazon basin: Fertilizer, manure and charcoal\ramendments. *Plant Soil* **2003**, *249*, 343–357. [CrossRef]
18. Xu, C.Y.; Hosseini-Bai, S.; Hao, Y.; Rachaputi, R.C.N.; Wang, H.; Xu, Z.; Wallace, H. Effect of biochar amendment on yield and photosynthesis of peanut on two types of soils. *Environ. Sci. Pollut. Res.* **2015**, *22*, 6112–6125. [CrossRef] [PubMed]
19. Rogovska, N.; Laird, D.A.; Rathke, S.J.; Karlen, D.L. Biochar impact on Midwestern Mollisols and maize nutrient availability. *Geoderma* **2014**, *230–231*, 34–347. [CrossRef]

20. Elmer, W.H.; Pignatello, J.J. Effect of Biochar Amendments on Mycorrhizal Associations and Fusarium Crown and Root Rot of Asparagus in Replant Soils. *Plant Dis.* **2011**, *95*, 960–966. [CrossRef]
21. van Zwieten, L.; Kimber, S.; Morris, S.; Chan, K.Y.; Downie, A.; Rust, J.; Joseph, S.; Cowie, A. Effects of biochar from slow pyrolysis of papermill waste on agronomic performance and soil fertility. *Plant Soil* **2010**, *327*, 235–246. [CrossRef]
22. Woldetsadik, D.; Drechsel, P.; Marschner, B.; Itanna, F.; Gebrekidan, H. Effect of biochar derived from faecal matter on yield and nutrient content of lettuce (*Lactuca sativa*) in two contrasting soils. *Environ. Syst. Res.* **2018**, *6*. [CrossRef]
23. Uzoma, K.C.; Inoue, M.; Andry, H.; Fujimaki, H.; Zahoor, A.; Nishihara, E. Effect of cow manure biochar on maize productivity under sandy soil condition. *Soil Use Manag.* **2011**, *27*, 205–212. [CrossRef]
24. Haefele, S.M.; Konboon, Y.; Wongboon, W.; Amarante, S.; Maarifat, A.A.; Pfeiffer, E.M.; Knoblauch, C. Effects and fate of biochar from rice residues in rice-based systems. *Field Crops Res.* **2011**, *121*, 430–440. [CrossRef]
25. Zhang, A.; Cui, L.; Pan, G.; Li, L.; Hussain, Q.; Zhang, X.; Zheng, J.; Crowley, D. Effect of biochar amendment on yield and methane and nitrous oxide emissions from a rice paddy from Tai Lake plain, China. *Agric. Ecosyst. Environ.* **2010**, *139*, 469–475. [CrossRef]
26. Ali, K.; Arif, M.; Shah, F.; Shehzad, A.; Munsif, F.; Mian, I.A.; Mian, A.A. Improvement in maize (*Zea mays* L.) growth and quality through integrated use of biochar. *Pak. J. Bot.* **2017**, *49*, 85–94.
27. Varela Milla, E.B.; Rivera, W.-J.; Huang, C.-C.; Chien, Y.-M.W. Agronomic properties and characterization of rice husk and wood biochars and their effect on the growth of water spinach in a field tes. *J. Soil Sci. Plant Nutr.* **2013**, *13*, 251–266. [CrossRef]
28. Alburquerque, J.A.; Salazar, P.; Barrón, V.; Torrent, J.; Del Campillo, M.D.C.; Gallardo, A.; Villar, R. Enhanced wheat yield by biochar addition under different mineral fertilization levels. *Agron. Sustain. Dev.* **2013**, *33*, 475–484. [CrossRef]
29. Baronti, S.; Alberti, G.; Vedove, G.D.; di Gennaro, F.; Fellet, G.; Genesio, L.; Miglietta, F.; Peressotti, A.; Vaccari, F.P. The biochar option to improve plant yields: First results from some field and pot experiments in Italy. *Ital. J. Agron.* **2010**, *5*, 3–11. [CrossRef]
30. Cornelissen, G.; Nurida, N.L.; Hale, S.E.; Martinsen, V.; Silvani, L.; Mulder, J. Science of the Total Environment Fading positive effect of biochar on crop yield and soil acidity during five growth seasons in an Indonesian Ultisol. *Sci. Total Environ.* **2018**, *634*, 561–568. [CrossRef]
31. Steiner, C.; Teixeira, W.G.; Lehmann, J.; Nehls, T.; de Macêdo, J.L.V.; Blum, W.E.H.; Zech, W. Long term effects of manure, charcoal and mineral fertilization on crop production and fertility on a highly weathered Central Amazonian upland soil. *Plant Soil* **2014**, *291*, 275–290. [CrossRef]
32. Carter, S.; Shackley, S.; Sohi, S.; Suy, T.; Haefele, S. The Impact of Biochar Application on Soil Properties and Plant Growth of Pot Grown Lettuce (Lactuca sativa) and Cabbage (Brassica chinensis). *Agronomy* **2013**, *3*, 404–418. [CrossRef]
33. Jin, Z.; Chen, C.; Chen, X.; Jiang, F.; Hopkins, I.; Zhang, X.; Han, Z.; Billy, G.; Benavides, J. Field Crops Research Soil acidity, available phosphorus content, and optimal biochar and nitrogen fertilizer application rates: A five-year field trial in upland red soil, China. *Field Crops Res.* **2019**, *232*, 77–87. [CrossRef]
34. Mukherjee, A.; Zimmerman, A.R.; Hamdan, R.; Cooper, W.T. Physicochemical changes in pyrogenic organic matter (biochar) after 15 months of field aging. *Solid Earth* **2014**, *5*, 693–704. [CrossRef]
35. Furtado, G.F.; Chaves, L.H.G.; Lima, G.S.; Andrade, E.M.G.; Souza, L.P. Growth of sesame in function with NPK and poultry litter biochar. *Int. J. Curr. Res.* **2016**, *8*, 38499–38504.
36. Ndor, E.; Jayeoba, O.; Asadu, C. Effect of Biochar Soil Amendment on Soil Properties and Yield of Sesame Varieties in Lafia, Nigeria. *Am. J. Exp. Agric.* **2015**, *9*, 1–8. [CrossRef]
37. Nurhayati, D.R. The effect of coconut shell charcoal on sesame (*Sesamum indicum* L.) yield grown on coastal sandy land area in bantul, indonesia. *Int. Res. J. Eng. Technol.* **2017**, *4*, 1035–1041.39.
38. Hussain, M.; Farooq, M.; Nawaz, A.; Al-Sadi, A.M.; Solaiman, Z.M.; Alghamdi, S.S.; Ammara, U.; Ok, Y.S.; Siddique, K.H.M. Biochar for crop production: Potential benefits and risks. *J. Soils Sediments* **2017**, *17*, 685–716. [CrossRef]
39. FAO/IIASA/ISRIC/ISS-CAS/JRC. *Harmonized World Soil Database (Version 1.2)*; FAO: Rome, Italy; IIASA: Laxenburg, Austria, 2012.
40. Truog, E. The determination of the readily available phosphorous of soils. *Agron. J.* **1930**, *22*, 874–882. [CrossRef]

41. Chapman, H.D. Cation–exchange capacity. In *Methods of Soil Analysis–Chemical and Microbiological Properties*; Black, C.A., Ed.; American Society of Agronomy: Madison, WI, USA, 1965; pp. 891–901.
42. Estefan, G.; Sommer, R.; Ryan, J. *Methods of Soil, Plant, and Water Analysis: A Manual for the West Asia and North*; International Center for Agricultural Research in the Dry Area (ICARDA): Beirut, Lebanon, 2013.
43. Mitchell, G.A.; Bingham, F.T.; Yermanos, D.M. Growth, mineral composition and seed characteristics of sesame as affected by nitrogen, phosphorus and potassium nutrition. *Soil Sci. Am. Proc.* **1974**, *38*, 925–931. [CrossRef]
44. Downie, A.; Crosky, A.; Munroe, P. Physical properties of biochar. In *Biochar for Environmental Management: Science and Technology*; Lehmann, J., Joseph, S., Eds.; Earthscan: London, UK, 2009; pp. 13–32.
45. Reeves, D.W. The role of soil organic matter in maintaining soil quality in continuous cropping systems. *Soil Tillage Res.* **1997**, *43*, 131–167. [CrossRef]
46. Alburquerque, J.A.; Calero, J.M.; Barrón, V.; Torrent, J.; del Campillo, M.C.; Gallardo, A.; Villar, R. Effects of biochars produced from different feedstocks on soil properties and sunflower growth. *J. Plant Nutr. Soil Sci.* **2014**, *177*, 16–25. [CrossRef]
47. Wang, Y.; Yin, R.; Liu, R. Characterization of biochar from fast pyrolysis and its effect on chemical properties of the tea garden soil. *J. Anal. Appl. Pyrolysis* **2014**, *110*, 375–381. [CrossRef]
48. Sollins, P.; Robertson, G.P.; Uehara, G. Nutrient mobility in variable- and permanent- charge soils. *Biogeochemistry* **1988**, *6*, 181–199. [CrossRef]
49. Bargmann, I.; Rillig, M.C.; Kruse, A.; Greef, J.M.; Kücke, M. Effects of hydrochar application on the dynamics of soluble nitrogen in soils and on plant availability. *J. Plant Nutr. Soil Sci.* **2014**, *177*, 48–58. [CrossRef]
50. Biederman, L.A.; Harpole, W.S. Biochar and its effects on plant productivity and nutrient cycling: A meta-analysis. *GCB Bioenergy* **2013**, *5*, 202–214. [CrossRef]
51. Limwikran, T.; Kheoruenromne, I.; Suddhiprakarn, A.; Prakongkep, N.; Gilkes, R.J. Dissolution of K, Ca, and P from biochar grains in tropical soils. *Geoderma* **2018**, *312*, 139–150. [CrossRef]
52. Abrishamkesh, S.; Gorji, M.; Asadi, H.; Bagheri-Marandi, G.H.; Pourbabaee, A.A. Effects of rice husk biochar application on the properties of alkaline soil and lentil growth. *Plant Soil Environ.* **2015**, *62*, 475–482. [CrossRef]
53. Atkinson, C.J.; Fitzgerald, J.D.; Hipps, N.A. Potential mechanisms for achieving agricultural benefits from biochar application to temperate soils: A review. *Plant Soil* **2010**, *337*, 1–18. [CrossRef]
54. Laird, D.A.; Fleming, P.; Davis, D.D.; Horton, R.; Wang, B.; Karlen, D.L. Impact of biochar amendments on the quality of a typical Midwestern agricultural soil. *Geoderma* **2010**, *158*, 443–449. [CrossRef]
55. Jien, S.H.; Wang, C.S. Effects of biochar on soil properties and erosion potential in a highly weathered soil. *Catena* **2013**, *110*, 225–233. [CrossRef]
56. Masulili, A.; Utomo, W.H.; MS, S. Rice Husk Biochar for Rice Based Cropping System in Acid Soil 1. The Characteristics of Rice Husk Biochar and Its Influence on the Properties of Acid Sulfate Soils and Rice Growth in West Kalimantan, Indonesia. *J. Agric. Sci.* **2010**, *2*. [CrossRef]
57. Si, L.; Xie, Y.; Ma, Q. The Short-Term Effects of Rice Straw Biochar, Nitrogen and Phosphorus Fertilizer on Rice Yield and Soil Properties in a Cold Waterlogged Paddy Field. *Sustainability* **2018**, *10*, 537. [CrossRef]
58. Hasanuzzaman, M.; Bhuyan, M.H.M.B.; Nahar, K.; Awal, A.; Masud, C.; Fujita, M. Potassium: A Vital Regulator of Plant Responses and Tolerance to Abiotic Stresses. *Agronomy* **2018**, *8*, 31. [CrossRef]
59. Koyama, S.; Hayashi, H. Rice yield and soil carbon dynamics over three years of applying rice husk charcoal to an Andosol paddy field. *Plant Prod. Sci.* **2017**, *20*, 176–182. [CrossRef]
60. Syuhada, A.B.; Shamshuddin, J.; Fauziah, C.I.; Rosenani, A.B.; Arifin, A. Biochar as soil amendment: Impact on chemical properties and corn nutrient uptake in a Podzol. *Can. J. Soil Sci.* **2016**, *412*, 1–13. [CrossRef]
61. Weil, R.R.; Brady, N.C. *The Nature and Properties of Soils*, 15th ed.; Pearson Education Ltd.: London, UK, 2016; p. 381.
62. Zemanová, V.; Břendová, K.; Pavlíková, D.; Kubátová, P.; Tlustoš, P. Effect of biochar application on the content of nutrients(Ca, Fe, K, Mg, Na, P) and amino acids in subsequently growing spinach and mustard. *Plant Soil Environ.* **2017**, *63*, 322–327. [CrossRef]
63. Butnan, S.; Deenik, J.L.; Toomsan, B.; Antal, M.J.; Vityakon, P. Biochar characteristics and application rates affecting corn growth and properties of soils contrasting in texture and mineralogy. *Geoderma* **2015**, *237*, 105–116. [CrossRef]

64. Khan, I.A.A. Biochar application and shoot cutting duration (days) influenced growth, yield and yield contributing parameters of *Brassica napus* L. *J. Biol. Agric. Healthc.* **2015**, *7*, 104–108.
65. Schulz, H.; Dunst, G.; Glaser, B. Positive effects of composted biochar on plant growth and soil fertility. *Agron. Sustain. Dev.* **2013**, *33*, 817–827. [CrossRef]
66. Kuppusamy, S.; Thavamani, P.; Megharaj, M.; Venkateswarlu, K.; Naidu, R. Agronomic and remedial benefits and risks of applying biochar to soil: Current knowledge and future research directions. *Environ. Int.* **2016**, *87*, 1–12. [CrossRef]
67. Persaud, T.; Homenauth, O.; Fredericks, D.; Hamer, S. Effect of Rice Husk Biochar as an Amendment on a Marginal Soil in Guyana. *J. World Environ.* **2018**, *8*, 20–25. [CrossRef]
68. Borchard, N.; Siemens, J.; Ladd, B.; Möller, A.; Amelung, W. Soil & Tillage Research Application of biochars to sandy and silty soil failed to increase maize yield under common agricultural practice. *Soil Tillage Res.* **2014**, *144*, 184–194.
69. Clough, T.; Condron, L.M.; Kammann, C.; Muller, C. A Review of Biochar and Soil Nitrogen Dynamics. *Agronomy* **2013**, *3*, 275–293. [CrossRef]
70. Haider, G.; Steffens, D.; Moser, G.; Müller, C.; Kammann, C.I. Biochar reduced nitrate leaching and improved soil moisture content without yield improvements in a four-year field study. *Agric. Ecosyst. Environ.* **2017**, *237*, 80–94. [CrossRef]
71. Sohi, S.P.; Krull, E.; Bol, R. A Review of Biochar and Its Use and Function in Soil. *Adv. Agron.* **2010**, *105*, 47–82.
72. Spokas, K.A.; Novak, J.M.; Masiello, C.A.; Johnson, M.G.; Colosky, E.C.; Ippolito, J.A.; Trigo, C. Physical Disintegration of Biochar: An Overlooked Process. *Environ. Sci. Technol. Lett.* **2014**, *1*, 326–332. [CrossRef]
73. Lehmann, J.; Rondon, M. Bio-Char Soil Management on Highly Weathered Soils in the Humid Tropics. In *Biological Approaches to Sustainable Soil Systems*; Uphoff, N., Ed.; CRC Press: Boca Ration, FL, USA, 2006; pp. 517–530.
74. Glaser, B.; Lehmann, J.; Zech, W. Ameliorating physical and chemical properties of highly weathered soils in the tropics with charcoal—A review. *Biol. Fertil. Soils* **2002**, *35*, 219–230. [CrossRef]
75. Hagemann, N.; Harter, J.; Kaldamukova, R.; Guzman-bustamante, I.; Ruser, R.; Graeff, S. Does soil aging affect the N$_2$O mitigation potential of biochar? A combined microcosm and field study. *GCB Bioenergy* **2017**, *9*, 953–964. [CrossRef]
76. Khorram, M.S.; Lin, D.; Zhang, Q.; Zheng, Y.; Fang, H.; Yu, Y. Effects of aging process on adsorption—desorption and bioavailability of fomesafen in an agricultural soil amended with rice hull biochar. *J. Environ. Sci.* **2016**, *56*, 180–191. [CrossRef]
77. Mahmoud, A.H.; Saleh, M.E.; Abdel-Salam, A.A. Effect of Rice Husk Biochar on Cadmium Immobilization in Soil and Uptake by Wheat Plant Grown on Lacustrine Soil. *Alex. J. Agric. Res.* **2011**, *56*, 117–125.
78. Woolf, D.; Amonette, J.E.; Street-perrott, F.A.; Lehmann, J.; Joseph, S. climate change. *Nat. Commun.* **2010**, *1*, 1–9. [CrossRef]
79. Clare, A.; Shackley, S.; Joseph, S.; Hammond, J. Competing uses for China's straw: The economic and carbon abatement potential of biochar. *GCB Bioenergy* **2015**, *7*, 1272–1282. [CrossRef]

© 2019 by the authors. Licensee MDPI, Basel, Switzerland. This article is an open access article distributed under the terms and conditions of the Creative Commons Attribution (CC BY) license (http://creativecommons.org/licenses/by/4.0/).

Brief Report

Characterization of Biochars Produced from Dairy Manure at High Pyrolysis Temperatures

Wen-Tien Tsai [1],*, Po-Cheng Huang [2] and Yu-Quan Lin [2]

[1] Graduate Institute of Bioresources, National Pingtung University of Science and Technology, Pingtung 912, Taiwan
[2] Department of Environmental Science and Engineering, National Pingtung University of Science and Technology, Pingtung 912, Taiwan; mike299123@gmail.com (P.-C.H.); wsx55222525@gmail.com (Y.-Q.L.)
* Correspondence: wttsai@mail.npust.edu.tw; Tel.: +886-8-770-3202

Received: 20 August 2019; Accepted: 9 October 2019; Published: 14 October 2019

Abstract: In this work, the thermochemical analyses of dairy manure (DM), including the proximate analysis, ultimate (elemental) analysis, calorific value, thermogravimetric analysis (TGA), and inorganic elements, were studied to evaluate its potential for producing DM-based char (DMC) with high porosity. The results showed that the biomass should be an available precursor for producing biochar materials based on its high contents of carbon (42.63%) and volatile matter (79.55%). In order to characterize their pore properties, the DMC products produced at high pyrolysis temperatures (500–900°C) were analyzed using surface area and porosity analyzer, pycnometer, and scanning electron microscopy-energy dispersive X-ray spectroscopy (SEM-EDS). The values of pore properties for the DMC products increased with an increase in pyrolysis temperature, leading to more pore development and condensed aromatic cluster at elevated temperatures. Because of the microporous and mesoporous structures from the N_2 adsorption–desorption isotherms with the hysteresis loops (H4 type), the Brunauer–Emmett–Teller (BET) surface area of the optimal biochar (DMC-900) was about 360 m^2/g, which was higher than the data reported in the literature. The highly porous structure was also seen from the SEM observations. More significantly, the cation exchange capacity (CEC) of the optimal DMC product showed a high value of 57.5 ± 16.1 cmol/kg. Based on the excellent pore and chemical properties, the DMC product could be used as an effective amendment and/or adsorbent for the removal of pollutants from the soil media and/or fluid streams.

Keywords: dairy manure; thermochemical property; pyrolysis; biochar; pore property

---

1. Introduction

Since the Kyoto Protocol adopted on 11 December 1997, humans are beginning to actively focus on mitigating greenhouse gas (GHG) emissions because it is irreversibly changing the planet's ecosystems via global warming. In this regard, the livestock sector plays a significant role in the globally anthropogenic emissions of GHG, including carbon dioxide (CO_2), methane (CH_4), and nitrous oxide (N_2O) [1]. According to the statistical data by the Food and Agriculture Organization [2], cattle-raising and its resulting manure are the most important contributors to the sector's GHG emissions, accounting for 65% of the livestock sector emissions. Regarding the main sources of emissions from the sector, the deposition of cattle manure on pastures and manure storage and processing generate substantial amounts of GHG. In order to mitigate GHG emissions and upgrade the recycling of nutrients and lignocellulosic sources, the thermochemical processes (e.g., pyrolysis) can convert cattle manure into renewable chemicals like char or biochar [3]. More importantly, this manure treatment can gain several positive benefits, including GHG emission reduction, biomass nutrients recycling, biofuels and carbon materials production, and waste management without public health concerns [4].

Among the thermochemical processes, pyrolysis is an available method to convert cattle manure into biofuels and/or chemicals. In the process, the lignocellulosic components of the biomass are broken down condensable hydrocarbon molecules (e.g., acetic acid), noncondensable gases (e.g., CO_2), and solid carbon as char or biochar under an inert atmosphere. Based on the heating rate, the pyrolysis technology may be divided into slow and fast processes [5]. The slow pyrolysis process ($\leq$10 °C/min) is used primarily for the production of biochar at higher temperatures ($\geq$400 °C), but the fast pyrolysis process ($\geq$100 °C/min) primarily for the production of bio-oils is at about 500 °C [6–8]. Furthermore, the resulting biochar not only provides a remarkable carbon sink for mitigating GHG emissions, but also possesses available pore structure for its potential use as an adsorbent and/or soil amendment [9].

Biochar is a rigid and porous carbon material. Its pore properties, however, vary over a wide range with its precursor type and production conditions [5]. Although there are many research studies for reusing dairy (cattle) manure (DM) as a feedstock for char production, few studies focused on the production of dairy manure-based char (DMC) at higher temperatures [3,10–17]. These studies almost produced biochars at a moderate temperature (300–700 °C). It is notable that their pore properties were not significant. For instance, Cantrell et al. [3] reported the physicochemical results for five manure-based biochars pyrolyzed at 350 °C and 700 °C, showing that the specific surface area (SSA) of DMC produced at 700 °C was 186.5 m^2/g. However, the SSA value of DMC produced at 700 °C in another study was only 74.0 m^2/g [17]. On the other hand, the biochars obtained at higher pyrolysis temperatures showed more stability in both abiotic and biotic incubations [18].

As reviewed above, there is very limited research focused on the production of biochar at higher pyrolysis temperatures (>700 °C) in the literature [19,20]. In the previous studies [21–23], the authors even investigated the thermochemical characterization of DM with relevance to its energy conversion and environmental implications. In order to evaluate DM as a precursor for producing biochars as solid fuels, the authors further prepared DMC products at 400–800 °C [21], suggesting that its resulting chars can be used as clean solid fuels based on their high values of carbon (60% by weight, or 60 wt%) and calorific value (22.3 MJ/kg), and low contents of nitrogen (0.5 wt%). In the preliminary evaluation [23], a biochar product was produced at 900 °C, showing its microporous/mesoporous textures with a specific surface area of about 300 m^2/g. In the present study, a series of DMC products, produced at higher charring temperatures (i.e., 500–900 °C), were used to see the structural changes based on their pore properties and true densities.

2. Materials and Methods

2.1. Materials

The dairy manure (DM) for producing biochar was obtained from the livestock farm at the National Pingtung University of Science and Technology (Pingtung, Taiwan). Details on the DM pretreatment were described in the previous studies [21–23].

2.2. Thermochemical Analysis of Oven-Dried DM

In this work, the thermochemical properties of DM, including proximate analysis, ultimate (elemental) analysis, calorific (heating) value, inorganic elements, and thermogravimetric analysis (TGA), were conducted to evaluate the potential for producing porous biochar at adequate pyrolysis conditions. Again, details on the thermochemical analysis of oven-dried DM were described previously [21–23].

2.3. Pyrolysis Experiments

Because the temperature has been shown to be the most important process parameter in the pyrolysis experiments [11,24], the DMC products were produced at higher pyrolysis temperatures (500–900 °C), moderate residence time (30 min), and low heating rate (10 °C/min) under an inert atmosphere by nitrogen flow. The pyrolysis conditions were similar to the previous study [21], except for the pyrolysis temperature. The yields of the DMC products were obtained by the difference between the weights of DM (5 g for each experiment) and DMC. The DMC products produced at 500, 600, 700, 800, and 900 °C were marked as DMC-500, DMC-600, DMC-700, DMC-800, and DMC-900, respectively.

2.4. Characterization of DMC

As mentioned above, the main purpose was to produce biochars with high pore properties (including specific surface area, pore volume, and porosity). Therefore, the physical characterization of DMC was mainly based on the nitrogen adsorption–desorption isotherms and true density by using a surface area and porosity analyzer (Model No.: ASAP 2020; Micromeritics Co., Norcross, GA, USA) and a gas pycnometer (Model No.: AccuPyc 1340; Micromeritics Co., Norcross, GA, USA), respectively. More details about the analytical conditions were determined previously [25,26]. On the other hand, the microstructural textures and elemental compositions on the surface of DM and DMC were measured by using a scanning electron microscopy (Model No.: S-3000N; Hitachi Co., Tokyo, Japan) coupled with an energy dispersive X-ray spectroscopy (Swift ED3000, Oxford Instruments, Abingdon, UK). Regarding the chemical characterization of DMC, it included proximate analysis, ultimate (elemental) analysis, and calorific value, which were similar to the thermochemical analysis of oven-dried DM in the Section 2.2.

3. Results and Discussion

3.1. Thermochemical Characterization of Dairy Manure (DM)

As shown in Table 1, there were high contents of carbon (C, 42.63%) and hydrogen (H, 6.43%) in the dried dairy manure (DM), which was consistent with the high value of volatile matter (79.55%) due to the undigested lignocellulose. The compositional characterization should be attributed to ruminant feed (or forage) like pangola grass and napier grass. Making a comparison between the molar ratios of DM ($C_6H_{10.9}$) and cellulose ($C_6H_{10}O_5$)/hemicellulose ($C_5H_8O_4$), their values were approximate to each other. Therefore, its calorific value was up to 18.4 MJ/kg-dry. The data in Table 1 were very close to those in the previous studies [25,26] and other reports [27–29]. On the other hand, the contents of inorganic elements in the biochar precursor (i.e., DM) will be important for various reasons, including soil fertility and contamination when reusing it (or resulting biochar) as an organic fertilizer, and slagging and fouling as it was burned in boilers. In this regard, the macro-nutrients (i.e., Ca, Mg, P, Si, K), micro-nutrients (i.e., Mn, Cu, Zn, Fe), and toxic metals (i.e., As, P, Cd, Cr) in the DM were determined in the present study based on the analyses of inductively coupled plasma-optical emission spectrometer (ICP-OES). These inorganic elements in the DM biomass will exist in the so-called ash, which is as high as 10.94% (Table 1). Also, the oxides or carbonates of alkali metals (i.e., K, Na) and alkaline earth metals (i.e., Ca, Mg, Sr, Ba) from the DM ash could evaporate at combustion temperature due to the relatively low melting points and subsequently condense on the down streams. As listed in Table 2, the primary inorganic elements included calcium (Ca), silicon (Si), phosphorus (P), magnesium (Mg), potassium (K), sodium (Na), and aluminum (Al). The contents of these ash-forming elements in DM were in accordance with those in grass biomass such as straw [30]. These inorganic elements could be present in the forms of oxide or carbonate like SiO_2, CaO, or $CaCO_3$. Regarding the concentrations of toxic metals in DM, the ash is almost free of them, including arsenic (As), cadmium (Cd), chromium (Cr), cobalt (Co), copper (Cu), nickel (Ni), and lead (Pb). However, it should be noted that the contents of zinc (Zn) and manganese (Mn) in Table 2 indicated small amounts (0.017% and 0.034%, respectively). Although these contents were significantly lower than the swine-based manure [31], their moderate toxicity may pose hazards to ecosystems.

Table 1. Proximate analysis, ultimate analysis, and calorific value of dairy manure (DM).

Property	Value [a]
Proximate analysis (wt%) [a]	
Volatile matter	79.55 ± 0.19
Ash content	10.94 ± 0.27
Fixed carbon [b]	9.51
Ultimate analysis (wt%) [a]	
Carbon (C)	42.63 ± 0.01
Hydrogen (H)	6.43 ± 0.08
Oxygen (O)	39.70 ± 0.03
Nitrogen (N)	2.01 ± 0.13
Sulfur (S)	0.42 ± 0.02
Calorific value (MJ/kg) [a]	18.40 ± 0.08

[a] On a dry basis; two or three measurements. [b] By difference.

Table 2. Contents of relevant trace elements of dairy manure (DM).

Inorganic Element	Value [a]	Method Detection Limit (ppm)
Ca (wt%)	2.140	8.4
Si (wt%)	1.130	11.3
P (wt%)	0.822	91.8
Mg (wt%)	0.642	5.4
K (wt%)	0.426	51.6
Na (wt%)	0.202	11.4
Al (wt%)	0.107	17.4
Fe (wt%)	0.081	6.6
Mn (wt%)	0.034	6.0
Zn (wt%)	0.017	4.2
Sr (wt%)	0.007	0.3
Ti (wt%)	0.006	1.2
Ba (wt%)	0.002	0.3
As (wt%)	ND [b]	0.5
Cd (wt%)	ND	2.4
Cr (wt%)	ND	7.8
Co (wt%)	ND	20.4
Cu (wt%)	ND	7.2
Ni (wt%)	ND	19.2
Pb (wt%)	ND	17.4

[a] On a dry basis (moisture free). [b] Not detectable.

In the measurement of thermogravimetric analysis (TGA), the TGA curve of DM (0.2 g), shown in Figure 1, was obtained at a heating rate of 10 °C/min under the nitrogen flow (50 cm^3/min). Obviously, there were three stages for the devolatilization and carbonization of biomass (i.e., DM) biopolymer constituents (i.e., hemicellulose, cellulose, and lignin) in different temperatures [5]. The first stage in the temperature range of 25–200 °C represented the thermal desorption of moisture attached and structural deformity of the biomass. The second stage ranging from 250 to 450 °C was the greatest change on the mass lose zone. From the derivative thermogravimetric (DTG) curve, the pyrolytic decompositions of hemicellulose and cellulose occurred at the peak temperatures of around 320 °C and 380 °C, respectively. This reaction zone was indicative of extensive decomposition of hemicellulose in the early stage due to its fragile structure chemically in comparison with cellulose and lignin. Subsequently, cellulose started to undergo decomposition reactions such as devolatilization and carbonization. The third stage starting from 430 °C was attributed to the rigorous decompositions of lignin and its resulting char, making it brittle and porous. In this regard, the pyrolysis temperature was adopted at above 500 °C in the charring experiments for producing porous biochars.

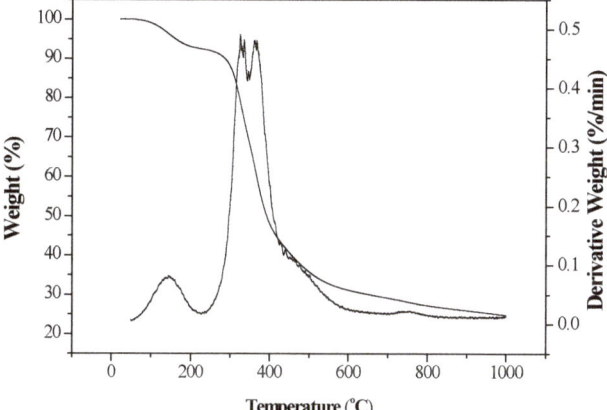

Figure 1. Thermogravimetric analysis (TGA) and derivative thermogravimetry (DTG) curves at heating rate of 10 °C/min for dairy manure (DM).

3.2. Yield and Pore Properties of DMC Products

For better utilization in the agricultural and environmental applications, the yields, densities, and pore properties of DMC products were determined in this work. As shown in Figure 2, the yields, ranging from 31.4% to 22.5%, indicated a decreasing trend as the pyrolysis temperature was increased from 500 °C to 900 °C. This variation should be attributed to the devolatilization and carbonization of solid char products during the progressive pyrolysis at higher temperatures, which was consistent with the TGA observations (Figure 1).

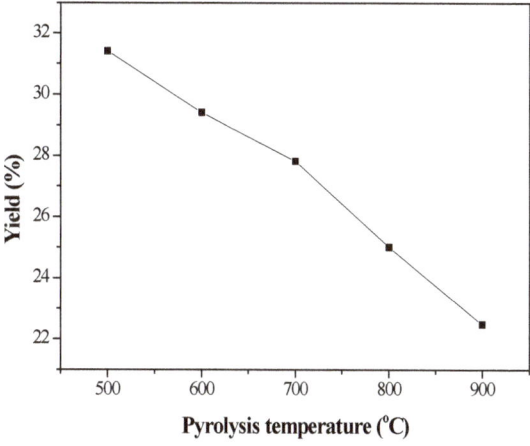

Figure 2. Yields of DM-derived biochar (DMC) as a function of pyrolysis temperature.

The pore properties of material involve the correlations between its specific surface area, pore volume, and density. Among them, specific surface area may be the most important parameter to indicate the adsorptive capacity and the quality of biochar. Based on the Brunauer–Emmett–Teller (BET) model, it was commonly measured by N_2 adsorption–desorption isotherms at −196 °C using the surface area and porosity analyzer. Table 3 summarized the pore properties of DMC products produced at various pyrolysis temperatures [32,33]. The true density is defined as the ratio of the biochar mass to the volume occupied by that mass, which was measured by a helium-displacement pycnometer. However, the particle density was estimated by the total pore volume and true density. Based on the definition of porosity in the porous particle, this property was further determined by the particle density and the true density. The data in Table 3 obviously indicated an increasing trend in pore properties, including the BET surface area, micropore area, total pore volume, micropore volume, true density (ρ_s), particle density (ρ_p), and porosity (ε_p). This result was in accordance with the observations of the average pore diameter (D_{ave}), DMC's yield, and the DM's TGA, suggesting that the rigorous carbonization (charring) at higher pyrolysis temperatures was favorable to the development of porous structure in the DMC products with aromatic carbon clusters [34]. Here, assuming that the pore is of cylindrical geometry, the data on D_{ave} were calculated from the ratio of the total pore volume (V_t) and the BET surface area (S_{BET}) and can be further validated in the pore size analysis. Consistently, more condensed aromatic structure of the charring materials can be formed at higher pyrolysis temperatures (>700 °C). Therefore, the optimal DMC product (i.e., DMC-900) with the maximal BET surface area of 360.6 m^2/g and true density of 2.2284 g/cm^3 was produced at 900 °C. Based on the suggestion by Keiluweit et al. [35], the DMC-900 product could be a turbostratic biochar.

Table 3. Pore properties and densities of DMC products at different temperatures (500–900 °C) held for 30 min.

Biochar Product	S_{BET} [a] (m^2/g)	S_{micro} [b] (m^2/g)	V_t [c] (cm^3/g)	V_{micro} [d] (cm^3/g)	D_{ave} [e] (Å)	ρ_s [f] (g/cm^3)	ρ_p [g] (g/cm^3)	ε_p [h] (-)
DMC-500	6.5	3.3	0.008	0.002	51.2	1.6838	1.6603	0.014
DMC-600	42.9	36.3	0.030	0.020	28.2	1.8741	1.7737	0.054
DMC-700	139.1	114.3	0.088	0.063	25.2	2.0108	1.7094	0.150
DMC-800	267.6	198.5	0.167	0.109	24.9	2.1245	1.5694	0.275
DMC-900	360.6	256.7	0.240	0.141	26.6	2.2824	1.4746	0.354

[a] BET surface area (S_{BET}) based on the relative pressure (P/P_0) ranging from 0.05 to 0.30. [b] Micropore area (S_{micro}) by t-plot method. [c] Total pore volume (V_t) obtained at relative pressure of about 0.95. [d] Micropore volume (V_{micro}) by t-plot method. [e] Calculated from the ratio of the total pore volume (V_t) and the BET surface area (S_{BET}) if the pore is of cylindrical geometry (i.e., average pore width = 4 × V_t/S_{BET}) [32,33]. [f] Measured by a pycnometer. [g] Estimated by the values of total pore volume (V_t) and true density (ρ_s) (i.e., $\rho_p = 1/[V_t + (1/\rho_s)]$) [32,33]. [h] Estimated by the values of particle density (ρ_p) and true density (ρ_s) (i.e., $\varepsilon_p = 1 - \rho_p/\rho_s$) [32,33].

As mentioned above, the nitrogen (N$_2$) adsorption–desorption isotherms is commonly used to analyze the porous characterization of carbon material. Furthermore, the pore size distributions of the resulting biochars at the desorption branch were obtained by the Barrett–Joyner–Halenda (BJH) method [36]. Figures 3 and 4 depicted the N$_2$ adsorption–desorption isotherms (at −196 °C) and pore size distribution curves of DMBC products, respectively. The isotherms in Figure 3 showed type I shape, as expected for microporous materials because of its high adsorption potential at relative pressure (P/P_0) of less than 0.05. Obviously, a sharp "knee" point near P/P_0 around 0.05 was observed in these isotherms, which corresponds to the monolayer capacity. However, the specific surface area was commonly calculated from the values of P/P_0 in the range of 0.05–0.30 [36]. As shown in Figure 3, the steep increase in the isotherm slope up to P/P_0 (above 0.95) can be explained by capillary condensation within the pores, followed by saturation as the pores become filled with nitrogen liquid. The significant increase in pore properties occurred when the pyrolysis temperature was between 600 and 900 °C which should be due to the rigorous charring and shrinking reactions, thus developing more porous structures in the DMC products (i.e., DMC-700, DMC-800, and DMC-900). On the other hand, the hysteresis loop observed in the isotherms (Figure 3) is a typical Type IV for mesoporous materials. According to the classification by the International Union of Pure and Applied Chemistry (IUPAC), the DMBC products should be classified as being of the H4 type [36], indicating that the resulting biochar products are complex carbon materials containing both micropores and mesopores. More consistently, their pore size distributions observed in Figure 4 indicated two narrow profiles for micropores (<2.0 nm) and mesopores (about 4.0 nm), respectively.

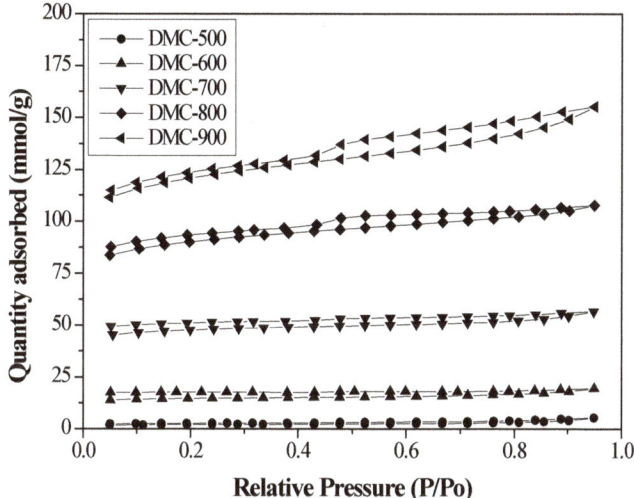

Figure 3. N$_2$ adsorption–desorption isotherms of DMC products (i.e., DMC-500, DMC-600, DMC-700, DMC-800, and DMC-900).

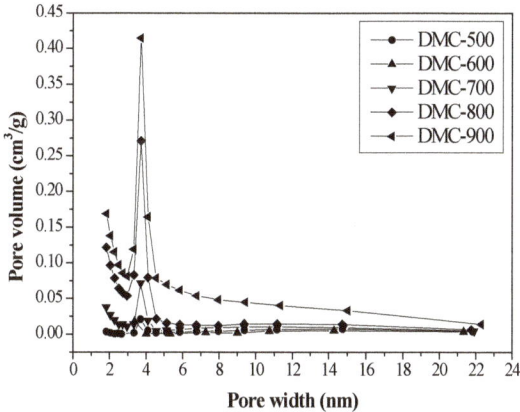

Figure 4. Pore size distributions of DMC products (i.e., DMC-500, DMC-600, DMC-700, DMC-800, and DMC-900).

In order to see the porous textures, the SEM observations were performed on the surface of DM and DMC products. Figure 5 showed the SEM micrographs (×1000) of DM and optimal DMC product (i.e., DMC-900). As seen in Figure 5a, the DM exhibited a rigid and compact matrix, which was indicative of its non-porous and rod-like features due to the lignocellulosic composition. By comparison, the optimal biochar product (DMC-900) displayed porous structures on the surface. The SEM observations were very consistent with their pore properties (Table 3). In addition, these findings were similar to the literature survey by Mukome and Parikh [37].

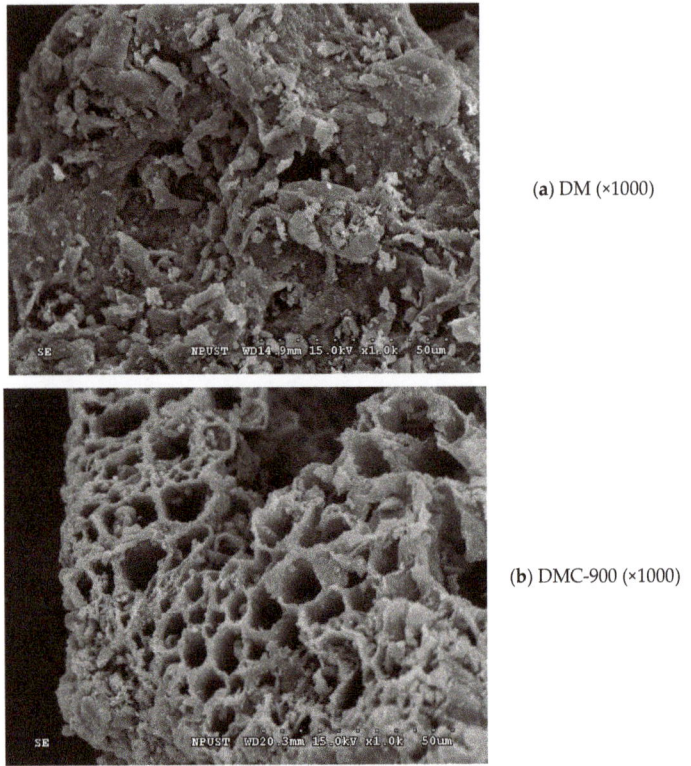

Figure 5. SEM images (×1000) of (**a**) dairy manure (DM), and (**b**) optimal biochar product (DMC-900).

3.3. Chemical Characterization of DMC Products

In order to correlate the chemical and agronomic properties of DMC products with pyrolysis temperature, Table 4 provided the data on the proximate analysis (ash content), ultimate (elemental) analysis (C/N), and calorific value. Obviously, the ash contents of DMC products indicated an upward trend, increasing from 25.26 wt% to 42.21 wt% as pyrolysis temperature increased. However, the calorific values of DMC products pyrolyzed from 500 °C to 700 °C slightly increased as their carbon contents increased, but they then decreased above 700 °C due to the decrease in the carbon content and/or the increase in the ash content. Compared to the data in Table 1, this can be further confirmed by the increase in the carbon content from 42.6 wt% to above 50 wt%, clarifying that the carbon content in the biochar products (i.e., DMC) significantly increased during the carbonization process. It should be noted that increasing the carbonization temperature from 800 °C to 900 °C will induce more gasification reaction, thus indicating a decrease in the carbon content of the resulting biochar product (Table 4). On the other hand, the nitrogen contents of the biochar products (i.e., DMC) indicated a decreasing trend. It can be ascribed to increase the emissions of N-containing pyrolytic gases at higher carbonization temperatures.

Figure 6 showed the elemental compositions on the surfaces of the optimal product DMC-900 by the energy dispersive X-ray spectroscopy (EDS). The contents of inorganic elements were consistent with the data in Table 2. It can be seen that the content of oxygen in the DMC-900 was still high, suggesting that the polar nature (i.e., hydrophilicity) enhanced by oxygen-containing functional groups on the surface will be more significant [33]. Furthermore, the cation exchange capacity (CEC) of the optimal product DMC-900 was determined in duplicate by the sodium acetate method to show a high value of 57.5 ± 16.1 cmol/kg probably due to its high specific surface area and the presence of

O-containing functional groups [37]. Therefore, the reuse of biochar as a soil amendment has been shown to increase soil CEC through electrostatic interaction due to its negatively charged surface [38]. In order to increase the biochar CEC by embedding more acidic oxygen functional groups on the surface, biochar was further treated with strong oxidants like hydrogen peroxide [39] and ozone [40]. Therefore, the removal of cationic pollutants (e.g., heavy metal ions, cationic dye) from aqueous solutions by using DM-based biochars has been widely studied in the literature [10–17].

Table 4. Chemical properties of DMC products at different temperatures (500–900 °C) held for 30 min.

Biochar Product	Proximate Analysis (wt%) [a]		Ultimate Analysis (wt%) [a]		Calorific Value [a] (MJ/kg)
	Ash	Combustible [b]	Carbon	Nitrogen	
DMC-500	25.26 ± 0.03	74.74	50.69 ± 0.75	1.99 ± 0.02	19.38 ± 0.11
DMC-600	27.56 ± 0.31	72.44	52.52 ± 0.71	1.62 ± 0.02	19.92 ± 0.10
DMC-700	29.48 ± 0.32	70.52	53.33 ± 0.71	1.43 ± 0.04	20.38 ± 1.04
DMC-800	32.65 ± 0.42	67.35	53.61 ± 0.06	1.25 ± 0.07	19.67 ± 0.03
DMC-900	42.21 ± 0.06	57.79	50.51 ± 0.09	0.78 ± 0.02	19.72 ± 0.32

[a] On a dry basis; two measurements. [b] Including volatile matter and fixed carbon. It was calculated by difference.

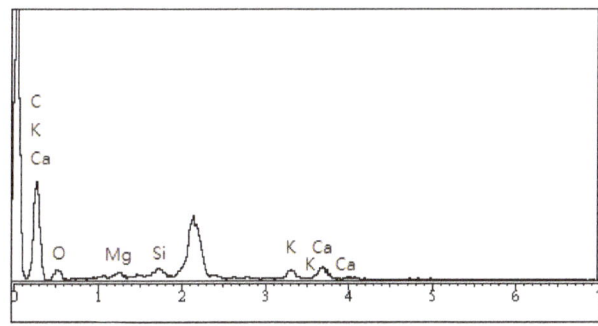

Element	Weight %	Atomic %
Carbon	71.91	81.88
Oxygen	15.50	13.25
Calcium	5.75	1.96
Potassium	3.71	1.30
Silicon	1.91	0.93
Magnesium	1.22	0.68

Figure 6. Energy-dispersive X-ray spectrometry (EDS) analyses of optimal biochar product (i.e., DMC-900).

4. Conclusions

A series of porous biochar products (DMC) were prepared from dried dairy manure (DM) at high pyrolysis temperatures (500–900 °C). The following conclusions were summarized:

1. According to the data on the thermogravimetric analysis (TGA) of DM, it is better to produce highly porous DMC at temperature above 500 °C because of the intense devolatilization of its lignocellulosic compositions and the increase in the aromaticity.
2. The optimal DMC product produced at 900 °C showed its BET surface area of 361 m^2/g and total pore volume of 0.24 cm^3/g.

3. From the N_2 adsorption–desorption isotherms with the hysteresis loops (H4 type), the DMC products are complex carbon materials, which contained both micropores (Type I) and mesopores (Type IV).
4. The carbon contents of the biochar products (i.e., DMC) significantly increased to above 50% during the carbonization process.
5. The cation exchange capacity (CEC) of the optimal DMC product showed a high value of 57.5 ± 16.1 cmol/kg.

Author Contributions: Conceptualization, W.-T.T.; methodology, P.-C.H.; validation, P.-C.H.; data curation, Y.-Q.L.; formal analysis, Y.-Q.L.; writing—original draft preparation, W.-T.T.; writing—review and editing, W.-T.T.

Funding: This research received no external funding.

Acknowledgments: The authors express sincere appreciations to the Instrument Centers of National Chung Hsing University, National Tsing-Hua University and National Pingtung University of Science & Technology for their supports in the analyses of organic elements, inorganic elements and scanning electron microscope- energy dispersive X-ray spectroscopy (SEM-EDS), respectively.

Conflicts of Interest: The authors declare no conflict of interest.

References

1. Intergovernmental Panel on Climate Change (IPCC). *2006-IPCC Guidelines for National Greenhouse Gases Inventories*; IPCC: Geneva, Switzerland, 2006.
2. Gerber, P.J.; Steinfeld, H.; Henderson, B.; Mottet, A.; Opio, C.; Dijkman, J.; Falcucci, A.; Tempio, G. *Tracking Climate Change Through Livestock—A Global Assessment of Emissions and Mitigation Opportunities*; Food and Agriculture Organization: Rome, Italy, 2013.
3. Cantrell, K.B.; Hunt, P.G.; Uchimiya, M.; Novak, J.M.; Ro, K.S. Impact of pyrolysis temperature and manure source on physicochemical characteristics of biochar. *Bioresour. Technol.* **2012**, *107*, 419–428. [CrossRef] [PubMed]
4. Cantrell, K.B.; Ducey, T.; Ro, K.S.; Hunt, P.G. Livestock waste-to-bioenergy generation opportunities. *Bioresour. Technol.* **2008**, *99*, 7941–7953. [CrossRef] [PubMed]
5. Basu, P. *Biomass Gasification, Pyrolysis and Torrefaction*, 2nd ed.; Academic Press: San Diego, CA, USA, 2013.
6. Tsai, W.T.; Chang, J.H.; Hsien, K.J.; Chang, Y.M. Production of pyrolytic liquids from industrial sewage sludges in an induction-heating reactor. *Bioresour. Technol.* **2009**, *100*, 406–412. [CrossRef] [PubMed]
7. Tsai, W.T.; Lee, M.K.; Chang, J.H.; Su, T.Y.; Chang, Y.M. Characterization of bio-oil from induction-heating pyrolysis of food-processing sewage using in an induction-heating reactor. *Bioresour. Technol.* **2009**, *100*, 2650–2654. [CrossRef] [PubMed]
8. Lee, M.K.; Tsai, W.T.; Tsai, Y.L.; Lin, S.H. Pyrolysis of napier grass in an induction-heating reactor. *J. Anal. Appl. Pyrolysis* **2010**, *88*, 110–116. [CrossRef]
9. Lehmann, J.; Joseph, S. Biochar for environmental management: An introduction. In *Biochar for Environmental Management*, 2nd ed.; Lehmann, J., Joseph, S., Eds.; Routledge: New York, NY, USA, 2015; pp. 1–13.
10. Cao, X.; Ma, L.; Gao, B.; Harris, W. Dairy-manure derived biochar effectively sorbs lead and atrazine. *Environ. Sci. Technol.* **2009**, *43*, 3285–3291. [CrossRef] [PubMed]
11. Cao, X.; Harris, W. Properties of dairy-manure-derived biochar pertinent to its potential use in remediation. *Bioresour. Technol.* **2010**, *101*, 5222–5228. [CrossRef] [PubMed]
12. Uchimiya, M.; Cantrell, K.B.; Hunt, P.G.; Novak, J.M.; Chang, S. Retention of heavy metals in a Typic Kandiudult amended with different manure-based biochars. *J. Environ. Qual.* **2012**, *41*, 1138–1149. [CrossRef]
13. Qian, L.B.; Chen, B.L. Duel role of biochars as adsorbents for aluminum: The effects of oxygen-containing organic components and the scattering of silicate particles. *Environ. Sci. Technol.* **2013**, *47*, 8759–8768.
14. Xu, X.Y.; Cao, X.D.; Zhao, L.; Wang, H.L.; Yu, H.R.; Gao, B. Removal of Cu, Zn, and Cd from aqueous solutions by the dairy manure-derived biochar. *Environ. Sci. Pollut. Res.* **2013**, *20*, 358–368. [CrossRef]

15. Zhu, Y.; Yi, B.J.; Yuan, Q.X.; Wang, M.; Yan, S.P. Removal of methylene blue from aqueous solution by cattle manure-derived low temperature biochar. *RSC Adv.* **2018**, *8*, 19917–19929. [CrossRef]
16. Chen, Z.L.; Zhang, J.Q.; Huang, L.; Yuan, Z.H.; Li, Z.J.; Liu, M.C. Removal of Cd and Pb with biochar made from dairy manure at low temperature. *J. Integr. Agric.* **2019**, *18*, 201–210. [CrossRef]
17. Zhao, B.; Xu, H.; Ma, F.; Zhang, T.; Nan, X. Effects of dairy manure biochar on adsorption of sulfate onto light sierozem and its mechanisms. *RSC Adv.* **2019**, *9*, 5218–5223. [CrossRef]
18. Jindo, K.; Sonoki, T. Comparative assessment of biochar stability using multiple indicators. *Agronomy* **2019**, *9*, 254. [CrossRef]
19. Fu, P.; Hu, S.; Xiang, J.; Sun, L.; Su, S.; Wang, J. Evaluation of the porous structure development of chars from pyrolysis of rice straw: Effects of pyrolysis temperature and heating rate. *J. Anal. Appl. Pyrolysis* **2012**, *98*, 177–183. [CrossRef]
20. Park, J.H.; Wang, J.J.; Meng, Y.L.; Wei, Z.; DeLaune, R.D.; Seo, D.C. Adsorption/desorption behavior of cationic and anionic dyes by biochars prepared at normal and high pyrolysis temperatures. *Colloids Surf. A* **2019**, *572*, 274–282. [CrossRef]
21. Liu, S.C.; Tsai, W.T. Thermochemical characteristics of dairy manure and its derived biochars from a fixed-bed pyrolysis. *Int. J. Green Energy* **2016**, *13*, 963–968. [CrossRef]
22. Tsai, W.T.; Liu, S.C. Thermochemical characterization of cattle manure relevant to its energy conversion and environmental implications. *Biomass Convers. Biorefin.* **2016**, *6*, 71–77. [CrossRef]
23. Tsai, T.W.; Hsu, C.H.; Lin, Y.Q. Highly porous and nutrients-rich biochar derived from dairy cattle manure and its potential for removal of cationic compound from water. *Agriculture* **2019**, *9*, 114. [CrossRef]
24. Steiner, C. Considerations in biochar characterization. In *Agricultural and Environmental Applications of Biochar: Advances and Barriers*; Guo, M., He, Z., Uchimiya, S.M., Eds.; Soil Science Society of America: Madison, WI, USA, 2016; pp. 87–100.
25. Hung, C.Y.; Tsai, W.T.; Chen, J.W.; Lin, Y.Q.; Chang, Y.M. Characterization of biochar prepared from biogas digestate. *Waste Manag.* **2017**, *66*, 53–60. [CrossRef] [PubMed]
26. Tsai, W.T.; Huang, P.C. Characterization of acid-leaching cocoa pod husk (CPH) and its resulting activated carbon. *Biomass Convers. Biorefin.* **2018**, *8*, 521–528. [CrossRef]
27. Fan, L.T.; Chen, L.C.; Mehta, C.D.; Chen, Y.R. Energy and available energy contents of cattle manure and digester sludge. *Agric. Wastes* **1985**, *13*, 239–249. [CrossRef]
28. Wang, L.; Shahbazi, A.; Hanna, M.A. Characterization of corn stover, distiller grains and cattle manure for thermochemical conversion. *Biomass Bioenergy* **2011**, *35*, 171–178. [CrossRef]
29. Wu, H.; Hanna, M.A.; Jones, D.D. Thermogravimetric characterization of dairy manure as pyrolysis and combustion feedstocks. *Waste Manag. Res.* **2012**, *30*, 1066–1071. [CrossRef] [PubMed]
30. Baernthaler, G.; Zischka, M.; Haraldsson, C.; Obernberger, I. Determination of major and minor ash-forming elements in solid biofuels. *Biomass Bioenergy* **2006**, *30*, 983–997. [CrossRef]
31. Hsu, J.H.; Lo, S.L. Effect of composting on characterization of copper, manganese, and zinc from swine manure compost. *Environ. Pollut.* **2001**, *114*, 119–127. [CrossRef]
32. Smith, J.M. *Chemical Engineering Kinetics*, 3rd ed.; McGraw-Hill: New York, NY, USA, 1981.
33. Suzuki, M. *Adsorption Engineering*; Elsevier: Amsterdam, The Netherlands, 1990.
34. Lian, F.; Xing, B. Black carbon (biochar) in water/soil environments: Molecular structure, sorption, stability, and potential risk. *Environ. Sci. Technol.* **2017**, *51*, 13517–13532. [CrossRef]
35. Keiluweit, M.; Nico, P.S.; Johnson, M.G.; Kleber, M. Dynamic molecular structure of plant biomass-derived black carbon (biochar). *Environ. Sci. Technol.* **2010**, *44*, 1247–1253. [CrossRef]
36. Lowell, S.; Shields, J.E.; Thomas, M.A.; Thommes, M. *Characterization of Porous Solids and Powders: Surface Area, Pore Size and Density*; Springer: Dordrecht, The Netherlands, 2006.
37. Mukome, F.N.D.; Parikh, S.J. Chemical, physical, and surface characterization of biochar. In *Biochar: Production, Characterization, and Applications*; Ok, Y.K., Uchimiya, S.M., Chang, S.X., Bolan, N., Eds.; CRC Press: Boca Raton, FL, USA, 2016; pp. 67–96.
38. Liang, B.; Lemann, J.; Solomon, D.; Kinyangi, J.; Grossman, J.; O'Neill, B.; Skjemstad, J.O.; Thies, J.; Luizao, F.J.; Petersen, J.; et al. Black carbon increases cation exchange capacity in soils. *Soil Sci. Soc. Am. J.* **2006**, *70*, 1719–1730. [CrossRef]

39. Huff, M.D.; Lee, J.W. Biochar-surface oxygenation with hydrogen peroxide. *J. Environ. Manag.* **2016**, *165*, 17–21. [CrossRef]
40. Huff, M.D.; Marshall, S.; Saeed, H.A.; Lee, J.W. Surface oxygenation of biochar through ozonization for dramatically enhancing cation exchange capacity. *Bioresour. Bioprocess.* **2018**, *5*, 18. [CrossRef]

© 2019 by the authors. Licensee MDPI, Basel, Switzerland. This article is an open access article distributed under the terms and conditions of the Creative Commons Attribution (CC BY) license (http://creativecommons.org/licenses/by/4.0/).

MDPI
St. Alban-Anlage 66
4052 Basel
Switzerland
Tel. +41 61 683 77 34
Fax +41 61 302 89 18
www.mdpi.com

Agronomy Editorial Office
E-mail: agronomy@mdpi.com
www.mdpi.com/journal/agronomy

www.ingramcontent.com/pod-product-compliance
Lightning Source LLC
LaVergne TN
LVHW071942080526
838202LV00064B/6652